160 550893 1
AF741187

UNCONVENTIONAL AND COMMUNITY TRANSPORT IN THE UNITED KINGDOM

TRANSPORTATION STUDIES

A series edited by

NORMAN ASHFORD, Loughborough University of Technology

WILLIAM G. BELL, Florida State University

Titles Published in Transportation Studies book series:

Volume 1: E. W. Allanson – Car Ownership Forecasting

Volume 2: Edited by Norman Ashford, William G. Bell and Tom A. Rich – Mobility and Transport for Elderly and Handicapped Persons

Volume 3: B. S. Hoyle – Seaports and Development: The Experience of Kenya and Tanzania

Volume 4: K. J. Button and A. D. Pearman – Applied Transport Economics: A Practical Case Studies Approach

Volume 5: Edmund J. Gubbins – The Shipping Industry; The Technology and Economics of Specialisation

Volume 6: Dušan Teodorović – Transportation Networks: A Quantitative Approach

Volume 7: G. G. Lucas – Road Vehicle Performance: Methods of Measurement and Calculation

Volume 8: M. G. Lay – Handbook of Road Technology (2 volumes)

Volume 9: Wayne Kenneth Talley – Transport Carrier Costing

Volume 10: Dušan Teodorović – Airline Operations Research

Volume 11: John R. McLean – Two-Lane Highway Traffic: Theory and Practice

Volume 12: S. R. Tolofari – Open Registry Shipping: A Comparative Study of Costs and Freight Rates

Volume 13: C-E. Norrbom and A. Ståhl – Mobility and Transport for Elderly and Disabled Persons

Volume 14: Stephen D. Nutley – Unconventional and Community Transport in the United Kingdom

This book is part of a series. The publisher will accept continuation orders which may be cancelled at any time and which provide for automatic billing and shipping of each title in the series upon publication. Please write for details.

UNCONVENTIONAL AND COMMUNITY TRANSPORT IN THE UNITED KINGDOM

Stephen D. Nutley

University of Ulster, Coleraine, Northern Ireland

GORDON AND BREACH SCIENCE PUBLISHERS
New York · Philadelphia · London · Paris · Montreux · Tokyo · Melbourne

388.40942 N

Copyright © 1990 by OPA (Amsterdam) B.V. All rights reserved. Published under license by Gordon and Breach Science Publishers S.A.

Gordon and Breach Science Publishers

Post Office Box 786
Cooper Station
New York, New York 10276
United States of America

5301 Tacony Street,
Drawer 330,
Philadelphia, Pennsylvania 19137
United States of America

Post Office Box 197
London WC2E 9PX
UK

58, rue Lhomond
75005 Paris
France

Post Office Box 161
1820 Montreux 2
Switzerland

3–14–9, Okubo
Shinjuku-ku, Tokyo 169
Japan

Private Bag 8
Camberwell, Victoria 3124
Australia

Library of Congress Cataloging-in-Publication Data

Nutley, Stephen D.
Unconventional and community transport in the United Kingdom / Stephen D. Nutley.
p. cm. -- (Transportation studies, ISSN 0278-3819 : v. 14)
Includes bibliographical references and index.
ISBN 2-88124-764-4
1. Paratransit services--Great Britain. 2. Transportation, Automotive--Great Britain--Technological innovations. I. Title. II. Series.
HE311.G7N88 1990 90-47763
388.4'0941--dc20 CIP

No part of this book may be reproduced or utilized in any form or by any means, electronic or mechanical, including photocopying and recording, or by any information storage or retrieval system, without permission in writing from the publishers. Printed in Great Britain by Billing & Sons Ltd, Worcester

CARDIFF
17 MAR 1997

To the memory of my father, Denis Nutley

CONTENTS

INTRODUCTION TO THE SERIES

This broad-ranging series of books will cover many of the varied aspects of transportation. The subject area will be generally divided into two parts: the first deals with planning and technological aspects of transportation, the second with specialized transportation.

Within the context of this series, technology and planning will include the wide spectrum of various aspects of the design and planning of vehicles and infrastructure for the transport of freight and passengers, as well as operational and management considerations. The general aim of the planning and technology series is to inform readers of the state of the art and to summarize the status of transportation.

The second part of the series will seek to generate monographs dealing with improving the mobility of those groups in society increasingly characterized as the transportation disadvantaged, particularly, but not exclusively, the elderly, the disabled and families with low incomes. It is anticipated that the content of these books will be derived largely from research, policy analysis and documental field experience. The subject matter will include advances in the relevant technology, service and methods demonstration, improved planning and methodology, major or proposed changes in public policy and innovative proposals for the development or change of systems.

Occasionally, more specific monographs will be published, presenting the results of individual studies in areas of special interest to planners and technologists.

As with any monograph series, the emphasis is on current information, and the material will be of interest to the transport practitioner, the postgraduate student and the academic working in the field.

NORMAN ASHFORD
WILLIAM G. BELL

It is with great sadness that we record the death of Dr. William G. Bell in April 1990. Co-editor of this series, Dr. Bell, in a relatively short period of activity in the area of transportation, materially improved the access of elderly and handicapped persons to public and private transportation. We have all benefited from his life's work and there is no better memorial to his life.

NORMAN ASHFORD

LIST OF TABLES

LIST OF FIGURES

PREFACE

This book does not aim to fill a gap in research, but it was strongly felt there was a need for it. At one level, there is an enormous pile of literature already published on 'unconventional modes' (UCMs) and 'community transport' (CT). All this is, however, very highly fragmented and widely dispersed among academic, technical or popular magazine articles, local authority and government documents, consultants' reports, conference papers, etc., which are 'semi-published' or difficult to access, and too often focus narrowly on individual transport services. Anyone brave enough to plough through this morass will be further confused by the inconsistent use of terminology. It was felt therefore that there was an urgent need, in the light of continuing growth of the UCM/CT sector, to bring all this disparate material between the same pair of covers in an organized fashion. An overview of the whole field is needed, within the changing policy context, together with a coherent classificatory framework to sort out the terminology.

Although it is to some extent inevitable, this book will be out of date by the time it is published. On frequent occasions the author was aware of being overtaken by events, having to re-write sections of chapters already prepared in order to accommodate new developments. This is particularly true in the aftermath of the 1985 Transport Act. The book was written at a time of much uncertainty and speculation; after it is published, answers to questions posed in the book may have become clear.

The structure is intended to permit alternative usage. Anyone with a general understanding of what UCMs and CT are will obtain a review of their origins and development in chapter 1 and a discussion of evaluation and policy role in chapter 14. Those uncertain of the meaning of UCMs and CT and of individual modes will find them all classified and explained in chapter 2. This may suffice for those not needing detailed analyses of any individual modes. Personal interests in specific types may be pursued by choosing from the mode-by-mode analysis in chapters 3–13. Careful cross-referencing has been provided. Anyone attempting to read the book from cover to cover will encounter some repetition.

The work is overwhelmingly a review, as the long bibliography testifies. It does, however, include some original research of my own, and it is still necessary to point out that references to the analysis of RUTEX in Wales and to the 'South Radnor' accessibility study (mainly in chapter 14) derive from work done under contract to the Welsh Office in 1977–83. Despite the passage of years, this does not necessarily reflect the views of the Secretary of State for Wales, present or past. The author acknowledges a 'rural' bias! Gratitude is also due to staff of the Department of Environment Library, London, and of the Scottish Development Department, Edinburgh, for access to planning material; to the former Community Transport Services Ltd, Manchester, the Northern Ireland Council for Voluntary Action, Belfast, and the Rural Development Commission, London, for access to unpublished records; to professional colleagues Tony Moyes, University of Wales at Aberystwyth, and Irvine Lavery, University of Ulster at Jordanstown, for copies of unpublished papers; to David Kaye of Louth for use of his unpublished manuscript on community buses; to innumerable local authority officers up and down the country who answered letters and telephone calls on unpredictable topics; and lastly to Nigel McDowell, Mary McCamphill and Kilian McDaid of my own department who undertook the photography and production of tables and figures, respectively.

Stephen D. Nutley
April 1989.

Chapter 1

THE DEVELOPMENT OF 'UNCONVENTIONAL' AND 'COMMUNITY' TRANSPORT

1.1 THE LONG TERM POLICY CONTEXT

Transport policy in the United Kingdom has suffered from an almost total lack of consensus since 1945, such that periods of stability have been few and short, and investment and development have been frustrated by periodic reorganizations under political direction. Probably the single most important issue has been the public/private split, where decision-makers' perceptions or ideological preferences have neglected the needs of that section of society dependent on public transport, and exaggerated the necessary or justifiable provision for the private car. This can be seen as a conflict between needs-based planning and demand-led planning, and the overall trend is strongly in favour of the latter despite frequent reversals of policy corresponding to changes of government. The public sector has never enjoyed any lasting economic security or political confidence, and is constantly appraised by its cost rather than by its vital social role. It is ultimately these attitudes, more than the realities of social change, lifestyles, affluence, mobility and car ownership, which have endorsed the decline of public transport and the predicament of those people dependent on it. Further policy changes of the 1980s, culminating in the Transport Act 1985 which enforces deregulation and privatisation of the bus sector in Great Britain, may not be so revolutionary when seen in this context. Transport policy is firmly back on its long term course, reversing a dozen years (1968-79) of modest progress for the public sector. More specific policy components, and local or sectoral attempts to solve real problems, have all been very severely constrained by the official view of the public sector as a financial burden on the state rather than (as in most countries of western Europe) as an essential piece of social infrastructure.

While attitudes have varied among governments and local authorities, there has been little consistent challenge to the orthodox view that subsidy of public transport is undesirable - a view that is often contrasted with enthusiastic expenditure on road-building. The major response of the public sector to social changes and motor car competition has been contraction, with wholesale route closures (road and rail) primarily in rural areas but also some thinning out in cities, and persistent frequency reductions at off-peak times or service withdrawals in evenings and weekends. Other responses to economic problems have included market research and network rationalisation projects, which still produce a contracted system even if it is more cost-effective. Coordination and planning were tried for a short time. From 1972 local authorities were required to coordinate public transport and justify their revenue support levels when claiming transport supplementary grant from government. From 1978 they were required to produce plans demonstrating how subsidies were directed at satisfying social needs. With a change of government in 1979, these procedures did not last long enough to be properly tested, and passed to a new regime for whom high costs were, *ipso facto*, evidence of failure. In the 1980s, a 'radical' Conservative government has used its ideological belief in the free market and its crusade against public expenditure to undermine previous support mechanisms for public services, but this has had the effect of renewing uncertainty, loss of confidence and fear of decline.

Major policy responses to the economic problems of public transport in the UK have traditionally been of three types. (It should be noted that most of the following refers specifically to Great Britain, as transport in Northern Ireland has always been separately administered, although the same trends are valid in principle).

(a) Reorganisation of Institutional Structures
Amalgamation of private bus companies under the free market was rife in the inter-war period.

Statutory reorganisation might start from the takeover of certain bus companies by the railways from 1928, includes the nationalisation of the railways and their bus subsidiaries in 1947, and of other territorial bus companies in 1968, includes also the establishment of passenger transport executives in the conurbations in 1968, and concludes with the privatization of National Bus Company subsidiaries from 1985.

(b) **Regulation by the State**
Over the long term, transport problems have been met with more, rather than less, regulation. The comprehensive Road Traffic Act 1930 aimed at securing safety and stability for the emerging bus industry, rather than competition, and lasted virtually untouched until the deregulating legislation of 1980 and 1985. Regulation of the railways dates back much earlier, and has withdrawn considerably from a grossly excessive peak in the 1940s. In the road passenger sector, a realisation that 'problems' might be helped by liberalizing operating practices and demarcation lines led to a flurry of permissive legislation from 1977.

(c) **Public Financial Support**
Almost an automatic response to economic problems is the demand for state subsidy on the grounds of the social importance of transport services. The belated introduction of bus subsidies in 1968, administered by local authorities on a route-specific basis, contrasted with new arrangements for the railways in 1974 in which uneconomic but socially necessary sectors were supported by a public service obligation block grant as a 'contract' with government. Escalating bus subsidy demands led to attempts at control by linking them to planning mechanisms under Acts of 1972 and 1978. In the 1980s, financial control is exerted by centrally imposed 'cash limits' on local authority spending, and the state is empowered to put a ceiling on all local transport subsidies regardless of the source of income. Implementation of the 1985 Act has brought the new approach of tendering for 'non-commercial' bus services.

In addition there are other policy options which are not as persistent as the above but which may offer possibilities for problem-solving at certain times, under certain conditions, or in a relatively minor way.

(i) **Technology**
The opportunity is rarely offered to solve immediate problems with a significant switch of technologies. Examples are the removal of municipal trams and trolleybuses and the dieselisation of the railways in the 1950s, and increasing interest in urban rapid transit systems from about 1970. In general, however, new technology is applied in a small scale, incremental manner, and its impact is variable. Few problems are exclusively technical. If the 'problem' is defined as the need to reduce costs on poorly patronised services, then technology may be required to adopt a different role.

(ii) **"A more commercial approach"**
The traditional public service role and reliance on subsidies has produced accusations of management inertia. Market analysis projects undertaken by bus companies in the 1970s led to network rationalisation. Deregulation and privatization were driven by the revival of free market ideology in the 1980s and supposed to produce "new opportunities". Such a trend could be positive or negative, with new investment and marketing on the one hand, but abandonment of marginal services on the other.

(iii) **The voluntary sector**
On the fringes of national transport policies but doing a vital job at community level, is this highly disparate group of services alternatively known as the 'supplementary', 'informal' or 'third level' sector. They range from the purely charitable, through supplementing statutory welfare transport and provision of specialized facilities for persons with impaired mobility, to attempts at 'public transport' in the normal sense. Rarely acknowledged in national policies until recently, there are many constraints on their potential role and contribution to wider transport development.

1.2 THE EMERGENCE OF 'UNCONVENTIONAL MODES'

'Unconventional modes' (UCMs) of transport do not fit neatly into any one of the above policy themes, but contain elements of each; it is also doubtful whether they can be regarded as a policy in their own right. The inability to reconcile gradually declining demand for conventional public transport with the continuation of adequate service levels to satisfy the significant demand that remains is the crux of the problem which is familiar to governments and operators in many countries. In the UK the political/economic climate is such that any policy which deviates from the traditional approaches listed above would still have to be aimed at cutting costs. Hence the emergence of a line of thought favouring consideration of novel, innovative modes of operation and/or vehicle types as alternatives to conventional services. This is based on the expectation that such a development might provide significant operating economies and also present a more suitable or attractive product for the consumer.

In the UK a variety of unusual or ephemeral types of public transportation has been emerging steadily over the last two decades, albeit in a spasmodic and uncoordinated manner. This has now reached the point where numbers, distribution, and apparent success in many cases, seem to offer a realistic and 'new' option for policy makers. Although the origin of the term is not clear, such types are normally known collectively as 'unconventional modes' (UCMs). It can be regarded as equivalent to the American term 'paratransit' (Kirby et al 1975, Orski 1975, Transportation Research Board 1980, 1985, 1986). The generality of the British label reflects the impossibility of devising a more precise title for the immense variety of service types which it embraces. While recognized internationally, the term 'paratransit' and other American nomenclature is not appropriate and seldom used in Britain.

Insofar as the problems of public transport can be identified in terms of a mismatch of supply and demand, then the case for new modes rests on the apparent rigidity of the former, characterized as it is by fixed routes and timetables, relatively high capacity and formal operating procedures. The new modes were therefore developed as departures from 'conventional' practice. Briefly, unconventional features include: flexible routeing and timing, multi-purpose operation, use of vehicles not normally available to the public, non-commercial operation by community groups, new technology and alternative sources of finance. With one exception (the railbus) they are all road-based. There are now so many different types, each with numerous variations, and described by inconsistent and confusing terminology, that particular attention has to be given to definition and classification (chapter 2).

The great diversity of UCMs reflects their diversity of originating circumstances and sources of innovation. It is important to stress that the development of UCMs was not the result of any officially prescribed strategy or plan, but in retrospect was uncoordinated, spasmodic and localized. This continues to be the case despite recent government encouragement of a greater policy role for UCMs. The basic idea of reducing public transport costs or optimizing revenue in areas of weak traffic potential by departing from conventional means of operation is certainly not a new one. Comparisons and initiatives have been drawn from other countries or from the past. For example, the inception of postbuses in Britain was influenced by their existence in many countries of continental Europe; light rail or 'railbus' vehicles on rural lines have historic precedents in the British Isles and contemporary parallels elsewhere in Europe; 'dial-a-ride' systems have a substantial pedigree in North America. Institutional and legal impediments deterred early adoption of such ideas in Britain on anything other than a small scale. In practice, the nature and pattern of UCM development is strongly dependent upon the widely varying degrees of commitment and enthusiasm shown by transport operators, local authorities and community groups.

UCMs in Britain belong almost entirely to the non-commercial side of the market. The simplest generalization that can be made about their origins is to say that individual 'schemes', projects or experiments were set up as a local response to the perceived inadequacy of existing public transport services in the areas concerned. There are two quite different situations.

(a) Rural Areas

First, where the quantity of public transport supplied by the market is insufficient to meet the travel needs of the local population, responsible authorities have responded traditionally by subsidization of conventional services, but may turn to UCMs as alternative solutions if they are believed to be a more cost-effective use of public resources. This situation is most commonly brought about in the rural areas where public transport provision has been in continuous decline from the 1950s, due to the growth in private car ownership, rising bus and rail operating costs, and demographic and social changes (for a general review see Moseley 1979). In rural areas 20-35% of households and 45-65% of adult individuals are still without a car. Plentiful public transport is within most people's memory and the loss is keenly felt; public authorities are concerned to seek solutions. Considerable academic work has demonstrated poor accessibility to local facilities and significant hardship for rural inhabitants (Wilmers & Moseley 1975, Hillman et al 1976, White 1978, Oxley & Seaton 1979, Moseley 1979, Halsall & Turton 1979, Nutley 1983, Cloke 1985). This contrasts with mounting political pressure to reduce subsidies by cutting services. The appearance of UCMs is a partial response to this impasse.

It was always envisaged that their adoption would constitute purely *local* solutions, and be *supplementary* to existing transport. UCMs in Britain originated as a distinctly *rural* phenomenon, which is in striking contrast to the experience of most other countries, especially paratransit in the United States. However, increasing government interest in this sector and the progressive removal of institutional and legislative barriers have recently led to such innovations being proposed as a more general 'solution' throughout the UK.

(b) Handicapped Persons

The second situation where the shortcomings of conventional public transport lead to a consumer demand for new modes concerns the special needs of handicapped persons. Even where conventional public transport is available, certain groups of people may find it difficult or impossible to use through personal disability or generally impaired mobility (e.g. the elderly). By normal understanding, about one in ten of the UK population is disabled, but a broader definition including the blind, mentally handicapped, the general frailty of old age, and temporary disability would raise this to one in four. While statutory agencies provide essential transport for health and welfare purposes, this is supplemented to a great extent by the voluntary sector (for a review see Bailey & Layzell 1983). A diverse set of modes catering exclusively for the needs of elderly and disabled people has been in existence for some decades, and has come to be known as 'community transport' (CT) due to the high involvement of volunteer labour. Such services are provided by charities and local community welfare groups and have varying degrees of exclusivity. Some are used as a back-up to statutory welfare transport of the sort provided by health authorities or local authority social services departments for 'institutional' journeys, such as to hospitals, surgeries or day centres. Most importantly, CT is expanding to provide in addition services to enhance the general mobility of these groups, over and above the limited welfare needs (Bailey & Appleby 1986, Sutton 1987).

These two 'markets' for the UCM/CT sector are not easily distinguished; there is considerable overlap. The relationship can be explained in terms of social needs. Briefly, the rural origins of most UCMs were a response to needs in a *spatial* sense. i.e. the problem is that of non-car owning rural inhabitants seeking access to necessary facilities some distance away, where conventional public transport runs at low frequencies or is non-existent. Conversely, CT aims to meet needs in a purely *physical* sense, i.e. regardless of the quantity of available public transport, it is unsuitable for people with physical disabilities due to unsympathetic design or operational characteristics. The latter problem therefore has no particular geographical basis: the great majority of CT schemes are located in urban areas, simply reflecting population distribution.

Should CT be regarded as 'unconventional'? While a large part of this highly diverse sector has become familiar over a long period of time, some modes certainly have features which may appear unconventional, and there is much confusion of terminology between CT and UCMs. At its simplest, the relationship is as shown on figure 1.1, with UCMs characterised by their innovation and CT by its use of voluntary labour. It should also be emphasised that UCMs, as a 'public transport solution', are in most cases available to the general public, while most CT modes are exclusively targeted on groups of people with special needs and 'the general public' is ineligible. But again, there is overlap.

1.3 RECENT HISTORY AND THE POLICY CONTEXT

Very few unconventional experiments were initiated independently by bus companies from purely commercial motives. Almost all were designed for unprofitable environments (predominantly rural, especially in the early phase) and claiming revenue support from local authorities. These bodies (County Councils in England and Wales, Regional Councils in Scotland) have therefore played a central role in collaborating with bus companies, liaising with relevant institutions, and assessing claims of social need from the local population. Councils have varied in adopting either an interventionist, managerial role in setting up their own unconventional schemes, or else a promotional role, encouraging others to do so. The role of central government, apart from legislation and policy, has been to conduct

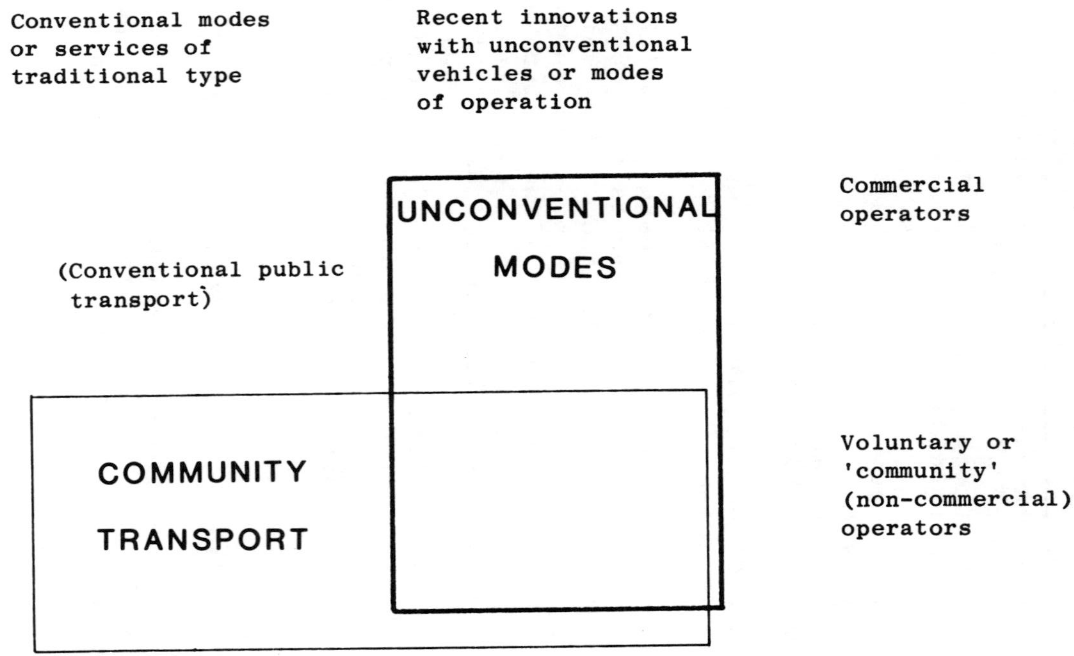

FIGURE 1.1 Unconventional Modes and Community Transport.

experiments or demonstration projects to test the feasibility and economics of alternative modes, to support and publish appropriate research through the Transport and Road Research Laboratory (TRRL), and occasionally to provide 'pump-priming' grants to get schemes off the ground.

Since the once-prosperous bus industry started to decline towards the end of the 1950s, familiar problems of rising costs, contracting services and mounting subsidy requests have beset successive governments of opposing ideologies. Transport in Britain has always been as much a political football as any other sector of the economy, fluctuating between Labour governments' preference for coordination, integration and planning, and Conservative preference for free market competition and deregulation, but despite this the policy role of UCMs has remained fairly consistent. Until recently, there was consensus that this role was to serve remote pockets of isolation, 'gap filling' between arterial routes, and not to compete with conventional buses. Growth over the last two decades has been steady but generally unobtrusive. It could be said that the 'profile' has been rising slowly, as governments have sought with increasing desperation to control bus costs and to consider alternatives. It must nevertheless be emphasised that the sector's contribution to policy has always been minor and subordinate to strategic objectives like 'coordination' or 'deregulation'.

Insofar as the origins of UCMs in Britain can be given a precise date, the earliest mention appears to be in the 'Jack Report' of 1961 - the first government enquiry into the economic problems of the bus industry (Ministry of Transport 1961). Amongst the more predictable economic and regulatory proposals, it considered the use of minibuses, the carriage of passengers on school contract buses and on local tradesmen's vehicles like the historic 'village carrier', conversely the carriage of goods on buses, and the postal bus. In the light of later developments it is interesting that the committee rejected postbuses on the lines of those already existing in Germany and Switzerland, and preferred the Post Office to place more mail contracts with bus companies.

Although the main conclusion of the report was (eventually) to lead to public subsidy for buses, interest in alternative solutions persisted. Surveys were done in six case study areas in 1963 (Ministry of Transport 1963). Apart from being amongst the earliest rural transport surveys, they are notable for seeking information on local mail van, voluntary car and ambulance services. These in turn led to a series of experiments in 1964/5 in four study areas, which attempted to implement new bus services aimed at solving problems exposed by the surveys (Ministry of Transport 1965). With the exception of a couple of school services, the experiments took the form of conventional buses on new routes, and most failed after a short time. What is probably more significant is that various unconventional options were considered - minibuses, car sharing/pooling, postbuses - but ruled out largely due to the rigidity of contemporary legislation. For further details see section 1.4 below.

Apart from some informal car sharing and a little public use of school buses, the credit for the first successful UCM to be implemented should go to the Post Office for initiating postbus services in 1967/8. Originally an experiment, the four prototypes - in Devon, Cumbria, Mid-Wales and SE Scotland - overcame early economic problems and all survive today, having laid the foundations for impressive expansion of this mode in the following decade.

The formative period of UCM development was the 1970s, when numerous services of different types were implemented around the country, initially on an experimental basis but many apparently succeeding and being copied elsewhere. Generally it was the local authorities who took the lead, under urgent pressure to control their transport budgets and aware that continued subsidy of conventional buses could not be sustained indefinitely. There was a growing feeling that something radically different had to be tried. For the same reasons bus companies were generally willing to cooperate with councils in experimenting with new modes, sensing mutual interest. It is impossible to date the origin of the idea that local communities themselves should accept some responsibility in providing services which a commercial bus company could no longer support, but the principle gained ground uncontentiously. New mode innovation showed little discernable pattern, being determined by no overall policy but merely "the art of the possible" under specific local conditions. Central government rather lagged behind in this activity.

Two modes completely new to the UK appeared in the early 1970s. The first British 'dial-a-ride' bus system commenced operations in Abingdon, Oxfordshire in 1972. This was the first attempt to introduce a mode proven successful in North America, and attractive at the time due to its 'technological' approach. This mode, however, contravenes the above generalizations on UCM origins in that it was introduced as a commercial venture by a public sector bus company, is confined to urban areas, and is not targeted on any well-defined 'problem' area. Four years later the number of dial-a-rides had risen to twelve, but this proved to be their peak and they were all withdrawn by the end of the decade. Compared with other UCMs, this mode is also anomalous in being the only one to fail (Oxley 1977, 1980).

The second of the most conspicuous new modes in the 1970s was the pioneering of the 'community bus' concept. First established in North Norfolk in 1975, the idea of handing over management and operation of all local services to volunteers from several village communities, but with the support of the public bus company and county council, was completely novel and does not seem to have precedents abroad. This project soon acquired national recognition, and a dozen similar schemes sprang up around the country by the end of the decade (Harris 1978, Banister 1982). Other forms of collaboration between local communities, public bus companies and local authorities resulted in individual examples of highly flexible, adaptable and responsive service types.

Modes with earlier roots expanded considerably during the 1970s. By far the most impressive growth rate was that achieved by postbuses, which multiplied ten-fold in six years (from 15 services in 1972 to 157 in 1978); the vast majority of this growth was in Scotland (Turnock 1977, Watts et al 1978a). Also, the potential for the wider use of private cars was beginning to be realised, with councils and local organizations seeking means of exploiting the rural tradition of lift-giving into a more widely available facility. Semi-formal arrangements for car sharing amongst members of local associations already existed, and some councils offered support to develop these into more organised services for people needing to make urgent trips. By 1975, Lincolnshire, Shropshire, Cumbria and Clwyd had 'social car' schemes (Rhys & Buxton 1974). Less novel perhaps, but arguably the most important mode as its potential is widespread, is the carriage of adult passengers on school contract buses. In many rural areas, these run greater mileage and are more frequent than stage carriage buses, and account for much greater public expenditure. Considerable potential for cost-cutting was seen in the integration of school and stage traffic, and most authorities devoted great effort towards re-casting services to this end, or using permits under the 1968 Act to carry commercial passengers on school buses (Awdas 1978).

A climate of increasing interest and discussion of UCMs was building up in the later 1970s. The TRRL held a 'state of the art' symposium in 1976 (Transport & Road Research Laboratory 1977a,b), but tended to concentrate on isolated examples of schemes in operation, and it was difficult to relate them to each other and to clarify their overall potential. It was accepted that there was insufficient practical experience of the various possible modes for confident planning. Comparative evaluation of diverse modes was recognized as a problem, perhaps not best achieved by normal quantitative methods (Martin 1978a, Tunbridge 1980). National Bus Company subsidiaries were cooperating with some UCM schemes such as community buses, but were also encouraged to experiment with more flexible service routes, park-and-ride, and the use of midibuses (Glassborow 1978): it is arguable whether the latter group is 'unconventional'. At the same time a nascent bandwagon effect was detectable amongst the community sector, with local conferences being organised to spur grassroots activity, and a stream of publicity and advice being issued by consumer groups (National Council for Social Service 1977, National Consumer Council 1978).

Two very important events were instrumental in consolidating these trends: the appearance of legislation aimed at relaxing regulatory barriers and the initiation of the government's RUTEX programme (see below), both starting in 1977. Following these, the Department of Transport joined the prevailing publicity drive by issuing a 'how to do it' manual for the aid of local organisations, called *A Guide to Community Transport* (Department of Transport 1978). Limited to social cars and community buses, it attempts to describe in simple language local demand assessment, planning, finance and licensing regulations, and directs interested groups to local authorities for assistance. The title of this publication, however, did expose the apparent lack of awareness of the traditional CT sector - the longstanding existence of many volunteer-based services devoted exclusively to the elderly and handicapped - and provoked claims that the 'community transport' label had been misappropriated (Sutton 1983). Since the reorganizations of local government and the health service in 1974, the demand for welfare transport had been increasing rapidly. This induced greater reliance on the voluntary sector to back up statutory services, and the number of local associations continued to grow, yet 'traditional' CT was still largely overlooked at national level.

For the potential of UCMs to be unleashed, or even tested, it was clear that new legislation would be needed to relax or remove the barriers to innovation that had become entrenched. Government recognition arrived in the 1977 White Paper, *Transport Policy*, which in its section on rural areas laid some emphasis on support for "simple low cost arrangements", citing specifically minibuses, postbuses, private cars with fare-paying passengers, and shared hire-cars (Department of Transport 1977). The necessary legislation was promised. In December 1975 an announcement had been made of the intention to conduct experiments into the feasibility and economics of alternative modes. Accordingly, the Passenger Vehicles (Experimental Areas) Act 1977 allowed the minister to relax the normal licensing regulations in certain designated areas so that these could take place.

The Rural Transport Experiments (RUTEX) were sponsored by the Department of Transport, Welsh Office and Scottish Office, and planned and monitored by the Transport and Road Research Laboratory with the aid of local working groups. Four experimental areas were chosen - in Devon, North Yorkshire, Strathclyde (Scotland) and Dyfed (Wales) - located on figure 1.2. Sixteen experimental transport services were designed, and fifteen actually carried out, starting in mid-1977 and running for up to two years. They can be classified various ways but the basic approaches were as follows.

(i) Carrying passengers on vehicles making regular journeys for other purposes, such as postbuses and school buses.
(ii) Allowing buses to operate on flexible routes according to pre-booked demand.
(iii) Making maximum use of voluntary effort by local residents in organising and running transport suited to their needs, such as one community bus and a number of schemes using private cars.
(iv) Arranging for the sharing of hire-cars.

All but two schemes were based on small vehicles. A more detailed listing of the modes involved can be found in section 1.4 below.

Progress and results of individual experiments have been described in a dozen TRRL reports, and some attempts at synthesis and inter-modal comparison can be found in the mid-term review (Balcombe 1979) and the final symposium proceedings (Transport & Road Research Laboratory 1980). Details are also available from numerous other sources, e.g. Milefanti (1978), Planning Exchange (1979), Welsh Office (1978, 1979). As might be expected, results of these experiments were highly variable, with most shared hire-cars and car services to hospitals proving unsuccessful. Otherwise, on termination of the experimental period those schemes which required public funding were adopted by the appropriate county council, and many survive today, although rarely in an unmodified form. Among possible criticisms of the programme (section 1.4), the main point is that RUTEX was motivated mainly by a desire to demonstrate means of providing rural transport *at low cost*, and the importance of maintaining service levels, 'accessibility', and satisfying 'social needs' were neglected (Nutley 1981, 1983).

No doubt the demonstration experience and publicity of RUTEX was valuable, but the extent of its influence is not easy to judge. Due to the change of government in 1979 there was no official 'verdict' on the experiments nor any kind of follow-up programme. The Select Committee on Nationalized Industries (1977-78), in an enquiry into the National Bus Company, heard evidence from a number of local representatives including community bus operators. While encouraging NBC support for community buses, it deferred judgement on RUTEX until after conclusion of the experiments. But the official response to the committee (Department of Transport 1979) came after the election and, ignoring RUTEX, merely spelled out the very different philosophy of the incoming administration.

Certainly, since RUTEX there has been a powerful expansion in the number of UCMs and continuing innovation of types. This is attributable to much more than the normal processes of innovation diffusion. After the demonstration effects of specific experimental projects, both state-sponsored and locally generated, other important influences have been the recent sequence of permissive legislation and government financial cuts of the 1980s.

Between 1977 and 1985 there was an Act of Parliament on land passenger transport in every year but one. Most of these reflected the Conservative government's determination to reduce public expenditure by disposing of nationalized industries and curbing the powers of local authorities. A considerable proportion of this legislation was devoted to encouraging UCMs, by progressively relaxing the regulations on aspects of road transport operation and removing bureaucratic impediments to new initiative. It is significant that this trend overlapped the change of government in 1979. First, the 1977 Act to reduce licensing regulations in experimental areas has already been mentioned. In the same year the Minibus Act set up a permit system to facilitate minibus operation by non-profit community groups. The 1978 Transport Act, *inter alia*, introduced community bus licences and eased the conditions for social cars. The 1980 Transport Act, whose main purpose was to deregulate express coach services, included sections to further liberalise car sharing schemes. In 1981 the Public Passenger Vehicles Act 'consolidated' all recent legislation on road passenger transport including UCMs. Finally, the 1985 Transport Act, whose main purpose was the deregulation of all stage bus services, contained sections to legalize the sharing of taxis and hire-cars at separate fares and further amended regulations on community use of small buses. See appendix 1 for details.

This sequence therefore shows the same permissive, deregulating trend that overrides the sharp ideological switch in policies for the overall transport sector. Governments of both parties have found it expedient to encourage UCMs. Thus, in the 1978 Transport Act, the Labour government welcomed informal, non-profit, 'community' or 'welfare' modes as compatible with its objective of meeting social needs under a framework of planning and coordination, while in the 1980 and 1985 Acts such modes were also compatible with the Conservative government's preoccupation with bus service deregulation and the supposed virtues of local enterprise. One irony is that while UCM and CT operators are now able to do many things previously out of bounds, there are still complaints that regulations are just as complex as before!

Associated with this was a re-definition of 'public transport' to encourage the integration of multifarious services already in existence but which were not accustomed to carrying the general public. Government circulars in 1978 included within the definition, for planning and coordination purposes: school transport, contracted or hired buses, hospital and welfare services, taxis and hire-cars, and any UCMs which may exist such as community buses and social cars. Reiterated in 1985, this was extended to cover services provided wholly or mainly for elderly or disabled people, belatedly recognizing the growth in CT projects. Presumably, such a re-definition was intended to draw attention to previously overlooked resources in the belief that these 'supplementary' services had enough spare capacity to be incorporated into the traditional public transport network. While there is much truth in this belief, in some cases it is rather naive.

The most depressing feature of the 1980s has been the atmosphere of retrenchment and austerity measures imposed on many sectors of the economy by the monetarist policies of the Conservative government. Progressive cutbacks in public expenditure at central and local levels, reinforced by new legal restrictions, have put intense pressure on the nationalized transport industries and severely curtailed the ability of local authorities to subsidize loss-making but socially necessary services (Whitelegg 1984). This has demolished the 1970s approach of building planning and financial mechanisms to provide for the social needs of each county, and has instead produced a climate of hurried and often arbitrary service cuts and an urgent search for stop-gap measures. Not surprisingly perhaps, local authorities' attentions turn to UCMs as low cost alternatives. As many commentators had feared (Banister 1982), modes ideally suited for a 'supplementary' role were now pressed into service as 'substitutes' for conventional buses which had to be withdrawn through lack of revenue support.

Present government attitudes are also exploiting the current growth trend in the 'self-help movement'. This highly diverse and uncoordinated movement seeks local improvements through community action (Wilmers et al 1981), and embraces voluntary or cooperative tasks such as local job creation, community housing and shops, conservation, etc. Of course, charities assisting with medical and social service needs have existed for a long time, and transport provision on a similar basis is a logical extension (Bailey & Layzell 1983, Sutton 1985). However, in transport the voluntary sector is continually being encouraged to accept a greater workload. It is often claimed that local communities usually contain a reservoir of talent, initiative and goodwill, waiting to be harnessed for the common good. This may be so, but 'self-help' is originally a defensive reaction to official indifference to local problems. The current emphasis on the voluntary sector could be interpreted as a cynical move by a right-wing government to proclaim the virtues of self-reliance and enterprise while at the same time divesting itself of traditional state responsibility for public service provision. In the context of deregulation, privatization and the general emphasis on individual enterprise, this could be seen as part of a wider trend from public support to 'self-help' (Banister 1981).

Through the later 1980s, the overriding policy thrust in road public passenger transport is the deregulation of bus services consequent upon the Transport Act 1985. This is linked in principle to the parallel policy of privatizing National Bus Company subsidiaries. For a readable guide to the complexities of the Act, see Open Tech Transport Project (1986). The arguments surrounding this highly contentious statute have been propounded at great length already (Farrington 1985, Knowles 1985, Gwilliam et al

1985, Beesley & Glaister 1985) and need not be repeated here. As the government's case for deregulation is founded primarily on the cost of subsidy (£558 million in 1982/3, three-quarters of which was due to metropolitan areas), and the supposed stifling of local enterprise under the old system, then it was predictable that UCMs would receive favourable emphasis. The White Paper, *Buses*, which preceded the Act, urged "the better use, particularly in rural areas, of services run on behalf of education, health and social services authorities, the Post Office, private firms and others", and was also most enthusiastic about shared taxis (Department of Transport 1984).

The crucial reform is that services able to run without subsidy (the 'commercial' sector) no longer require a road service licence and may operate without restrictions on numbers or routes. Very few UCMs are able to benefit from this: primarily, the new phenomenon of urban PSV minibuses (and it is arguable whether these are 'unconventional'), and possibly also the shared taxis and hire-cars (legalized by the same Act). In the 'non-commercial' sector, which is likely to include the rural areas and most UCMs, operators submit tenders for subsidy to the local authority. While acceptable in general terms, this procedure caused alarm among local authorities and community groups who feared that UCMs diligently fostered over the years would be undercut by new arrivals. In recognition of this, postbuses and community buses were granted exemption from tendering until 1991.

Specific measures incorporated in the 1985 Act were the legalization of taxi/hire-car sharing, a simplified licensing system for taxi operators to run small buses, and regulatory changes for minibus permits and community buses. Most notable was the establishment of a Rural Transport Innovation Grant (RTIG) to provide pump-priming funds for setting up new "innovatory" transport projects in the countryside. A budget specifically for this purpose is a completely novel event, although it must be pointed out that the total value of £1.27 million is puny compared with the run-down of local authority subsidy in recent years. Also, to compensate for loss of cross-subsidy, a transitional Rural Bus Grant was allowed, originally at 6 pence per mile, payable by the DTp directly to operators.

Detailed discussion of post-deregulation changes and the current status and future role of the UCM/CT sector will be deferred to the conclusion of the book (chapter 14). Early information on the effects of deregulation tended to be fragmented, anecdotal, or confused with political rhetoric. Half a dozen research projects are in progress (mid-1988) and have yet to report in full, but some preliminary information is summarised in Balcombe et al (1987, 1988) and Rickard et al (1988). In the new circumstances, there are two sources of data on UCMs. The TRRL maintains a database on post-deregulation services, from which one might extract those appearing 'unconventional'. Consequently, Finch et al (1986) were able to identify services run by taxi operators and commercial minibuses. In November 1986, there were over 50 semi-stage shared taxi or taxibus services, and almost 200 locations known to have mini- or midibuses. This is updated further in Finch (1988) and Balcombe et al (1988). Both have expanded considerably, the former due to legal changes and the latter following recent commercial trends (Turner & White 1987). Also, the RTIG takes a broad view of what is "innovative", and is able to assist many schemes which do not necessarily qualify as UCMs, such as the appointment of local transport advisers and (again) commercial minibuses. Apart from these, of the UCMs which had appeared by the end of 1987, the most impressive was the inauguration of fifteen new community buses. It should also be noted that these benefits vary widely in type and location, with many areas - especially in Wales and Scotland - being under-represented.

A final trend that needs to be pointed out is that continuing expansion and diversification of UCMs has led to a broadening of their role, away from their rural origins. With a greater awareness of the CT sector, which mainly operates in urban areas, and the recent rise of other 'town' modes such as commercial minibuses and shared taxis, consideration of UCMs is no longer associated solely with the 'rural transport problem'. They now tend to be regarded as a legitimate policy option for the country as a whole, and the question then arises as to their future role in a period of intense change after deregulation. Some people have expressed fears about the complete demise of traditional public transport in unprofitable environments, such as the rural areas, with UCMs merely a transitional stage in this decline (e.g. Banister 1982). Nevertheless, current trends indicate that the UCM/CT sector might be expected to *replace* conventional transport, at least in some areas, exceeding their former 'supplementary' role. Is this possible? The major issue at present must be to define this role, especially the part played by the voluntary sector, in the context of a coherent policy for public transport as a whole, and then to support it with the necessary powers and resources.

1.4 GOVERNMENT EXPERIMENTS

This section briefly summarises the experimental services inaugurated under the two rounds of official experiments in 1964/5 and 1977-9. In the case of RUTEX, operational characteristics and economic performance are analysed in detail under the appropriate modal headings in chapters 4 to 12.

(a) Ministry of Transport Experiments 1964/5

The four case study areas are located on figure 1.2. The new services attempted in each are listed below, but as they are now historic and proved to have little influence no further details are given. If necessary, these can be traced in Ministry of Transport (1965).

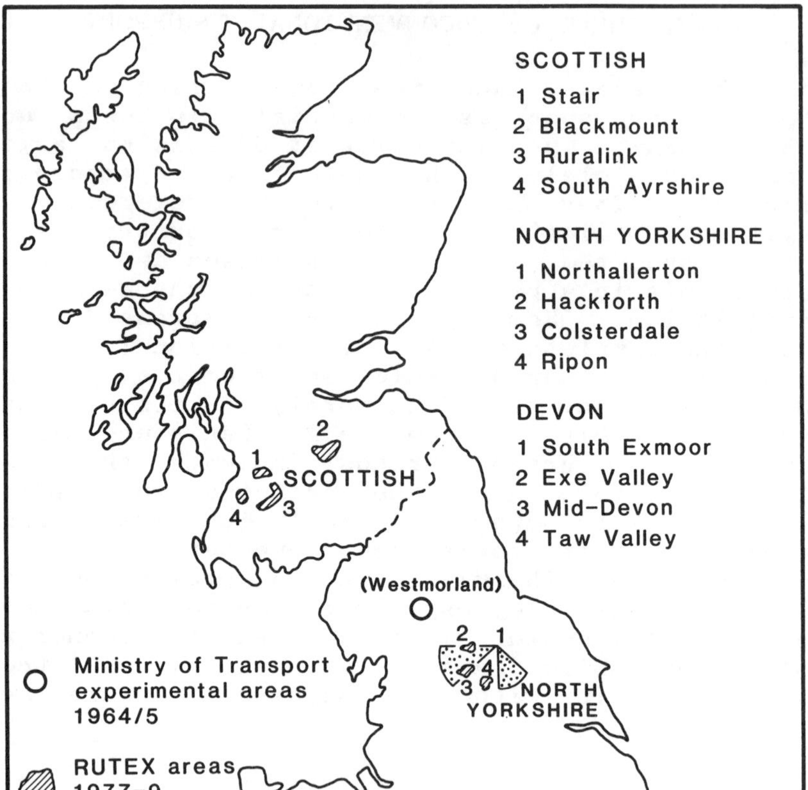

FIGURE 1.2 Experimental areas 1964-5 and 1977-9.

(i) **Lincolnshire.** North West of Horncastle.
New weekday service run by 'territorial' bus company: discontinued after 13 weeks.
Two school contract buses run by 'territorial' bus company authorized to carry fare-paying passengers under road service licence: continued indefinitely.
(ii) **Montgomeryshire** (now Powys). West of Welshpool.
Reinstated market day service run by independent operator: discontinued after 13 weeks.
(iii) **Northamptonshire.** North of Brackley.
New once-weekly service by independent operator: experiment ran for 26 weeks; evening service on same route discontinued after five weeks.
Five extensions to existing services either to reach new localities or to improve frequencies: most discontinued after eight weeks.
(iv) **Westmorland** (now Cumbria). South East of Appleby.
No new services implemented: only minor alterations to schedules.

(b) **Rural Transport Experiments (RUTEX) 1977-9**
The four designated areas are located on figure 1.2, but in all cases except Wales individual experimental services were conducted in separate sub-areas. Although there are some cases of UCMs being coordinated with existing conventional transport, only in the Welsh study area were the various UCMs coordinated with each other. Various classifications are possible, but initially the schemes are listed by area.

(i) **North Yorkshire.** Four experiments.

$1. *Ripon Flexibus.* A conventional bus operated by a NBC company on a circular route west of Ripon, partly replacing pre-existing services, to give rural dwellers access to town, also with a tourist role. Three optional diversions from the fixed route, made in response to passenger demand (North Yorkshire Rutex 1979).
$2. *Northallerton Hospital Transport Service.* A shared hire-car service for the purpose of conveying out-patients and visitors to Northallerton DGH from sectors of the rural hinterland without convenient bus routes. Service provided by three commercial hire-car firms, with

advance bookings coordinated centrally. These arrangements terminated in June 1979 (North Yorkshire Rutex 1980a).
$3. *Colsterdale Car Service.* A shared hire-car service to provide residents of Colsterdale (west of Masham) with connections to local buses (North Yorkshire Rutex 1980b).
$4. *Hackforth and District car sharing service.* A social car scheme using private cars and volunteer drivers with a local coordinating committee. Operating in three parishes south of Catterick, it aimed to provide access to local towns for general purposes (North Yorkshire Rutex 1980c).

Plans for another car scheme north-west of Northallerton were not proceeded with.

(ii) Devon. Four experiments.

$1. *Exe Valley Market Bus.* A community bus depending on local volunteer drivers and management, using a vehicle provided by TRRL, with maintenance and driver training undertaken by the local NBC subsidiary. Replacing conventional bus services, the community bus operates in an area north-west of Tiverton, providing mainly shopping journeys with demand-responsive routeing, plus hires and excursions (Devon Rutex 1978).
$2. *Taw Valley car service.* A shared hire-car service operating in a large diffuse rural area centred on Chulmleigh. Divided into sub-zones and served by numerous hire-car operators, possible journeys were pre-bookable and very diverse. Services in one sub-zone did not commence due to the lack of a suitable operator, and those in another zone terminated in June 1979 (Devon Rutex 1980a).
$3. *South Exmoor car service.* A social car scheme using private cars and volunteer drivers with a local coordinating committee. Operating in five parishes between South Molton and Dulverton, it aimed to provide access to these towns, supplementing existing buses, and to hospitals further afield (Devon Rutex 1980b).
$4. *Mid-Devon lift-giving scheme.* A special authorization issued under contemporary

legislation to allow private motorists to accept payment for lifts. Applied to ten parishes between Crediton and Tiverton. No organization or advertising. This arrangement was soon made redundant by the 1978 Transport Act which extended such provisions throughout the country (Devon Rutex 1979).

(iii) Scottish (southern parts of Strathclyde Region). Four experiments.

$1. *Ruralink.* A new midibus service run by a Scottish Bus Group subsidiary in the Dalmellington/Cumnock area, designed to connect several small towns and two hospitals. Three diversions were possible, in response to demand. Also carried small goods packages (Scottish Rutex 1981a).
$2. *"Blackmount services".* An attempt to provide new integrated services in an area of Lanark District, to replace conventional buses withdrawn. One existing postbus was extended (Biggar-Carnwath) and one new postbus commenced (Biggar-Lanark). These were complemented by a new minibus service (the Medwyn Gypsy), run by an independent operator, which also included two demand-responsive diversions. In phase II of the experiment, the latter was extended northwards on a demand-responsive, variable-route basis (Scottish Rutex 1979).
$3. *Stair car service.* A shared hire-car service operating in an area east of Ayr, intended to act as a feeder to local bus routes. The scheme ceased in June 1979 (Scottish Rutex 1980).
$4. *South Ayrshire Hospital Transport Service.* A social car scheme using private cars, volunteer drivers and local organisers. The objective was to transport people from a designated area south-east of Maybole to hospitals in Ayr and elsewhere. This scheme ceased in June 1979 (Scottish Rutex 1981b).

For all schemes see also Planning Exchange (1979).

(iv) **Welsh** (Dinefwr District in the County of Dyfed). Three experiments.

$1. *Postbus.* A new postbus service operated two routes, one to the north and one to the south of Llandovery, allowing frequent access to the town.
$2. *School buses.* Four school contract services provided by an independent operator were issued permits to carry the general public. They all connected rural areas with the town of Llandovery, but only during term-time.
$3. *Voluntary Essential Transport Scheme (VETS).* A social car scheme using private cars and volunteer drivers, organised by the WRVS. Limited to 'essential' journeys, it was available in a wide area around Llandovery and Llandeilo.

These three experiments were intended to be complementary, and are described in Dinefwr Rutex (1979), Welsh Office (1978, 1979).

Unless stated otherwise, the new services continued after the experimental period under the aegis of the county/regional council involved. Inevitably, most underwent further changes in later years, with some disappearing (like the Flexibus) and others expanding (like the Dinefwr social cars).

The programme was designed to incorporate a wide range of UCMs within the four areas, but also to have more than one example of each, so that it would be possible to compare modes with each other and also to see how any one mode performs in different circumstances. The individual service descriptions given above might more usefully be classified in a different way, to demonstrate the 'unconventional' approach and to facilitate modal comparisons.

(i) Carrying passengers on vehicles making regular journeys for other purposes (dual-purpose transport), thereby rationalising traffic: postbuses in Blackmount (Scotland) and Llandovery (Wales); school buses also in Llandovery area.
(ii) Allowing buses to operate on flexible routes and schedules according to pre-booked demand, thereby achieving a better match of supply and demand in geographical terms: Flexibus (Yorkshire), Ruralink and Medwyn Gypsy (Scotland);

the Exe Valley community bus (Devon); all car schemes also have this facility.
(iii) Making use of voluntary effort by local residents in organising and running their own transport, thereby achieving lower costs, local identity and understanding of community needs: Exe Valley community bus (Devon); social car schemes in Hackforth (Yorkshire), South Exmoor (Devon), South Ayrshire for hospitals (Scotland), Dinefwr VETS (Wales); and without any organisation, the mid-Devon lift-giving scheme.
(iv) Arranging for the sharing of commercial hire-cars, thereby achieving flexible service with lower cost to the passenger: Northallerton hospital service, and in Colsterdale (Yorkshire), Taw Valley (Devon) and Stair (Scotland).
(v) Using small vehicles, to cater for low demand, narrow roads, and (possibly) to reduce costs: all schemes except Ripon Flexibus and Llandovery school buses.
(After Balcombe 1979, amended).

The diagnostic features of each experiment are also summarised in table 1.1.

Apart from the fact that the 1979 change of government prevented any coherent follow-up to RUTEX, the main reservations that need to be expressed about the programme refer to its evaluation. Specifically, published reports demonstrate a great emphasis on cost factors and passenger loadings, and appear to neglect the social character of the problems which the programme was designed to tackle. First, it has never been revealed exactly how the RUTEX areas were chosen, or whether this selection was based on any prior socio-economic analysis, but it is known that the areas were supposed to be 'typical' and not 'extreme'. This is arguable: in the Welsh case, retrospective analysis suggests the study area was not typical of rural Wales (Nutley 1983). Decisions on specific sites for experiment were based on discussions among the local working groups and with other organizations (Milefanti 1978). Household surveys were done immediately prior to commencement of the experiments, not to provide data for identifying problems and planning operations, but to establish contemporary travel

TABLE 1.1 Rural Transport Experiments (RUTEX) 1977-9 (source: Balcombe 1979).

Scheme	Vehicle	Operator	Service type
Ripon Flexibus (N.Yorks) Ruralink (Strathclyde) Medwyn Gypsy and postbuses (Strathclyde)	Standard bus Midibus (27seats) Minibus, in conjunction with 2 postbuses (minibuses)	NBC SBG private operator Post Office	Fixed routes and schedules, with diversions on demand
Llandovery Postbus (Dyfed) Llandovery School Buses (Dyfed)	Minibus 4 Standard buses	Post Office Private operators contracted to County Council	Fixed routes and schedules
Exe Valley Market Bus (Devon)	Minibus	volunteer drivers, local committee	Fixed schedule; route determined by advance booking from passengers
Colsterdale Car Service (N. Yorks) Taw Valley Car Service(Devon) Stair Car Service(Strathclyde)	Shared hire-cars	private operators	pre-booked feeder services connecting with buses/ trains in local villages
Northallerton Hospital Service (N. Yorks) South Ayrshire Hospital Transport Service(Strathclyde)	Shared hire cars Private cars	private operators private motorists, local committee	pre-booked services to hospitals for outpatients and visitors
Voluntary Essential Transport Scheme (Dyfed)	Private cars	private motorists, WRVS	pre-booked lifts to doctors, hospitals, etc.
South Exmoor Car Service(Devon) Hackforth & District Car Service (N. Yorks)	Private cars	private motorists, local committee	pre-booked lifts to local towns
Mid-Devon lift-giving scheme (Devon)	Private cars	no organisation	lifts arranged on informal basis

patterns in order to assess the effects of the new services. The only review of these survey data (Coe & Fairhead 1980) emphasises the very low levels of demand, as expressed by actual trips, and concludes that the 'market' for public transport cannot be successfully estimated from such data.

Originally, the aims of RUTEX were laudably wide, embracing not only the practicality of UCMs, in operational and economic terms, for meeting the demands of low density areas, and the type and extent of necessary legislative changes, but also the assessment of transport usage relative to needs, and means of selecting modal types for particular areas (Milefanti 1978, Balcombe 1979). However, TRRL reports on the experiments tend to be starkly factual, uncritical, with little depth of analysis, and are unclear on whether aims were successfully achieved. The unavoidable impression is that the prime motivation of RUTEX was to demonstrate how rural transport could be provided *at low cost*, and the 'problem' has therefore been interpreted from the traditional operator's or public authority's point of view. Reports contain no analysis of local problems, and no attempts to relate the operational results to the impact they may have had upon those problems.

This was disappointing in that RUTEX coincided with a trend amongst academics towards interpreting the problem in terms of people's *accessibility* to necessary local facilities (shops, doctors, etc), and developing ways of measuring this. Under this concept, the role of transport is to provide a linkage between the locations of demand (e.g. a village) and supply (e.g. a country town). The degree to which necessary linkages can be satisfied is the prevailing level of accessibility; a methodology to quantify this had recently been published (Moseley et al 1977). Implicit in this is some degree of social *need* for transport (Koutsopoulos 1980). Unsatisfied needs (lack of accessibility) can be identified, so that transport planning could be directed towards filling the gaps. In fact, an approach similar in principle was urged on local authorities by the contemporary Transport

Act 1978. Also disappointing in TRRL reports is the emphasis on low demand (i.e. patronage) experienced by the new services, for the implication is that need is low and there is no latent or hidden demand. Further, there is no real assessment of the social benefits of the new services, as these seem to be understood only in terms of passenger loadings. Consequently, measures of cost-effectiveness are limited to simplistic indices such as cost per trip. A valid alternative would be to express 'benefits' in terms of accessibility procured, which gives a better appreciation of the project's social value (Nutley 1981, 1983).

There are indications that government was aware of the difficulty of evaluating the RUTEX programme. Conventional cost-benefit analysis was recognised to be unsuitable, as normal methods of putting values on the benefits could not be applied. The value of travel time saved, for example, is irrelevant in a rural area which previously had no transport at all, where there is no choice of alternatives, and where the main real and perceived benefit is a newly acquired mobility and access to services (Milefanti 1978). While the need for more appropriate evaluation methods was recognized, at the time of the project's inception no methodology was agreed upon, and was felt to be outside the scope of the programme. Assessment was therefore limited to the familiar account of costs, passengers and revenues. Despite the optimistic tone of official comment, the main criticism of RUTEX must be its failure to define objectives in terms of social problems and to take a 'problem-solving' approach.

Chapter 2

CLASSIFICATION OF MODES

2.1 WHAT IS 'UNCONVENTIONAL'?

The foregoing review of UCM development in the UK, and the policy context, was deliberately brief. Before proceeding to analyse in detail the performance of individual modes, it is essential to devote some attention to defining the field of interest. Although obviously not a precise term, one should attempt to lay down some principles or guidelines as a preliminary definition of 'unconventional modes'. Following this, the next objectives must be to clarify precisely which modal types are included within this umbrella term, to define these individually, and to construct a meaningful classification which reflects their character and relationships. Enough has been said above to suggest the complexity of such a task, and this complexity is increasing all the time as new 'types' appear, transgressing previous boundaries. The main criteria are as follows.

(i) Definition of the UCM sector must be counterposed to what is 'conventional' under UK conditions. What is 'unconventional' in Britain may be 'normal' in other countries, and may have been influenced by other forms of organisation abroad, e.g. postbuses were encouraged by their existence in many states of continental Europe. Also, some types may resemble 'appropriate' transport technologies in the Third World, e.g. the sharing of taxis. Therefore, definition is dependent on identifiable departures from various aspects of conventional UK operating practice, together with some degree of innovation or 'newness'. Relevant divergences from the norm will be itemised below.

(ii) Modes adopting unconventional features are generally aimed at unprofitable environments. They are united by their orientation towards a common problem. That is, they are stimulated by the problem of costs persistently exceeding revenues, in circumstances where passenger demand is generally low, either in absolute terms as in the countryside, or due to the special needs of minority groups such as the infirm. Some cases abroad do not fit this criterion, such as urban commuter 'ridesharing' in the United States, which has different motivations (Transportation Research Board 1986). This criterion also has the advantage of excluding modes which appear 'unconventional' because they are the first to adopt new technology, such as urban rapid transit systems, which belong to a quite different category!

(iii) The *modus operandi* is more crucial than vehicle type or ownership. Generally, it will be the operating characteristics which constitute the main attempt to cater for low and scattered demands, given that the indivisibility of the single vehicle is a common part of the problem. New vehicle designs, or use of unfamiliar vehicle types, is usually secondary. The main exception is where the only impediment to the use of conventional transport by certain people is design standards which are unsympathetic to those with physical handicaps.

(iv) They must be available to the general public. As a putative solution to problems of uneconomic, scarce or non-existent public transport, UCMs must be available to all persons likely to experience such problems. Ideally, they should be open to everyone without the use of a car, and not discriminating on grounds of 'essential trips' or 'urgent cases'. In practice this is extremely problematic, in that some UCM types owe their *raison d'etre* to deliberate targetting of groups with special needs, or were designed on the assumption that only essential cases could be handled. There is no clear dividing line, e.g. some social car schemes have fairly general eligibility criteria while others are very restrictive. Most, but not all, of the CT sector is specifically aimed at the 'special needs' population.

This last point indicates the need to retain flexibility by acknowledging the diverse views on the range and types of modes to be included.

There is no intention to be dogmatic, or to impose hard-and-fast rules on what is and what is not 'unconventional'. It has to be remembered that other labels are used, such as 'informal', 'third level' or 'supplementary' transport, which may have slightly different connotations.

The relationship between the UCM and CT sectors was illustrated on figure 1.1, which suggested that the voluntary sector alone is not intrinsically 'unconventional'. It may be thought so either if it is operating services of an innovatory nature, or if it is attempting to operate transport for the general public. On the other hand, the voluntary sector fulfilling familiar welfare or charitable roles and offering transport services of a type traditionally associated with those roles, would be 'conventional'. This still leaves much scope for argument. Figure 2.1 complements figure 1.1 by articulating the passenger eligibility dimension. This defines UCMs and part of the CT sector by their availability for the general public; the latter case would cover community buses, some social cars and various lift-giving schemes. These aside, the great majority of CT services are dedicated exclusively to groups with special transport needs such as the elderly and disabled, and may therefore appear to be more closely associated with statutory welfare transport provided by health authorities or social services departments. Alternative terminology also confuses the issue, as the welfare/community sector is sometimes referred to as 'special', 'accessible' or 'needs-based' transport, encompassing both statutory and voluntary provision. However, the more recent expansion of CT into the provision of services for the general mobility (i.e. everyday purposes) of its clients means that it can claim to be playing the role of 'public transport' for people who cannot use a conventional bus. Again, rigid definitions would not be appropriate.

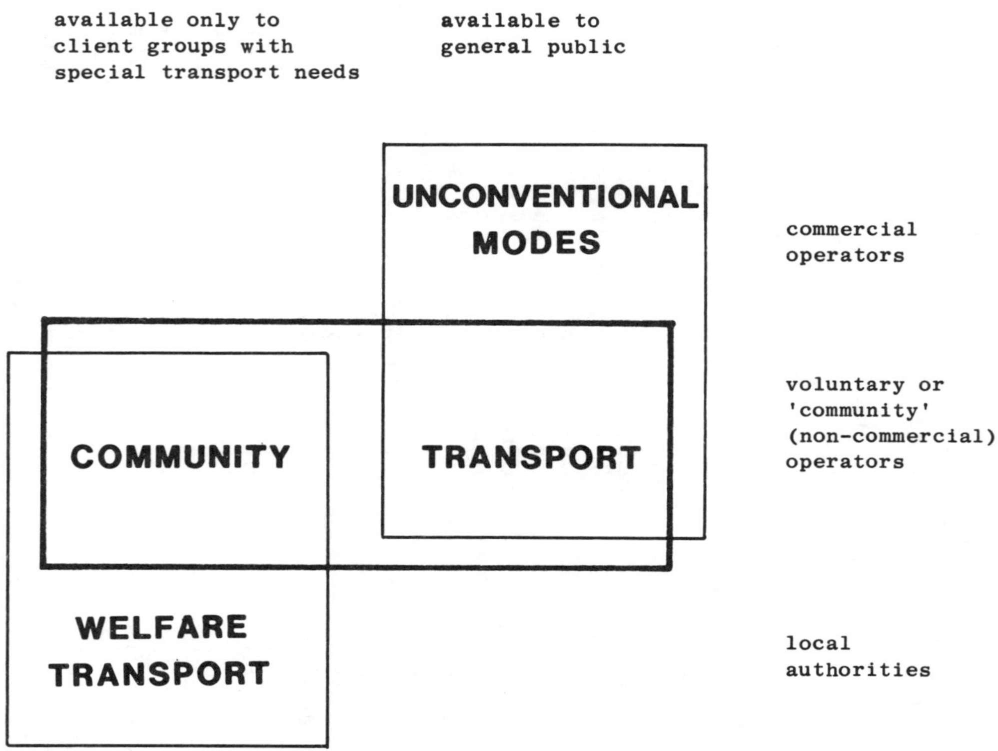

FIGURE 2.1 Unconventional Modes and related sectors.

2.2 TOWARDS A CLASSIFICATION

Having already counselled flexibility in defining the overall field, it follows that there is considerable room for argument over which modes should be included. It would be consistent with the above to consider all modes regarded in some quarters as 'unconventional', and to discuss their diagnostic features in relation to a sensible classificatory framework. The importance of classification at the present time is due to the accumulating impression that recent expansion of UCM numbers and broadening diversity of types have produced a situation of forbidding complexity. This urgently needs to be clarified. Renewed emphasis on definition and classification is also necessary for other reasons.

(i) The lack of any official or legal definition. The uncoordinated and/or informal origins of UCMs have prevented any official definition. Possibly the only mode to have a clear-cut *legal* definition is the community bus (originally in the 1978 Transport Act), and even this was applied *post facto*.

(ii) The inadequacy of existing classifications. In the current literature the usual approach is simply to list each mode with a separate description, but such lists tend to be incomplete or unstructured (e.g. Moseley 1979, Banister 1985). There have been few attempts at systematic classification of the whole field, and these are unclear or over-simplified (e.g. in Banister 1985, Sutton 1985). A complex classification of American paratransit appears in Kirby et al (1975) but is not applicable to the UK. The attempt by the European Conference of Ministers of Transport (Bovy & Krayenbuhl 1979) shows strong American influence, albeit with some French terminology. A list of experimental services produced by the OECD (Organisation for Economic Cooperation & Development 1979) has only a crude breakdown by type.

(iii) The excessive fragmentation of existing literature on the topic. A review of sources of information is given below (section 2.4) but, although ostensibly very wide, this literature is dominated by detailed descriptions of individual services which tend to obscure the overall view. Classification provides a framework to which this kind of work can be related.

(iv) The inadequacy of data sources. Following a classification, one can then attempt to discover the number of services in each modal category, and their distribution around the country. Unfortunately, very few UCMs have their own separate licensing procedure and there is no official listing or database which might give numbers in each category (NB. after deregulation, the TRRL's Bus Registration Index partly fulfils this role). Data from a miscellany of sources (section 2.4) is easier to interpret with reference to a classification scheme.

(v) The confusion of terminology. British nomenclature is quite different from American paratransit usage, although the latter appears to be more internationally acceptable. Terminology is inconsistent, vague, and shows duplication (section 2.3). The scheme below will attempt to achieve rationalisation and consistency.

(vi) Finally, type differentiation is also essential for planning or developmental purposes, to establish the potential of each mode to satisfy local travel needs, to clarify relative economic performance and cost/benefit effectiveness relative to each other and to conventional transport.

As explained in the previous section, the 'unconventional' sector is defined primarily by the adoption of features which diverge significantly from accustomed operational practice in the UK. Consequently, the classification of individual modes depends on identifiable departures from this 'conventional' norm. Such an approach is in line with UCM origins in that contemporary problems were seen as stemming from the operational rigidity of the traditional bus or train, and its unsuitability for the dispersed demand in rural areas and for the handicapped.

At the risk of labouring the obvious, it might be convenient to state the attributes of conventional operation in the UK. It is available to the general public, i.e. there are no eligibility restrictions, although some elderly people, the handicapped, and young children may experience difficulties in using it. It is available to individuals making separate journeys at separate fares, and is quite distinct from group travel between the same origin and destination on a private hire or excursion basis. There is no normal requirement for pre-booking.

Perhaps the most conspicuous feature of conventional transport that appears rigid, or at least unsuited to certain social or physical environments, is the maintenance of a network of fixed routes operated according to a fixed timetable. Such inflexibility is strongly at variance with conditions where demand is dispersed in space and time; this is the case in rural areas, but might also apply to minority groups in cities. Registration of routes and timetables with an application for a road service licence, under the licensing system prevailing up to the 1985 Transport Act, did not make it easy for operators to vary their routes and schedules in response to demand. Similarly, boarding and alighting points are normally fixed, especially at termini and within towns, although in many rural areas drivers have traditionally picked up and dropped off passengers at any point along the route.

Also 'conventional' is the exclusive allocation of a vehicle to one specific mode of operation or type of business, the most common being stage carriage service. Exceptions to this are the practice of interchanging passenger service vehicles amongst stage carriage, excursions and private hire work, and the use of PSVs for school transport. However, there is much further scope for dual- or multi-purpose operation such that different traffics can be handled by the same vehicle, either simultaneously or consecutively.

It is normal for passenger services to be

provided by a public or private sector bus company, with a duty to operate commercially or as efficiently as possible even if they do not actually succeed in covering their costs. Vehicles are normally owned by the companies which run them, and driven by professionally trained employees of these companies. One-man-operation can now be regarded as 'conventional'. Finance is normally obtained from a combination of fares, excursions and hires revenue, and if applicable, also by internal cross-subsidy from more profitable routes or business sectors, or by external subsidy from the local authority.

Finally, the vehicles themselves are of a 'conventional' design believed to be best suited to meet the above requirements - i.e. the familiar 45-50 seat single-deck bus or coach plus 70-80 seat double-deckers in urban areas. Unconventional modes of operation frequently demand new designs or technologies to accommodate them, while 'technology' itself may be a direct means of reducing costs on orthodox services. Rather problematic is the case of smaller passenger vehicles including minibuses (9-16 seats) and midibuses or small buses (17-35 seats). These can hardly be regarded as new technology, and their use on stage services was traditionally very rare until a recent revival. This is a good example of the earlier point that 'unconventional' status depends on mode of operation rather than vehicle type.

Attempts to mitigate the operational rigidity of conventional transport followed three main lines of enquiry:

(i) reduce costs, such as by the use of volunteer labour; and/or

(ii) increase the supply of vehicles, such as by using those not normally available to the general public; and/or

(iii) match supply more closely to demand, such as by operating on a demand-responsive basis.

More specifically, one can identify the following unconventional features which all constitute significant departures from established practice.

(a) Circumscribing the eligibility of passengers and/or destinations.
(b) Adopting flexible routeing and timing.
(c) Rationalising different traffics by various types of multi-purpose operation.
(d) Encouraging 'community' management and/or operation and the use of private vehicles.
(e) Obtaining finance from alternative sources.
(f) Applying new technology.

Simplicity is confounded by the inevitable fact that most UCMs qualify with more than one of these features. Nevertheless, this is the basis for an initial classification of types, which is given below (section 2.5).

2.3 THE CONFUSION OF TERMINOLOGY

The uncoordinated nature of UCM development in Britain is reflected in the idiosyncratic nomenclature adopted by local authorities and operators for the various modal types. This owes little to international experience, and differs significantly from American-influenced 'paratransit' terminology. In Europe, Bovy & Krayenbuhl (1979) report on the difficulty of finding suitable terms in French and German, and regard the American as international standard. "Transports semi-collectifs", as a translation of "paratransit", does not seem appropriate for it implies merely an intermediate field between public and individual transport. In the UK alone, this confusion means that different names are often applied to the same mode, or conversely that the same name is given to two or more modes exhibiting significant differences.

For example, local schemes to enable private cars to carry other persons with urgent travel needs are variously described as 'social car', 'voluntary car', 'essential car', 'country car', etc. The most widely accepted generic term is 'social car', which will be used below for all such schemes. Terms like ''car sharing' and 'lift giving' are too vague: in some cases they are social car schemes, in others they are different.

These terms could refer to any car carrying more than one person, such as a family outing!

Similarly, the common term 'dial-a-ride' merely implies some form of telephone communication to obtain transport, and it could therefore apply to various modes including ambulances and car schemes. A limited improvement is to change it to 'dial-a-bus'. Three different modes have been described as 'dial-a-ride' - initially, public service vehicles operating on a demand-responsive basis with the use of 'hi-tech' communications equipment; second, pre-bookable door-to-door transport for people with physical handicaps; third, less commonly, an otherwise conventional PSV with demand-responsive diversions from a fixed route. These need to be carefully distinguished.

Despite the legal definition, the term 'community bus' remains imprecise, being applied to a range of local bus projects with varying degrees of voluntary labour input. Moreover, any bus might claim to be 'serving the community'! Strictly, community representatives should at least operate the bus, and ideally, manage it too. However, the term is sometimes used where the community merely advises a commercial bus company, or where the community has part-time use of the vehicle. Most confusion surrounds the label 'community transport'. It was mentioned above how the traditional CT sector was overlooked, in favour of volunteer-run rural schemes such as community buses and social cars, as was exposed by the Department of Transport's (1978) self-help manual called *A Guide to Community Transport*. A 'community bus' could also be a volunteer-run dial-a-ride service for the handicapped, or simply a minibus available for hire by community groups. In a TRRL survey on the use of minibuses, Walmsley (1985) found it necessary to advise that 'community bus' should not be confused with 'community transport'! In the text below, a community bus is a rural scheme, operated and managed by local volunteers, and available to the general public.

The opening up of the private sector after deregulation has produced a plethora of new service names. A 'taxibus' could be either a bus or a taxi or a minibus service run by a taxi firm. Most new names are merely the product of imaginative marketing, so that 'market bus', 'village bus', 'shoppers bus', etc, are definitely *not* new modes.

The classification scheme in section 2.5 below will endeavour to explain terms in current use in relation to the unconventional features of the modes concerned, and where necessary, to achieve a more consistent terminology.

2.4 SOURCES OF INFORMATION

Available information on UCMs is as wide and diverse as the modes themselves. The main problem is that the vast majority of the literature is devoted to detailed descriptions of individual 'schemes' in specific areas, and is therefore very localized almost to the point of parochialism. Although the sum total of written work is very large, it is fragmentary and scattered among many sources, such as local authority, planning and consultants' reports, conference papers, academic articles, transport enthusiasts' journals, etc. Lengthy bibliographies can be found in Moseley et al (1977, vol 2), White (1978), Bovy & Krayenbuhl (1979) and Banister (1985), although UCMs are usually included within 'rural transport'. Rather strangely, there is no general overview text on UCMs in Britain, the nearest approach being the TRRL's deliberations over RUTEX (Transport & Road Research Laboratory 1977a,b, 1980, Balcombe 1979). Short reviews have been attempted by Nutley (1988a,b). On the CT sector, the most comprehensive review of 'special' transport, statutory and voluntary, for the elderly and disabled is Bailey & Layzell (1983). Shorter reviews are Bailey & Appleby (1986) and Sutton (1987).

International comparisons are a complex and perilous undertaking. Information on the United

States is easiest to find: the most comprehensive still seems to be Kirby et al (1975), together with collections of papers such as Orski (1975) and proceedings of the Transportation Research Board (1980, 1985, 1986). A comprehensive international review would be an impossible task. Bendixson's (1977) popular, journalistic approach reflects the optimism of the early years. The Organisation for Economic Cooperation & Development (1979) lists experimental services in member states. The Economic Research Centre (1979) of the ECMT discusses many European and some Third World examples within an American framework. Rimmer (1980) also tries to bridge the gap between Western countries and the Third World, but is too brief and selective.

Due to this diversity of sources, the effort of collating the large number of items produces an assemblage which is very incomplete in its coverage of relevant modes throughout the UK. It has already been pointed out that with very few exceptions there is no separate licensing procedure for each mode, and therefore no official list of numbers in each category. Holders of community bus licences may be identified from Traffic Commissioners' records. Since deregulation, bus services run by taxi operators require a special restricted operator's licence which would identify these. The latter, together with PSV minibuses, can be extracted from the TRRL's Bus Registration Index. Recent UCM schemes assisted through the Rural Transport Innovation Grant are listed by the fund administrators. This still leaves a very large number of UCMs originating up to 1985. The only other systematic source for these is the Scottish Postbus Timetable, which contains most of the country's postbuses. Nevertheless, these sources still omit the great majority of UCMs, especially the less formal ones. It is impossible therefore to answer the apparently simple question: how many UCMs exist in the UK?

In the CT sector, attempting to compile a comprehensive list of operators, service types and locations has so far proved impossible, and may remain so for some time. The sheer size and variety of the field make it difficult to decide on which types should be included. Very many services are so informal that they may be unknown outside their own local area. Only two or three authorities have been able to produce a complete list of 'elderly and disabled' transport in their areas. There are two attempts to compile national lists: Community Transport Services Ltd keeps a register of CT services known to it, currently standing at 1700; the Department of Transport is aiming to compile information on 2000. The national total is unknown, but may approach five times these figures.

In the circumstances, the nearest approach to a nationwide inventory of UCMs was thought to be the annual transport planning and policy documents produced by local authorities. In England and Wales, county coucils had been obliged under the 1978 Transport Act to produce Public Transport Plans (PTPs). Although this obligation was removed by the 1985 Act with local authorities no longer having coordination powers, a series of plans remains for the years 1979-86 as an extremely useful body of data at county level. In Scotland there are no PTPs, but equivalent information can be derived from regional councils' Transport Policies and Programmes (TPPs), especially from the 'comprehensive' versions produced every four years. The TPP is a roads and transport budgetary document required from all councils in Great Britain since 1974, but which in Scotland has expanded its public transport policy role in the absence of PTPs. Unfortunately there is no equivalent in Northern Ireland as local authorities in the province have no transport responsibilities.

A search of PTPs and TPPs can give a revealing insight into local authority attitudes and policies, the role of UCMs within these, and preferences for one mode over another. However, some councils have a much greater commitment to public transport planning and coordination than others, and consequently there can be very wide variation in the quality and information content of these documents. As far as UCMs are concerned, some councils like to draw attention to their activities in this field, and active projects may receive disproportionate publicity; others mention the existence of projects without giving enough operational details to classify them! Some informal schemes may be unknown to the council. Other problems with this source are that councils may not report schemes which are not in receipt of subsidy, and also that, having announced the establishment of a new service, subsequent annual plans fail to report whether it is continuing to operate successfully.

Such a survey inevitably has many gaps, some of which may be cross-checked against any of the other sources mentioned above. It is thought unlikely that the resulting data, however imperfect, could have been bettered by a questionnaire survey to local authorities. From this mixture of sources, the text below attempts to establish the range of UCM types currently existing in Britain, and the numbers and distribution of each.

2.5 CLASSIFICATION BY UNCONVENTIONAL FEATURES

Following the definition of unconventional features - departures from normal UK operating practice - that was given in section 2.2, UCMs are now described and classified under the six headings enumerated therein. It is also attempted to represent the classification in an appropriate tabular or graphical form, for ease of reference and comparison. It must be stressed that relationships are so complex that they cannot be expressed by any single diagram or table, but need several complementary figures to do them full justice.

The simplest approach is to enumerate all modes which appear to exhibit unconventional characteristics, and to identify their diagnostic features within the agreed classification. Accordingly, figure 2.2 lists eighteen modes. This list cannot of course be absolutely definitive due to uncertainties of definition, confusion over nomenclature, and type mutations.

MODE \ U/C FEATURES	Restricted eligibility	Flexible routeing & timing (or D/R)	Multi-purpose transport (vehicles or operations)	Community management and/or operation	Alternative finance	New design, technology
D/R diversion (Flexibus)		✓				
Multiple service bus		‡	✓			
Contract bus				‡	✓	
Subscription bus				✓	✓	
Free shoppers bus					✓	
School bus			✓			
Postbus			✓			
Community bus		✓	✓	✓	‡	
PSV dial-a-bus		✓				‡
'welfare' dial-a-bus	✓	✓		‡	‡	‡
Social car	‡	✓		✓	‡	
Hospital car	✓	✓		‡	‡	
Lift-giving scheme		✓		✓	‡	
Car pooling		✓		✓	‡	
Shared taxi/ hire-car		✓	‡			
Passenger/ freight service (Courier)		‡	✓			
Demountable vehicle			✓			‡
Railbus						✓

✓ Primary unconventional feature

‡ Secondary unconventional feature or important in some cases only

FIGURE 2.2 Classification of modes by their 'unconventional' features.

Disagreement is inevitable, with the possible exclusion of candidate modes believed by others to be 'unconventional', although it is thought less likely that any of the eighteen listed will be contested.

Modes tabulated will all be defined and briefly discussed in this section, while specific service details and operational data are left to chapters 3-13. Concise definitions of these modes are given in appendix 2. A few qualifications are in order at this stage. While most type labels may be familiar, one - the 'multiple service bus' - was invented by the author to embrace several existing services which did not seem to have a collective name (section 2.5(c)). Of the possible omissions from the list, the most contentious is likely to be the minibus used as a commercial PSV, which is regarded by many as a brilliant innovation. For the moment, suffice it to say that these do not qualify under the classification scheme, unless perhaps they are demand-responsive (section 2.5(g)). Despite this, their novelty and their importance earns them a place in chapter 3.

It is also intended to establish the numbers of services within each modal category that are currently in existence. For present purposes, it is convenient to take the year 1985 as a basis for comparison, so that one can appreciate the situation on the eve of deregulation. Relevant numbers are estimated under the appropriate heading below. The classification has also been discussed in Nutley (1988b), and the 1985 distribution of modes in Nutley (1988a) within a rural context.

(a) Passenger Eligibility

Ideally, one should insist that all UCMs must still be available to the general public without restriction. In practice, there are many situations where transport can only be provided at acceptable cost if the service frequency is kept to very low levels, and these can only be achieved by restricting the types of people eligible to use it. This is the case for most schemes which rely upon voluntary labour, especially in rural areas where demand is low and dispersed. Two restrictions seem reasonable in the circumstances: first, that aspiring passengers do not have the use of a car, and second, that their intended journey cannot be made by existing conventional public transport nor realistically by taxi.

Potential problems arise when restrictions go further than this. Transport provided by local voluntary groups with very weak funding is seen as a 'last resort' mode, and as such eligibility is often restricted to 'urgent' or 'essential' trips by persons with special needs. This is usually interpreted to mean the disabled, infirm or elderly seeking access to medical or welfare facilities. Some services provide transport only to specific doctors' surgeries or hospitals. Certain UCMs can therefore be classified by passenger eligibility and their permitted destinations, as shown on figure 2.3 with reference to car-based schemes.

A social car scheme (as explained above, other names are used) is one which relies on car owners using their own private vehicles to convey other members of the community to destinations which they could not reach by any other means. A local organising committee, often drawn from an existing community group, assembles a pool of volunteer drivers and arranges suitably timed journeys in response to travel requests from eligible persons. Drivers' costs are reimbursed through the committee, and the whole scheme is usually financed by the county council. Social car services emerged in the mid-1970s - early innovators included Lincolnshire, Shropshire, Cumbria, Clwyd and Dyfed (Rhys & Buxton 1974, Clwyd 1978, Fearnside 1980). It is impossible to count numbers in this case, but most counties now have some schemes of various types.

There are many type variations, but the main issue here is eligibility. Most social cars are 'restricted' to some degree (figure 2.3). A 'hospital car' may be either a social car exclusively serving a hospital, or a volunteer service organised by the health authority for the same purpose. No clear dividing line can be drawn

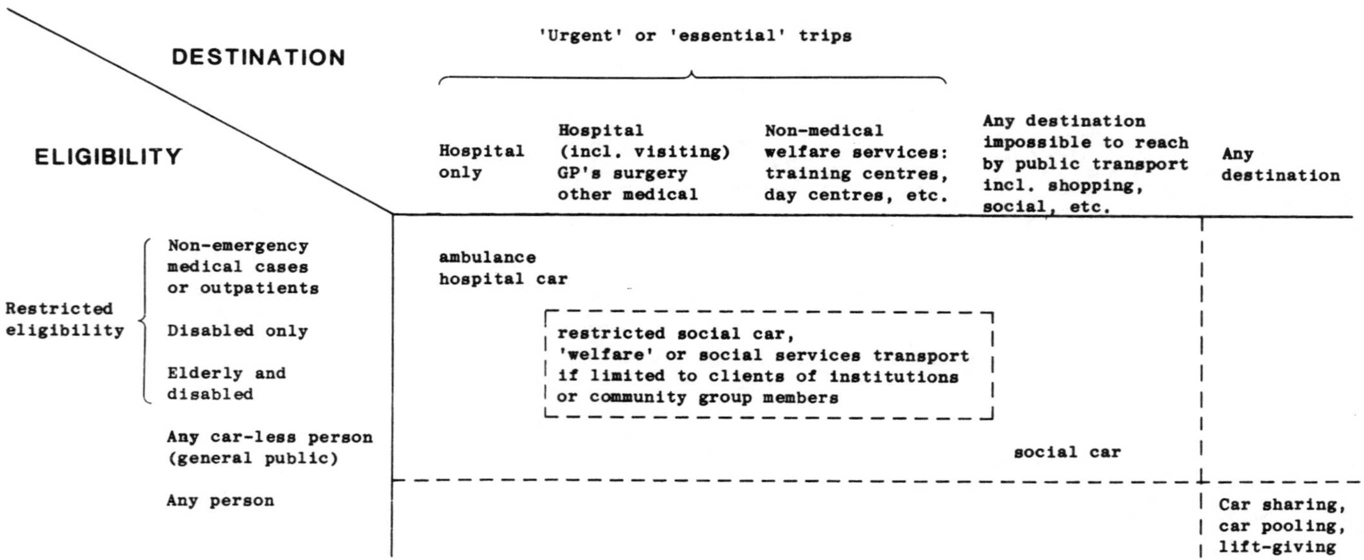

FIGURE 2.3 Restrictions on passengers and destinations: car modes.

between the definition of UCMs as supplementary services for the general public, and medical or welfare transport for the benefit of exclusive client groups. Many county PTPs agree on the definition of 'welfare transport' as that provided by specific institutions solely for their own clients or patients. This would include: ambulances and hospital cars provided by health authorities, 'social services transport' owned or hired by local authority social services departments to permit access to training centres, day centres, etc, and 'voluntary' or 'community' transport which backs up statutory provision by covering a more diverse range of medical, welfare and social needs, primarily for the elderly and disabled (Bailey 1979, Bailey & Layzell 1983). The latter group necessarily restricts passenger eligibility, and is not orientated towards the type of problem which originally stimulated the growth of UCMs. The range of voluntary car schemes depicted in figure 2.3 straddle the boundaries of the three sectors shown on figure 2.1.

Eligibility restrictions can lead to difficult problems, such as the possibility that in some areas with no conventional public transport, the elderly and disabled may have more transport opportunities than non-car owning able-bodied persons. Statutory health and welfare services are under great pressure to accept cases only on strict criteria of need, with 'non-urgent' trips being diverted to the CT sector; this may worsen transport problems in areas where CT modes do not exist. There is also the question of whether a social car coordinator would refuse a request for an 'inessential' trip such as shopping. It would seem that such arrangements do not pretend to sustain the mobility levels that are necessary to pursue a normal lifestyle.

There are however other ways of using privately owned vehicles (cars and minibuses) which do not impose such restrictions on passengers. These are dealt with under (d) below.

(b) Flexible Routeing and Timing

In environments without high densities of population adjacent to convenient corridor routes, conventional bus operation is likely to suffer from a frustrating inability to match indivisible supply with dispersed demand. Dispersal of passenger demand may apply in both geographical and temporal senses, and the problem can therefore be seen as one of time-space coordination, particularly relevant to rural areas but also to low density suburbs. Accordingly, the optimization of passengers and revenue relative to operating costs suggests the need for a demand-responsive (D/R), or 'demand-activated' mode. Numerous UCMs have been developed with the aim of achieving flexibility in routeing and/or timing, and may thus be classified according to space and time parameters (figure 2.4). Others have proposed a 2x2 classification, identifying modes with a variable route at fixed times and vice versa (Bovy & Krayenbuhl 1979, Banister 1985), but this is felt to be too simple. Some modes are loosely timetabled allowing some flexibility and diversions on demand, and thus the route is variable between the same start and finish points (figure 2.4). True D/R operation embraces any origin and any destination at times governed entirely by demand.

A minimum provision for a conventional bus to be more responsive to demand would be to authorise D/R diversions, e.g. from a major road to serve villages a few kilometres away. This would have implications for timetabling, and would require some form of pre-booking for passengers to board at an outlying point (Gallop 1975). An alternative approach is to operate over different routes on different days of the week, e.g. connecting different groups of villages with the same town. On figures 2.2 and 2.4 the term 'Flexibus' is suggested as a general label for such modes, the name deriving from a similar service inaugurated under RUTEX (North Yorkshire Rutex 1979).

The ideal D/R operation is 'many-to-many', not constrained by fixed routes or timetables, and able to respond immediately to telephone requests

TIME SPACE	Fixed timing	Flexible timing	Demand-responsive timing
Fixed origin /destination fixed route	Conventional bus railway ferry		Shuttle bus Shuttle ferry Optional ferry Shared taxi/hire-car Jitney (not in UK)
Fixed O/D variable route		D/R diversion (Flexibus) School bus Postbus Community bus Passenger/freight service (Courier) Shared taxi/hire-car Social car Many-to-one/few dial-a-bus	Shared taxi/hire-car Social car
Variable O/D, Variable route	Car sharing for regular trips Car pooling for journey to work		Taxi Hire-car Lift-giving scheme Social car Hospital car Many-to-many dial-a-bus (not in UK?)

FIGURE 2.4 Modes classified by time and space flexibility.

for transport. A theoretical 'dial-a-bus' system requires vehicles with radio control and an operations centre with computer facilities to optimize bus despatching and scheduling. This is a complex, technological solution which has found favour in North America but which, after a number of trials, has not persisted beyond the experimental phase in the UK. During the 1970s twelve 'dial-a-ride' bus systems were tried out in England, with much publicity and optimism. In common with American experience, only urban areas were found suitable for this type of operation. Service configurations varied among 'many-to-one', 'many-to-few' and diversions from a fixed route. However, all these projects had been discontinued by 1980, primarily due to high costs, and usually replaced by conventional buses (Watts 1976a,b, Oxley 1977, 1980). This contrasts with continuing successful operation of advanced systems in North America and other European countries.

With the failure of the 1970s operations, the present trend seems to be towards a convergence with the Flexibus-type workings described above. The result is a service partially demand-responsive, such as diversions from a fixed route, often with the use of minibuses.

Also as explained above, the term 'dial-a-ride' has undergone a shift in meaning, and is now applied to door-to-door CT services for the elderly and handicapped, using small buses with adaptations such as wheelchair lifts. While these claim to offer a 'many-to-many' service, this is only possible if passengers give sufficient advance notice. The two types of dial-a-ride are sometimes distinguished as '1970s' and '1980s' versions (Bailey 1984), but on figure 2.2 they are labelled 'PSV dial-a-bus' and 'welfare dial-a-bus' respectively.

Apart from the private car, the most flexible of existing modes are the taxi, which can ply for hire, and the hire-car, which must be pre-booked. Dating back to the horse-cab era, the taxi is certainly 'conventional'. The impression has been gaining ground in official circles that taxis and hire-cars constitute an under-exploited transport resource, in view of their D/R operation and the subjective advantages of small vehicles. The rationale for their unconventional use is rather different. This is based on the desire to make them available to a wider market, which means reducing cost to the passenger. On a commercial basis, this can only be done by allowing people to share at separate fares. But taxi sharing was illegal in Britain until implementation of the 1985 Transport Act.

In Northern Ireland, however, taxi sharing started earlier and is now widespread, although there are continuing doubts about its legality as the 1985 Act does not apply. The earliest and by far the largest taxi sharing operation in the UK is that of the 'Black Taxis' of Belfast. These run fixed radial routes into the city centre whenever there is a full load of passengers (Lavery 1986a,b). Despite official disfavour, they must be regarded as a success due to high passenger demand, and are attracting renewed attention for possible application elsewhere in the UK.

In Great Britain shared taxi and hire-car schemes, requiring special legal dispensation before 1986, are extremely diverse in location, character and success rate. Sharing requires at least a semi-fixed route, plus some advertising; some services are loosely timetabled while others are more responsive to demand. The 'Taxibus' service in Stourport, for example, runs on a fixed route at any time according to demand, with fares dependent on numbers of people sharing (Coe 1985). This and the Belfast operations seem to be the UK's nearest approach to the jitney, common in other parts of the world. An alternative mode of operation, as in Dent (Cumbria), is to advertise the number of spare seats on journeys booked in advance by other persons. Town services seem more likely to succeed: shared hire-cars introduced under RUTEX were not generally successful, and many rural schemes have expired since. Excluding Northern Ireland, by 1985 there were about 20-30 schemes in existence, but they have expanded greatly since legalization (Finch 1988). The irony here is that 'unconventional' use of taxis means *reducing* the flexibility of traditional D/R operation in order to run 'semi-stage' services.

The social car, described above, is a typical D/R mode (although a few are timetabled). For the sake of completion, figure 2.4 includes buses run on a 'shuttle' basis such as may be found between city centre railway stations, for access to airports and to some hotels. They may wait to fill with passengers before departing, perhaps subject to a maximum time interval, and could thus be described as D/R. Some ferries operate in a similar way, at least during times of peak demand. There is also the case, common in the Scottish Highlands, where small ferries have a notional timetable to cross estuaries or to islands, but do not depart unless an intending passenger telephones beforehand. These are described as 'optional ferries'.

(c) Multi-Purpose Operation

Dual-purpose or multi-purpose transport aims to secure more efficient utilisation of resources by arranging for any vehicle and its driver to handle more than one type of business or client group. There are two ways of doing this. First, the two (rarely more) types of traffic may be handled by a single vehicle simultaneously, with the mode of operation designed to suit both as far as possible. Second, a single vehicle may convey different traffics consecutively, using suitably diverse modes of operation at different times of day or days of the week. For want of a better term, the latter is described below, as on figure 2.2, as a 'multiple service bus'.

A persistent issue has been how best to combine schools traffic with fare-paying adult passengers. There are three ways by which local education authorities can meet their statutory obligations to transport eligible pupils to school: (i) by paying a local bus operator to carry them on existing stage services; (ii) by contracting an operator to provide a bus exclusively for school journeys; (iii) by exclusive use of the LEA's own vehicles. The first option constitutes dual-purpose transport in that pupils share with ordinary passengers, and has been normal practice since the 1944 Education Act. With the other two modes, sharing is possible either by obtaining a permit under the 1968 Transport Act or by licensing as a public service bus, and it is these which are regarded as 'unconventional'. Alternatively, both traffics could be provided for on a 'consecutive' basis by the 'extended school contract', in which school contract buses are also used for public services in the off-peak at marginal cost. It is probably true to say that all local authorities have made considerable efforts to integrate school and public services by various combinations of the above modes (Awdas 1978). School buses may therefore have a claim to be the most widespread of UCMs, if they can still be regarded as such.

The best known dual-purpose mode is the postbus, which is arguably the most conspicuous and popular of all UCMs. Their origins in Britain were as direct replacements of pre-existing mailvan collection and delivery rounds, with new vehicles to accommodate passengers: small adapted vans are the most common, but Land Rovers and cars are used in remoter areas. British postbuses therefore differ substantially from those in continental Europe (e.g. Holding 1983 on Switzerland) in that the latter have been historically the main providers of rural road passenger services, and there is full integration, in some cases without the need for door-to-door mail delivery. As recounted in chapter 1, numbers of services expanded dramatically in the 1970s, and by 1985 there were 173 in Britain, of which 131 were in Scotland (Turnock 1977, Watts et al 1978a,b).

There has in fact been little expansion in the 1980s for institutional and financial reasons, the difficulty of finding suitable routes, and the loss of new bus grant (Watts 1984). Their success in Scotland is reflected in their acceptance by regional councils as permanent fixtures and no longer 'unconventional'. On the other hand, a number of authorities, especially in England, are reappraising their postbuses and some have recently been withdrawn. The overall UK total is generally stable.

In view of the tremendous number of public and private small goods vehicles in use, there should be some potential in the combination of freight carriage with passenger services. Despite some traditional carriage of parcels on bus services in remoter areas, this is an underdeveloped field. The only significant project of this sort is the Border Courier system in the Borders Region of Scotland. Starting in 1979, this uses multi-purpose vehicles whose main job is to convey medical supplies between doctors' surgeries and the regional hospital, but will also carry sundry packages for various council departments, as well as passengers; it is partly demand-responsive (Borders 1981). The Border Courier is still unique in the UK, although other types of freight-carrying buses are being tried out in Strathclyde Region.

A further example of dual-purpose transport is the conveyance of local authority social services clients on the same buses as commercial passengers, thus achieving a rationalisation of specialized council vehicles and also of bus subsidies. This is done under the East Sussex Coordinated Rural Transport (ESCORT) project. The vehicles also take schoolchildren, have a wheelchair lift, and adopt D/R flexible routeing (Watts 1985). Here too, there seems to be scope for replication elsewhere, although local authorities report legal obstructions to the use of social services department vehicles for off-peak public service.

Multi-purpose transport may also be implemented in a 'consecutive' manner. This may be done where a single vehicle is scheduled in such a way that it operates (for example) a commuter or railway feeder service on weekday mornings, market day shopping services to distant towns, an off-peak doctor's surgery link on alternate days, and if suitably equipped special services for the disabled to day centres or leisure centres. With such versatility, a 'multiple service bus' should certainly be an efficient use of resources. Examples are the 'St Ives midibus' which formerly operated in Cambridgeshire and Northamptonshire, and the so-called 'Bassetlaw community bus' in Nottinghamshire (Cambridgeshire 1977, Childs 1978). The latter is a case of misplaced terminology!

(d) 'Community' Operation and Private Vehicles

Community involvement and the use of volunteer labour in public transport is encouraged primarily to reduce costs, in view of the oft-quoted figure of 70% bus costs being due to drivers' wages. Of equal importance is the aim to expand transport resources in terms of vehicles and personnel, in line with the 'self-help' movement. As explained above, the term 'community transport' could apply to any scheme with volunteer drivers and/or organisers, but those available to the general public are community buses, many social cars and various lift-giving arrangements. It should be reiterated that the vast majority of CT schemes are reserved for the elderly and handicapped. These include: minibuses (perhaps with customized coachwork and special fittings) owned by local authorities or associations and available for hire by community groups, dial-a-ride vehicles (again with special fittings), ambulances, lorries for furniture removals, 'restricted' social cars (see above), and private cars run by charitable individuals. The ultimate UK total for this disparate collection is unknown. Very few authorities have attempted to produce comprehensive lists for their areas, but projecting these on a *per caput* basis gives an estimate of between 6000 and 10000 for the whole country.

Probably the most important element of the community bus ideal is that overall management responsibility lies with the local organising committee. The 'classic' community bus follows the model established by the first project, in North Norfolk from 1975, with the vehicle provided by a commercial bus company, volunteer drivers trained by them, and the committee scheduling a combination of stage services, D/R operations, excursions and hires, and the whole project being underwritten by the county council (Orriss 1976, Harris 1978). Other early innovations were in East Sussex, Clwyd, Northamptonshire and Devon

(Banister 1982, Southdown 1977, Clwyd 1978, Devon Rutex 1978). On a fairly strict definition, one could identify 39 community bus projects in Britain by 1985; a wider definition (see below) would extend this to over 50. Two years later, the Rural Transport Innovation Grant had stimulated the arrival of another fifteen.

These figures include many variations on the original model: some schemes are very much smaller, and many now reduce initial costs by obtaining their vehicle from another source. A notable case is Grampian Region, where the council makes over its own school minibuses to local voluntary groups who operate stage services in the off-peak; ten such routes were in existence (Grampian 1985). These could be regarded as part-time community buses, or 'consecutive' dual-purpose transport. As mentioned under the 'terminology' heading, there are many dubious claims to the 'community bus' identity, ranging from minibuses for group hire to professionally operated services like the Bassetlaw scheme.

Community participation can also be activated by making existing private vehicles available to other persons, thus increasing public transport resources and achieving more efficient vehicle use.

Social car schemes are obviously relevant here, but there is also a number of related concepts with potentially confusing terminology. Useful definitions have been provided by Green (1978). Informal lift-giving has been a sociable tradition for generations, especially in the countryside, but without any organization it cannot be called a 'scheme' as such. A 'lift-giving scheme' implies at least some mutual advertising between drivers and aspiring passengers, and payment has been legal since 1978. 'Car sharing' is often used as a generic term for all car-based arrangements, but more specifically means giving lifts on regular trips, which would normally involve some payment by the passenger. 'Car pooling' is an agreement among car owners to share one person's car, in turn, for regular trips to the same destination, especially for the journey to work. Terminology and reearch in this field is strongly influenced by American experience where car sharing has a different rationale - mainly to cut costs for urban commuter trips, and not necessarily to assist the car-less or the handicapped.

Similar attention has been paid to the potential of private minibuses being made available for wider use. Thirty-five percent of the country's 82,000 minibuses belong to the voluntary sector (Walmsley 1985), and it would be desirable if those owned by local groups or associations could be opened up to persons other than private passengers or club members. Minibus permits, introduced originally under the 1977 Minibus Act, were intended to help in this respect by allowing educational, religious and welfare bodies to run vehicles for hire or reward without public service licensing. Many 'welfare' CT schemes use this mechanism, but its possible application is much wider. About 12,000 permits have been issued, and permits are held by 35% of voluntary sector minibuses.

The wider use of cars and minibuses by such means is the most difficult of all modes to assess. Surveys have shown that car pooling is engaged in by 3% of car commuters (Vincent & Wood 1979), while 7% would consider using an organised car sharing system (Bonsall 1980). Urban work-related trips, as in the 'Yorkshare' experiment (Bonsall et al 1983), come closest to the American concept of 'ridesharing'. Van (minibus) pooling is extremely rare in Britain; legal and economic factors inhibit American-style usage (Greening & Jackson 1982). Equivalent commuter services in Britain mainly use vehicles owned or contracted by employers, and known as 'works buses'.

(e) Alternative Finance

This category refers to means of meeting transport costs from sources not normally invoked for this purpose. Volunteer-based welfare transport projects are able to call upon various local authority budgets not applicable to commercial buses. In addition, many are eligible to claim from central government funds such as the Urban Programme and schemes run by the Manpower Services Commission. Most CT operations include some element of local fund-raising.

Other public agencies would predictably be reluctant to consider financing bus services. Uniquely, there is a case in Surrey of a public service bus being operated by the health authority (normally only responsible for 'institutional' travel needs) on a route providing access to a hospital, where a conventional bus had been withdrawn. 'Peterbus', named after the hospital involved, is co-sponsored by the health authority and the county council (Surrey PTP 1984/5).

Bus services felt to be socially necessary may as a last resort be 'bought in' from a commercial operator by lower tier authorities such as district, town or parish councils out of their own meagre resources (these are the 'contract buses' referred to on figure 2.2). There is also the case of buses being paid for by local traders to bring consumers to a shopping centre from outlying areas, these being 'free' to the passenger. The most frequently quoted rural example is the Cromer market bus in Norfolk, dating from 1972 (Moseley et al 1977, vol 2). A search of PTPs reveals a number of others around the country, especially those serving large stores and hypermarkets such as Asda and Tesco, often in suburban locations.

Another variation on the financial theme is to form a club of potential bus users and to support the service from subscriptions revenue. Members then travel 'free' while others pay fares, and excursions may make a further contribution. The pioneer 'subscription bus' (or 'cooperative bus') was the Horncastle Bus Club in Lincolnshire, which commenced in 1976 and remained the only example in Britain. To add to the terminological confusion, this is sometimes called a community bus (Kaye 1986) due to the role of local organisers, although the vehicle is operated by an independent bus company. There is no apparent demand for urban commuter subscription bus services on the American model.

(f) New Technology
Within the diverse field of UCMs, the potential application of new technology is primarily to reduce costs, to derive a cheaper service more appropriate for traditionally unprofitable territories such as rural areas. However, the classification in figure 2.2 suggested that the technological factor, relative to other unconventional features, is of minor or secondary importance. There is little doubt that the operational factor, the institutional factor, and the legal factor have each been more significant in UCM development than technology alone.

As an aside, it is interesting to consider briefly the role of technology within the broad span of transport history, and to contrast this with its role in solving short term problems of the kind now under discussion. Texts on the economic history of transport predictably follow the 'cycles of technology' approach - turnpikes, canals, railways, electric tramways, motor buses, motor cars... As a gross generalisation, it could be said that successive leaps in technology were inspired by the need to improve performance, usually meaning increased speed or load capacity. Although incurring high costs in the short term, through successful adoption and diffusion these produced lower costs in the long term. In the context of contemporary problems to which UCMs are addressed, the urgent objective is to reduce costs *immediately*, even if performance is lower. This means implementation of a simpler technology, or even going back to an older technology. In this respect there is a resemblance to 'appropriate technology' of the Third World, and this may be a valid analogy if it is not taken too far (Nutley 1988c).

Some modes require technical fittings such as communications equipment for dial-a-rides or ramps and lifts for the conveyance of disabled persons, but these are secondary to other unconventional features such as D/R operation. New design is more common than technology, as seen in the modified vehicles required for postbuses, freight-carrying buses, and various minibus models.

The most distinctive technological innovation in the roads sector is the demountable vehicle. This concept has already been adopted by the haulage and construction industries, and a public transport version has more recently been developed by Strathclyde Passenger Transport Executive. The objective is to handle passenger and freight traffic by the use of substitutable bus and van bodies, either of which can be attached to a single chassis/cab unit. An experimental version has been built, popularly known as the 'swap-body bus', and has operated on the Isle of Arran since 1984 (Rowlands 1984). It could be pointed out, however, that this is really a dual-purpose mode, and the technology is secondary.

Finally, it is under this 'technology' heading that the railways have their only claim to a UCM. The previously intractable problem of uneconomic rural lines, suffering from high costs and obsolete equipment, and under persistent threat of closure, is being tackled by infrastructure renewal with low-cost technology. The need for lightweight rolling stock with low servicing costs and low track wear has resurrected the 'railbus' concept. Several experimental versions were based on a Leyland National bus body with a railway underframe (Hodge 1981). One of these was sold to Northern Ireland Railways, and has been in intermittent service on two branch lines. Although various versions underwent trials on problem rural lines, the production railbus which emerged in 1983 as the Class 141 was put into service only in urbanized West Yorkshire. From the viewpoint of British Rail management this is not an exclusively 'rural' problem, but is part of a broader policy to replace the ageing diesel multiple unit fleet on Provincial Sector routes, and thus a progression of vehicle types was developed. This 'classic' railbus has recently been upgraded to enhance passenger capacity, and later models have reverted to 'train' rather than 'bus' design (Ford 1986).

In this case the role of the UCM was a transitory stage of a cyclical development progression, from the old DMU through the railbus to modern light rail. The Class 141 is the only mode characterised exclusively by 'technology' as an unconventional feature, and moreover is a good example of 'appropriate' technology, as it employs one which is simplified and which harks back to an old idea.

(g) Others
The above sections have briefly described the eighteen modes within the terminology used in figure 2.2. Other claimants for the 'unconventional' label should be considered.

(i) **'Welfare' CT modes**
As shown on figures 1.1 and 2.1, these overlap with UCMs in a way which is virtually impossible to disentangle. Arguably, most of this sector could be excluded because (a) they are not, by definition, available to the general public, and (b) voluntary or charitable involvement in welfare transport is not new, but has a long history. On the other hand, the new 'welfare' dial-a-ride services certainly appear 'unconventional'. There is no intention to be dogmatic about this, and the whole CT sector will be discussed below.

(ii) **Urban PSV minibuses**
The other most significant candidate for inclusion in the list of UCMs is the use of minibuses on commercial stage carriage services. The first appeared in Glasgow in 1983, some on routes unsuitable for conventional buses, others on a flexible D/R basis. However, the first to assume an apparently 'conventional' role on high frequency urban routes were those introduced by the National Bus Company in Exeter in 1984, and were believed so successful that minibuses were rapidly implemented in other towns (Murrell & Newport 1985, Stonham 1985, Turner & White 1987). Despite many refinements in design, this is hardly new technology. They appear a novelty in that a common mode (the minibus) is used in an unaccustomed manner (as a high frequency public service) in unfamiliar environments (city centres). They would not normally qualify under the criteria (a)-(f) above, but may do so if they are run with a degree of demand-responsiveness.

Nevertheless, the 'minibus revolution' which has intensified in the wake of deregulation (Knowles et al 1988) is a major event and cannot be disregarded on narrow grounds of definition.

(iii) **PSV cars**
There is one example of the new-found preference for small vehicles among commercial operators going to the extreme of using cars. In rural Essex the 'Countrycar' service is run by the local bus company on a normal scheduled basis. This differs from all other car schemes in that it is neither community transport nor demand-responsive. On its potential UCM status, the arguments given with respect to minibuses would seem to apply here also.

(iv) **Recreational transport**
Transport provided for access to recreational or tourist sites may exhibit some unorthodox features (Moyes 1977, Halsall 1982) but are inevitably seasonal and not, by definition, orientated towards everyday travel needs. They are excluded.

A survey of county/regional PTPs and TPPs reveals variations in local authority perceptions of what is 'unconventional'. Many welcome new 'experimental services', by which is usually meant the trying out of new routes by ordinary buses. One must also be cautious of promotional campaigns which try to dress up conventional buses as an imaginative new product with attractive names and advertising: this is even more prevalent after deregulation. Sometimes associated with these is a 'hail stop' facility, by which passengers may be picked up and dropped off anywhere, away from regular bus stops. This has always been normal practice in the countryside, but is now sometimes permitted in the case of minibuses on town routes. Arguably, this is a D/R adaptation of a fixed scheduled route.

Local authorities often draw attention to infrastructural features which, although not 'modes', may facilitate a more imaginative and appropriate use of existing transport resources. Hence, some PTPs/TPPs applaud park-and-ride facilities in the same context as UCMs. Some authorities set much store on the appointment of 'local transport advisers', usually in villages, to liaise between community, bus companies and council over local service provision (e.g. Oxfordshire). A few others have appointed 'brokers' to coordinate the provision of public and voluntary sector transport in meeting trip demands from numerous agencies (e.g. Hampshire, East Sussex) (Davies 1987).

2.6 ALTERNATIVE VIEWPOINTS

The initial classification in figure 2.2 provides a framework for discussion by identifying deviations from normal practice which distinguish certain modes as 'unconventional'. No single tabulation, however, can do full justice to the complexity of relationships that exist in this field.

The six unconventional features highlighted in this figure and in the above section can perhaps be collapsed into four factors - the operational, institutional, legal and technological - which govern in more general terms the possibilities for transport development. Figure 2.5 has been drawn in such a way as to indicate the relative influence of these on the UCMs. The most important factors are seen to be the operational - the means by which the service is delivered - and the institutional - the agencies or organizations providing the service. Legal and technological factors are significant mainly in supportive roles - in allowing different operational and institutional arrangements, and in designing suitable equipment. Only permit minibuses, shared taxis, the demountable vehicle and the railbus owe their existence very largely to legal change and technology transfer.

Going further, it would be useful to place UCMs within the context of the overall public transport sector. Figure 2.6 is designed to show how conventional modes - both commercial and non-commercial - can be developed through the adoption of various unconventional features to

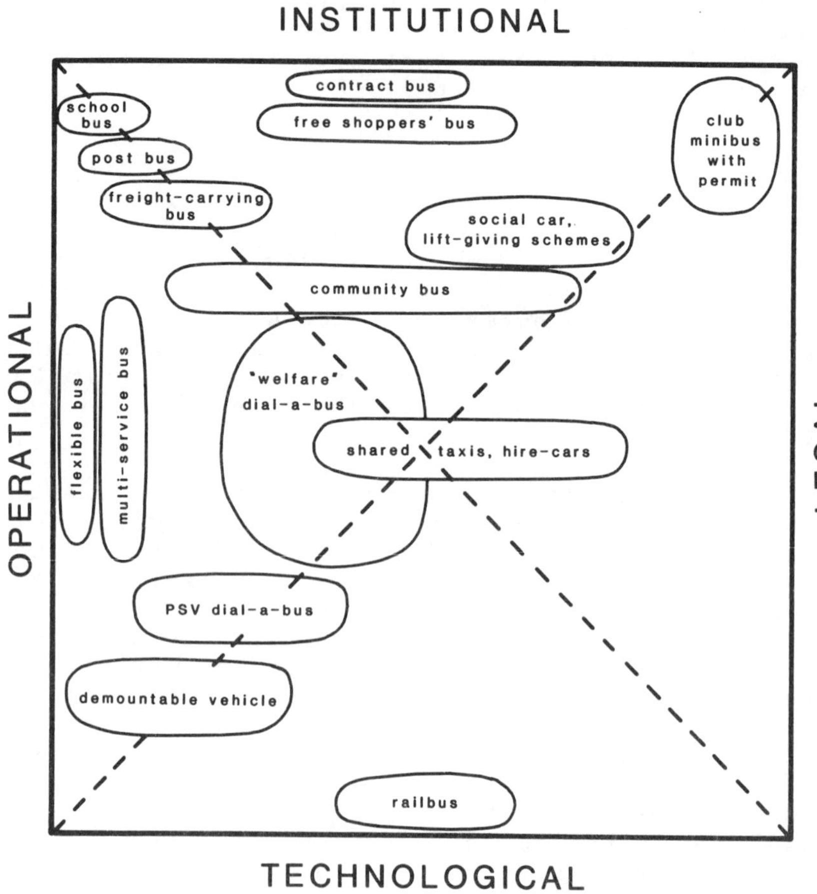

FIGURE 2.5 Relative importance of four basic dimensions.

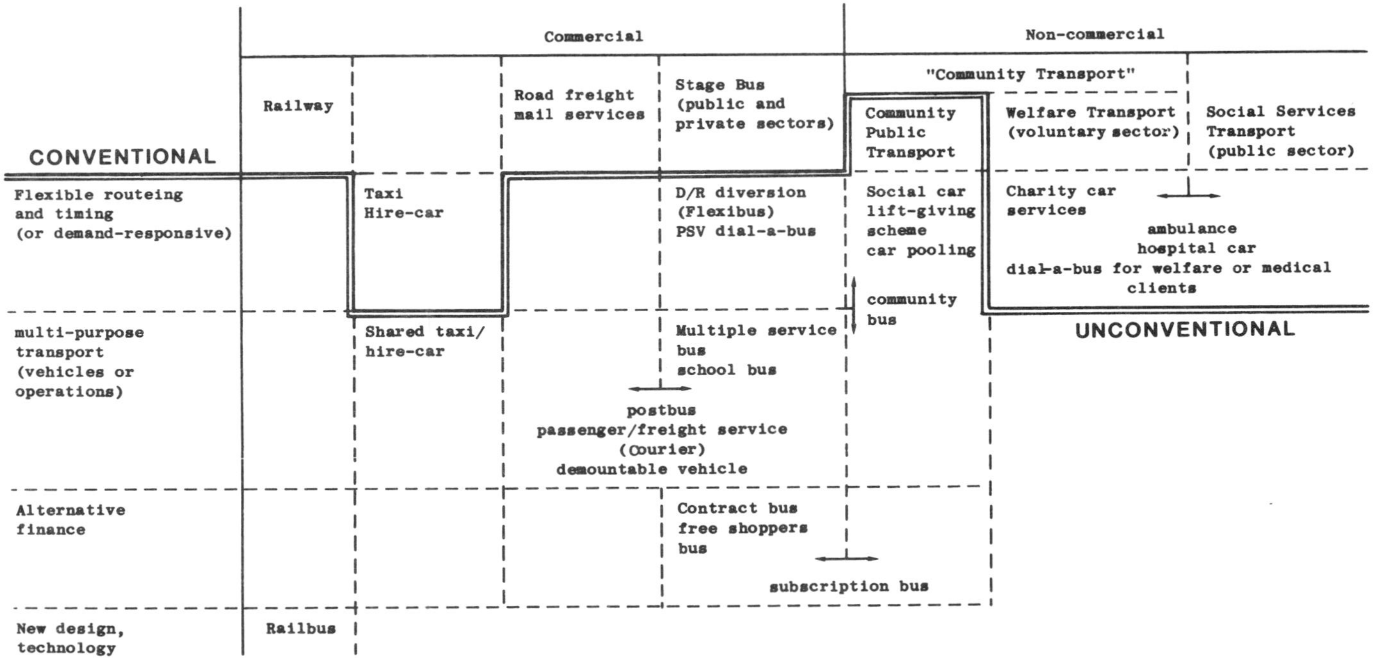

FIGURE 2.6 Conventional modes and ‘unconventional’ adaptations.

produce the set of UCMs already described. One problem with this approach is that it poses the question as to the point where 'conventional' gives way to 'unconventional', and the figure implies a rigid or arbitrary distinction. Note that in this case 'community' operations are shown not as an unconventional adaptation of the norm, but as an established non-commercial sector, within which only its attempts to run 'public transport' would be unorthodox.

An alternative approach is offered in figure 2.7, which is designed both as a taxonomy (complementing the classifications in figures 2.2, 2.5 and 2.6) and as a model of local transport resource development. Under a political/economic climate hostile to any increase in public expenditure, the two main options are (a) to make more apposite use of existing public service vehicles by exploiting their potential to adopt unconventional features, and (b) to increase the supply of available vehicles by using or adapting those not normally accustomed to conveying the general public. UCM examples are given under the appropriate heading (Nutley 1988b).

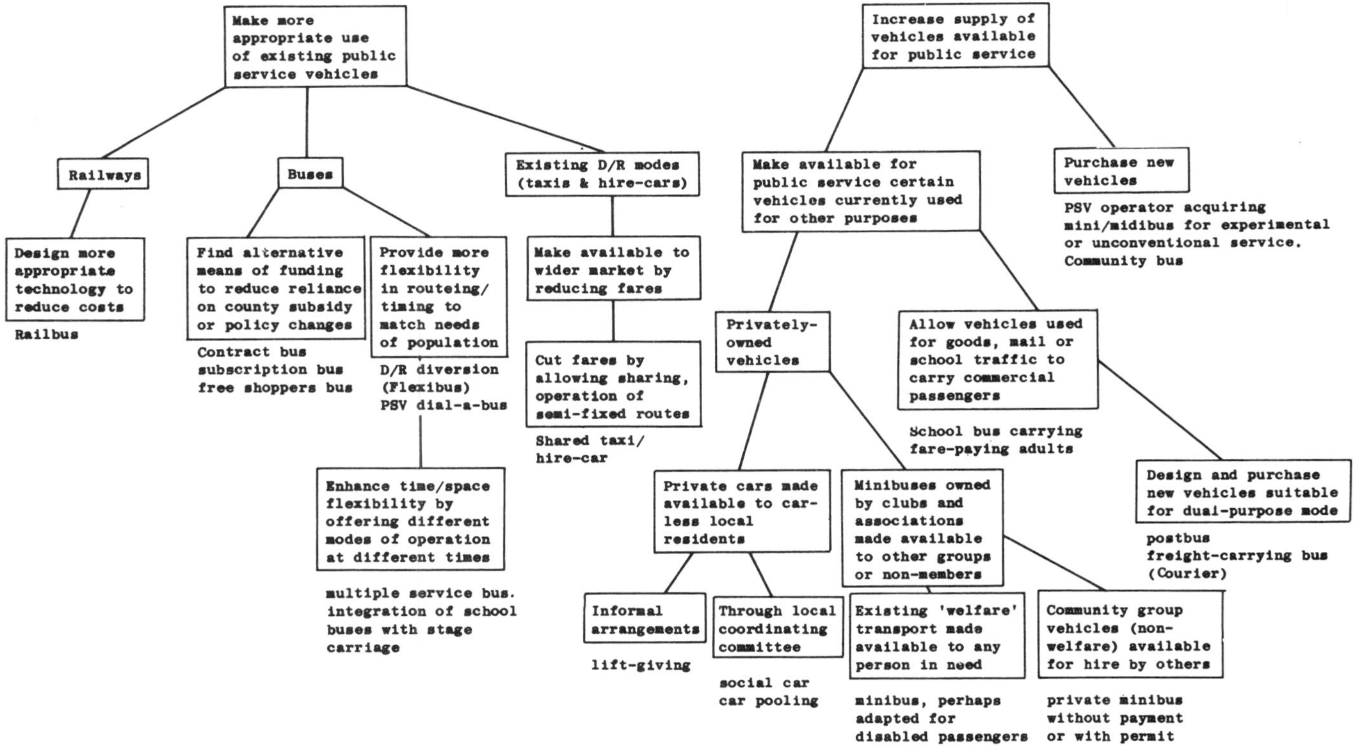

FIGURE 2.7 Transport resource development using unconventional modes.

Chapter 3

DIAL-A-RIDE IN THE 1970s

3.1 THE GENESIS OF DIAL-A-RIDE IN BRITAIN

(a) Concept

This mode refers to public service buses operating in a demand-responsive (D/R) manner with the aid of a significant 'technological' component. It was noted in chapter 2 that there is some ambiguity in the term 'dial-a-ride' in that, as it implies some form of telephone communication to summon transport, it might also apply to other modes, but in deference to accepted usage the term will be retained. Having said that, it should be reiterated that three distinct modes (classified differently in chapter 2) have variously been described as 'dial-a-ride'.

(i) A public service bus operating flexible routes and times on a demand-responsive basis, booked by telephone or other means, and controlled by 'hi-tech' communications equipment.
(ii) A minibus specially adapted for use by the elderly and handicapped, also operating on a D/R basis, most likely with a high input of volunteer labour (a CT mode).
(iii) A PSV with flexible routes or D/R diversions from a fixed route, but with no special technology.

Mode (i), the original form, is the concern here. Mode (ii) is dealt with under 'welfare' community transport in chapter 9, and mode (iii) is covered in chapter 4.

Dial-a-ride (DaR) is somewhat anomalous among the broad field of UCMs in that they were originally intended as commercial ventures (albeit experimental), were not targeted on any well-defined 'problem', were confined to urban areas, and most significantly, are the only UCM to be judged a failure, as all systems had expired by 1980. The following review is therefore largely of historic interest. Types (i) and (ii) above have sometimes been distinguished as 'DaRs of the 1970s' and 'DaRs of the 1980s' respectively (Bailey 1984, Sutton 1987).

Before describing the reality of British systems, some comment is needed on the nature of D/R operation. The basic types are represented in figure 3.1. The most idealistic form is 'many-to-many' (many origins to many destinations), at times determined by the consumer - complete flexibility in space and time, responding precisely to demand. In this way it resembles a taxi, but uses a larger vehicle and carries passengers at separate fares, thus aiming to reduce cost to the user. The immediate problem is reconciling true D/R operation with the need to rationalize the individual requirements of passengers travelling collectively. Inevitably, various forms of compromise emerge.

Attempts to run ideal 'many-to-many' services will encounter predictable problems. The first passenger is being transported from A to B. A second passenger requests, via the control centre, to be picked up at C and taken to D. A decision has to be made whether to divert the bus to C and then to take both passengers as efficiently as possible to destinations B and D; or to deliver the first passenger to B before moving on to C; or to despatch a second bus for the trip between C and D. Multiply this by the real-life demands of an urban area and the problems become formidable. Depending on the relative locations of trip origins and destinations demanded, distances, passenger journey times, vehicle running costs, and availability of buses, some complex calculations need to be made. Taking a devious route to pick up several passengers from different places and delivering them to different destinations might rationalize vehicle time and cost, but would also increase the journey time for each passenger, thus annulling the initial advantage of instant door-to-door service. An 'efficient' route for the bus would be an 'inefficient' route for each passenger, and vice versa.

Solving problems of this sort would require

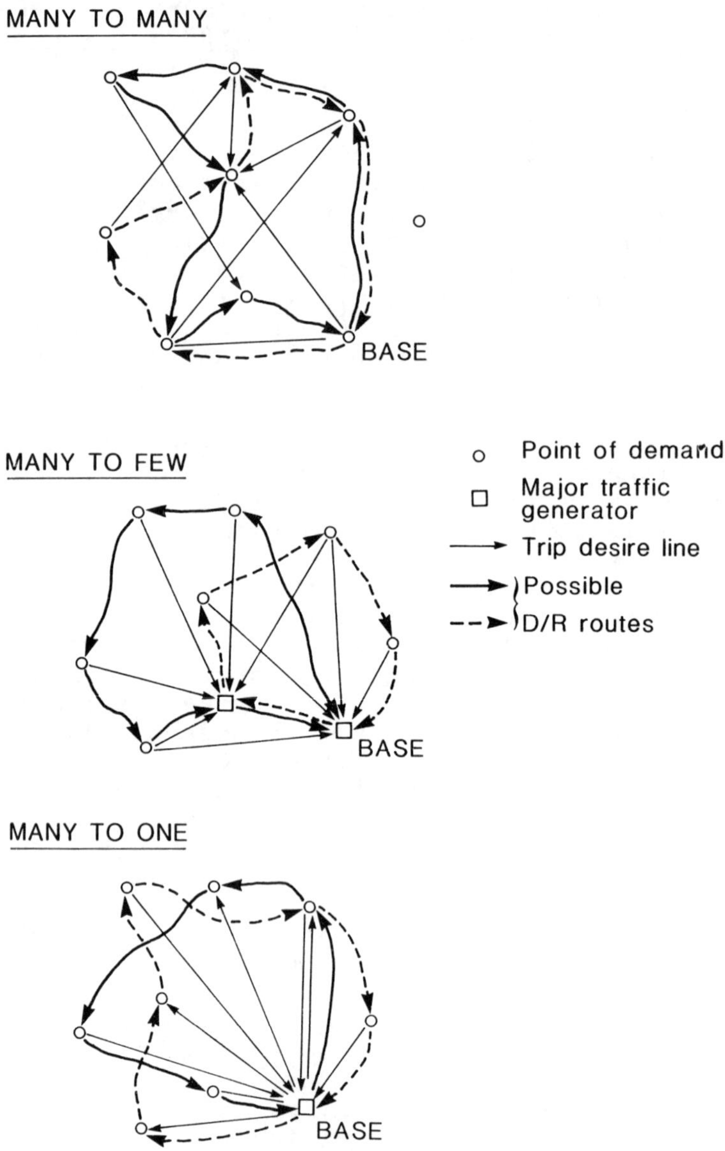

FIGURE 3.1 **Alternative dial-a-ride configurations.**

a computer, with an algorithm designed to minimize overall bus mileage/cost, or to minimize passenger wait times and/or journey times, or to optimize some kind of compromise specification. In response to telephone requests into the control centre, decisions would have to be made in real-time and radioed to the buses. Although DaR is ostensibly a technological solution, the cost and other implications of this technology have deterred full adoption and ruled out true many-to-many systems. Computer scheduling has been used in some large U.S. systems, but not in any of the British ones where less ambitious operation made them unnecessary (NB microcomputers were not available at this time). A common response has been to downgrade the theoretical D/R concept to 'many-to-few' (usually where two or three destinations are fairly close to each other, e.g. linking a residential area to a shopping centre, railway station and hospital) and 'many-to-one' (usually to a town centre). Some degree of scheduling and priority routeing is often introduced to make operations more practical. See figures 3.1 and 2.4 for combinations of routeing and timing options.

It is unlikely that any British system has ever offered an instantaneous many-to-many capability. Many of the new 'welfare' DaRs operate a door-to-door service for the mobility impaired, but in these cases immediate response to demand is impossible and pre-booking with 1-3 days notice is required.

Even for many-to-few services, planning needs to be very careful and sophisticated. Tunbridge & Mitchell (1977) demonstrate three alternative strategies for covering a given set of pick-up and set-down points within a service area from a base in the town centre (the main destination). See figure 3.2. In (a) 'sequential hold', the bus leaves town and sets down passengers at successive points in the service area until it reaches its hold point at the far end, from which at the scheduled time it returns to town picking up passengers en route. In (b) 'sequential tour', the bus again sets down passengers progressively, but in this case begins

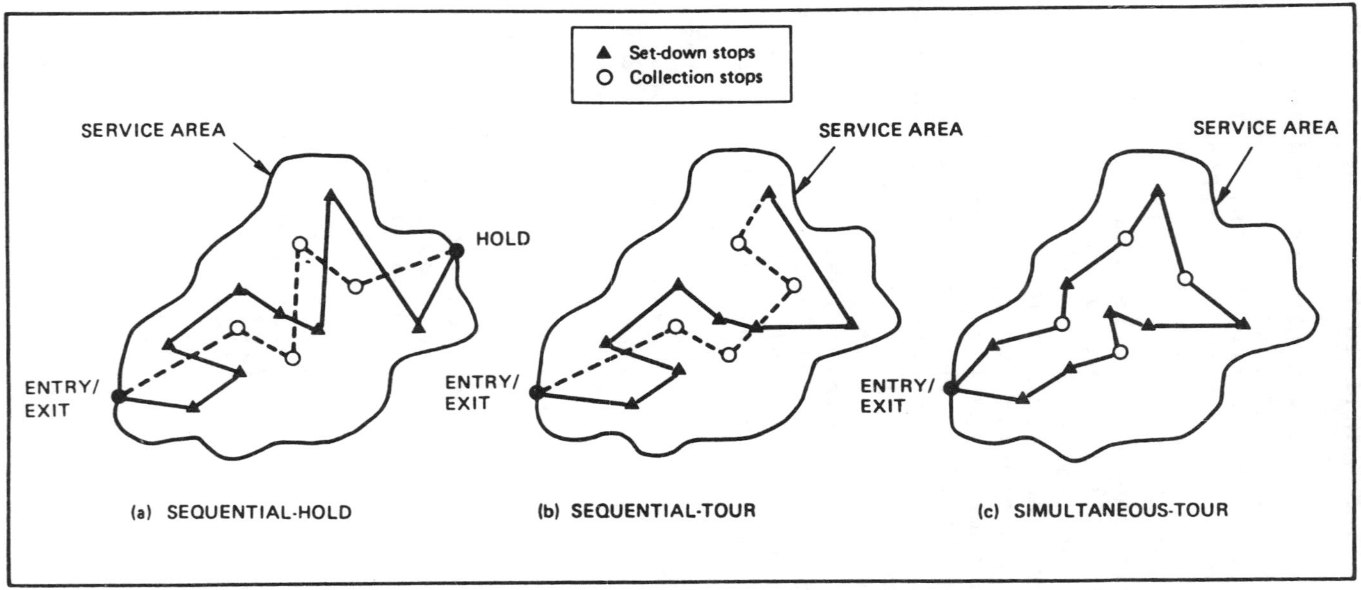

FIGURE 3.2 Theoretical strategies for many-to-few/one DaRs (Tunbridge & Mitchell 1977, (c) crown copyright).

its collection tour as soon as it is empty, without any hold point. In (c) 'simultaneous tour', the bus makes a single circuit of all pick-ups and set-downs in any order that minimizes overall distance. Of these, (a) is least flexible and may involve dead mileage and time, but is most suitable for timetabling; (b) minimizes individual passenger journey time but would probably increase vehicle time, and (c) vice versa. It is possible to calculate the travel distance for a given number of pick-up/set-down points in a given service area, and also the required vehicle capacity and productivity for an assumed level of demand. In reality, DaR planning was more pragmatic.

(b) Origins

The origins of DaR in Britain have been briefly surveyed by Grimmer (1979) and Oxley (1980). This is one of the few occasions where UCMs have been strongly influenced by predecessors in other countries, in this case the United States. In several ways this mode is a typical product of the 1960s, and arose through the coincidence of several factors. Bus decline became serious in Britain after about 1965 with conventional operations coming under severe financial pressure (a legal provision for state subsidy was granted in 1968). An early response was to seek ways of making public transport more attractive, and at the time 'technology' as a solution was widely acclaimed and politically popular.

Corporate and government research projects, mainly in the USA, were investigating 'personal rapid transit', monorail, maglev, etc., and despite high costs, it seemed important to sustain momentum by proving such technologies in real-world conditions. From the beginning, they were "solutions looking for problems." Powerful forces were pushing sophisticated technology - the aerospace industry, some American universities, and big companies such as Ford and General Motors. "Demand-responsive transit" (DRT) was part of this movement. To this powerful lobby, DaR was a potential market for unexploited technology. The first American DaR system commenced in 1969.

In Britain, contemporary opinion was receptive to these developments (Mitchell 1974). Bendixson (1977) encapsulates the enthusiasm of the time. DaR was seen as a possible solution to the problems of running buses in areas of lower population densities and catering for diverse and less predictable trip demands, without reducing the level of service. Accordingly, in 1972 the Department of the Environment started a research programme, mainly by the TRRL, into the potential of DaR systems for Britain. This also involved assistance towards the establishment of the Harlow DaR, as a deliberate experiment. It is symptomatic of the genesis of this new mode that the first plans for an operational system (at Chelmsford, but abortive) were laid by Ford of Europe, Marconi and the Eastern National bus company. The first systems to get off the ground, however, were independent commercial ventures by a combination of bus companies, taxi firms and local authorities. City of Oxford Motor Services was the first with the Abingdon DaR commencing in June 1972. The Maidstone DaR, two months later, was originally the idea of a local taxi operator (having radio control equipment), but subsequently taken over by the municipal bus company. Within four years of the start, twelve systems had been inaugurated (table 3.1 and figure 3.3).

While the logistical problems of D/R operation would be eased if demand was low as in a rural area, it was realised that the higher costs attributable to technology could only be met in a high demand environment. All but one of the British systems were therefore introduced into urban areas, although there was variation among suburbs, new towns, and inner metropolitan zones. The only rural example was the Sawtry & Yaxley DaR, a small scale operation about which little information has emerged. Hopes of inserting a rural DaR into the RUTEX programme did not materialize (the nearest approach was the Exe Valley Market Bus - see chapter 10). DaRs were seen as complementing the rest of the bus system, playing a local feeder or distributor role. There was even some hope in the early years that DaRs would encourage a switch away from the private car and, in the light of the 1973 energy crisis, have

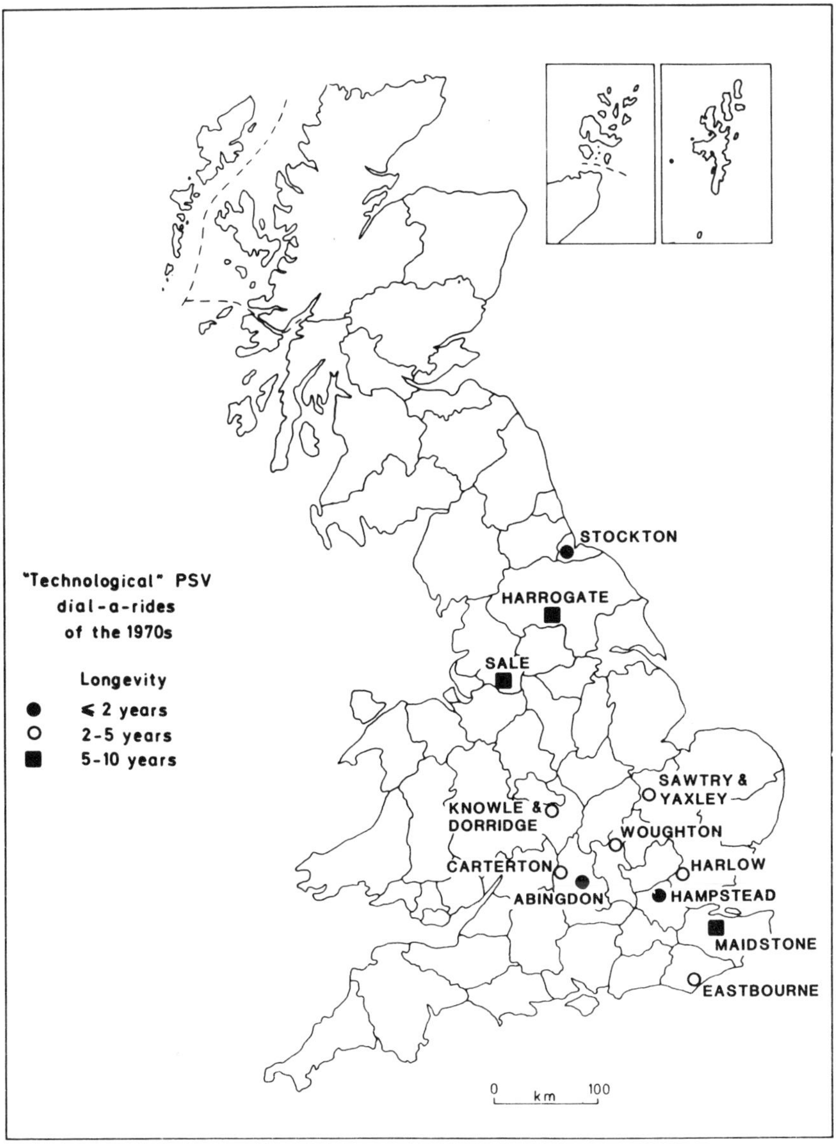

FIGURE 3.3 ‘Technological’ PSV dial-a-rides 1972-80.

TABLE 3.1 British dial-a-rides of the 1970s (source: Grimmer 1979, Oxley 1977, 1980).

	Start	Finish	Operator	No. of buses	Days operated	Bus Revenue Operation Hours (BROH) per week
Abingdon (Oxfordshire)	1972	1973	City of Oxford Motor Services (NBC)	1	Tue,Sun	20
Maidstone (Kent)	1972	1980	Maidstone Borough Council (originally Dennis Radio Taxis)	2	Mon-Sat	138
Harrogate (N.Yorkshire)	1972	1980	West Yorkshire Road Car Co. (NBC)	2	Mon-Sat	46
Eastbourne (E.Sussex)	1973	1976	Eastbourne Borough Council	1	Mon-Sat	60
Carterton (Oxfordshire)	1973	1976	City of Oxford Motor Services (NBC)	1	Mon-Sat	44
Harlow (Essex)	1974	1976	London Country Bus Services (NBC)	3	Mon-Sat	249
Hampstead (London)	1974	1976	London Transport	3	Mon-Sat	281
Sale (Manchester)	1974	1980	Dial-a-Ride Ltd. (a subsidiary of Greater Manchester PTE)	Up to 9	Mon-Sat	334
Woughton (Milton Keynes)	1975	1979	United Counties Omnibus Co. (NBC)	Up to 6	Mon-Sun	369
Knowle & Dorridge (W.Midlands)	1975	1978	West Midlands PTE	4	Mon-Sat	310
Stockton (Cleveland)	1976	1976	Cleveland Transit	n.a.	n.a.	n.a.
Sawtry & Yaxley (Cambridgeshire)	1976	1977(part) 1980	Eastern Counties Omnibus Co. (NBC)	1	Tue,Wed,Fri	7

some impact on aggregate fuel consumption and traffic congestion. This aim was soon refuted, and a new objective appeared which was to test the public acceptability of the 'superior' quality of service at higher fares.

3.2 DIAL-A-RIDE OPERATIONS

The most convenient source of detailed information on the operational characteristics and performance of each DaR system is Oxley (1977). Otherwise, results for the mode as a whole tend to be influenced by those services which have been most studied. The Harlow system, as a TRRL experiment, has generated most data with half a dozen TRRL reports (e.g. Speller & Mitchell 1975, Watts 1976b, Martin 1977). Useful reports on other systems are: Abingdon (Slevin & Cooper 1973), Harrogate 'Chauffeur Coach' (Crockford 1974), Carterton (Watts 1976a), Knowle & Dorridge (Lloyd 1977).

Despite the generalizations made in the previous section, there is considerable diversity among the operating environments and system organization. Table 3.2 indicates the size range among service areas ranging from small parts of a new town (Harlow) and an inner London suburb (Hampstead), to coverage of the whole of a medium size town (Harrogate). Usually services are confined to particular suburbs or housing estates, but in two other cases (Abingdon and Carterton) the whole area of a small town is covered. There is therefore wide variation in population density between 800-900 persons/sq km (the country town of Maidstone and part of Milton Keynes new town (Woughton) and 6000/sq km in Hampstead. The large size and high density of the Sale (Manchester) service area is also notable. Car ownership rates are high: all but three (Sale, Harrogate and Eastbourne) had rates higher than the national average for the mid-1970s (55% by households). The 84% rate in Knowle & Dorridge (outer suburbs of Birmingham) is extraordinary. Such areas would certainly be expected to be difficult territory

TABLE 3.2 PSV dial-a-rides: trip generation (source: Grimmer 1979, Oxley 1977, Watts 1976a).

	Catchment population	Service area: km^2	Population density persons/km^2	Households with car %	Av. weekly patronage	Trips per person per week	Bus Revenue Operation Hours (BROH) per week
Abingdon	17700	7	2500	65*	67	0.004	20
Maidstone	11500	14	800	77	662	0.06	138
Harrogate	59500	35	1700	53*	626	0.01	46
Eastbourne	6000	3	1800	42*	450	0.08	60
Carterton	12000	3	3700	79	422	0.04	44
Harlow	5500	1.8	3050	70	3281	0.60	249
Hampstead	9500	1.6	6000	78	2688	0.28	281
Sale	56000	15	3700	46	3054	0.06	334
Woughton	9000	10	900	68	3412	0.40	369
Knowle & Dorridge	12500	4	3000	84	6321	0.51	310

Data refers to maximum period of operation up to 1976
*1971

for conventional buses, but does raise questions about problem identification and service planning.

As seen from table 3.1, most DaRs were operated by public sector (municipal, Passenger Transport Executive or National Bus Company subsidiary) bus companies; in some cases this allowed the use of existing control facilities, and in others it permitted costing on a marginal basis. The main exception is Sale, where the system was designed by American consultants and started as a private commercial venture, later coming under the aegis of Greater Manchester PTE. In terms of the number of vehicles involved the early systems were all small, with the first five running only one or two buses. This unfortunately raises the impact of fixed overheads such as the cost of control equipment and staff; larger systems in America do achieve economies of scale in this respect. It is part of the DaR ethos that vehicles deployed are small: most used standard production 15-17 seater minibuses which were essential for penetrating housing estates. In retrospect, DaRs have played an important part in promoting the use of small buses as a step towards greater popularity in the 1980s.

For this type of mode it is not so appropriate to gauge service levels by mileage or frequency, and thus it is usual to relate performance to the bus revenue operation hours (BROH). Using this variable, one can see from table 3.1 that the second five systems were of a significantly larger scale than the first five. Oxley (1980) distinguishes 'first generation' systems, which were small in scale (one or two buses, 20-138 BROH per week representing 10-33 round trips per day) and serving a relatively specialised demand; 'second generation' systems provided a more intensive coverage (3-9 buses, 249-369 BROH per week representing 40-70 daily round trips).

As indicated, British DaR systems watered down the true D/R ethos for the sake of practical operation. Services were nearly all many-to-few or many-to-one, orientated towards the town centre where there might be two or three fixed stops.

The Harlow DaR served an older residential area 3 km from the new town centre, and thus differs in that the service area is remote from the eventual destination. In Hampstead the main destination was a London Underground rail station. Woughton, a residential suburb of Milton Keynes, was linked to two remote centres in the new city, at either end of a D/R zone. The Knowle & Dorridge DaR had a dominant commuting function, being focused on a suburban railway station.

DaRs had a formal 'service area' often divided into zones. Many followed a 'basic route' to cover each zone, often taking a radial form centred on the main destination, with diversions according to received travel requests. A limited number of fixed stops was normal, usually at the far end of each route and at the most popular intermediate places. Most had some sections of fixed route, at least in town centres - Sale and Knowle/Dorridge being the most flexible in this respect. Carterton was notable for being divided into four quadrants, each served by a circular fixed route. Within D/R zones, detailed routeing was left to the driver who had been given a list of notified pick-up points; Sale was exceptional with buses routed each stop ahead by the dispatcher over the radio. Several route configurations are illustrated (figures 3.4-3.6).

It follows from this that timing could not be completely D/R. The 'basic routes' were controlled by a schedule of departures from the central point and (loosely) at other fixed points. Telephone callers requesting a pick-up could be given a predicted time. The larger 'second generation' systems operated throughout the day, typically 0700-2300 or even later, at 15-minute intervals in the peak, while the smaller systems had shorter hours and/or lower frequencies. Sale is again exceptional, running a many-to-one ('gather') system into the town centre in the morning peak and the reverse one-to-many ('scatter') in the afternoon peak; in the off-peak was the only serious attempt among British DaRs to run an unscheduled 'many-to-many' service.

Telephone ownership might be a constraint

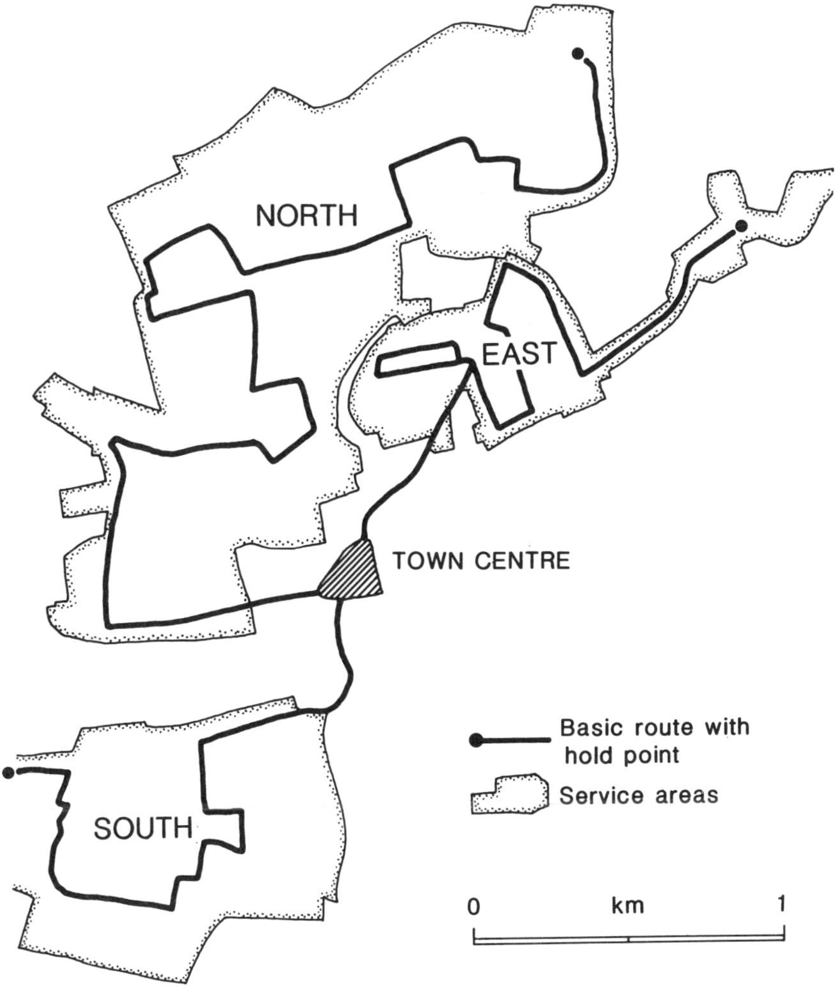

FIGURE 3.4 Dial-a-ride in Abingdon 1972-3 (after: Oxley 1977).

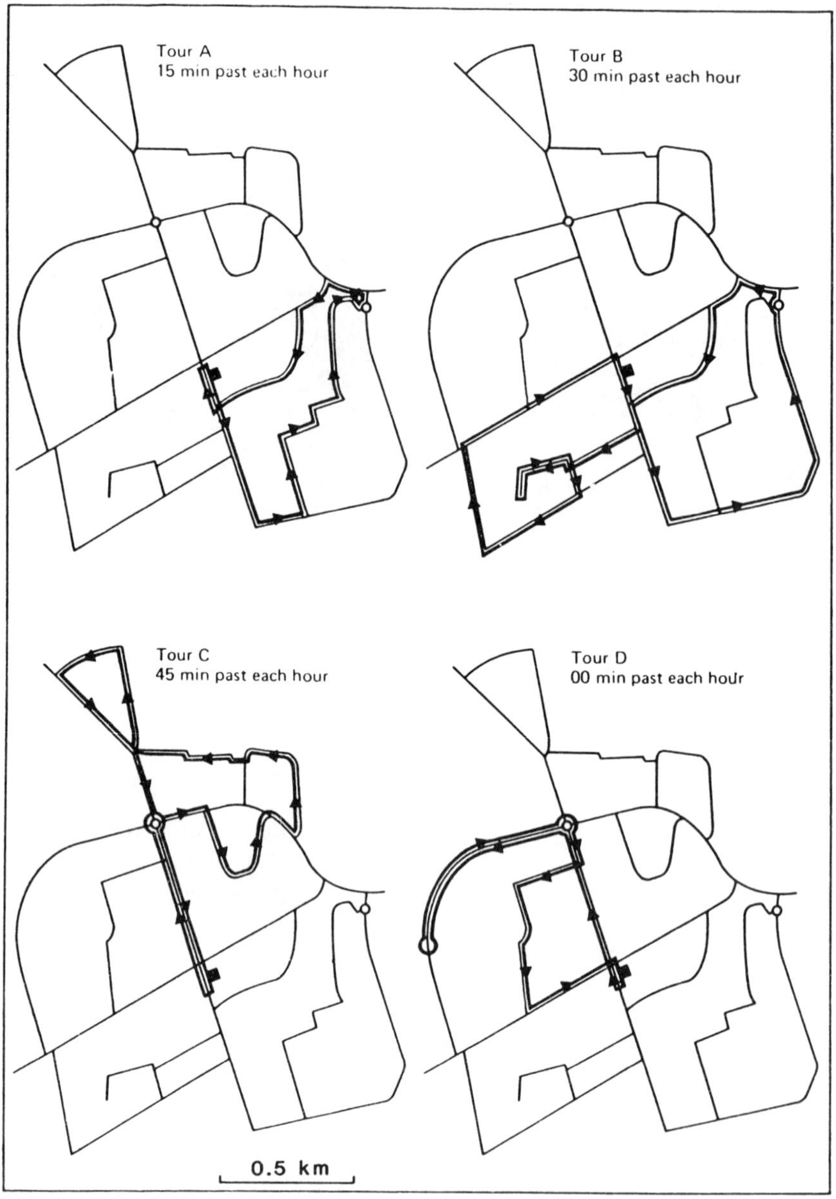

FIGURE 3.5 Dial-a-ride in Carterton 1973-6 (Watts 1976a, (c) crown copyright).

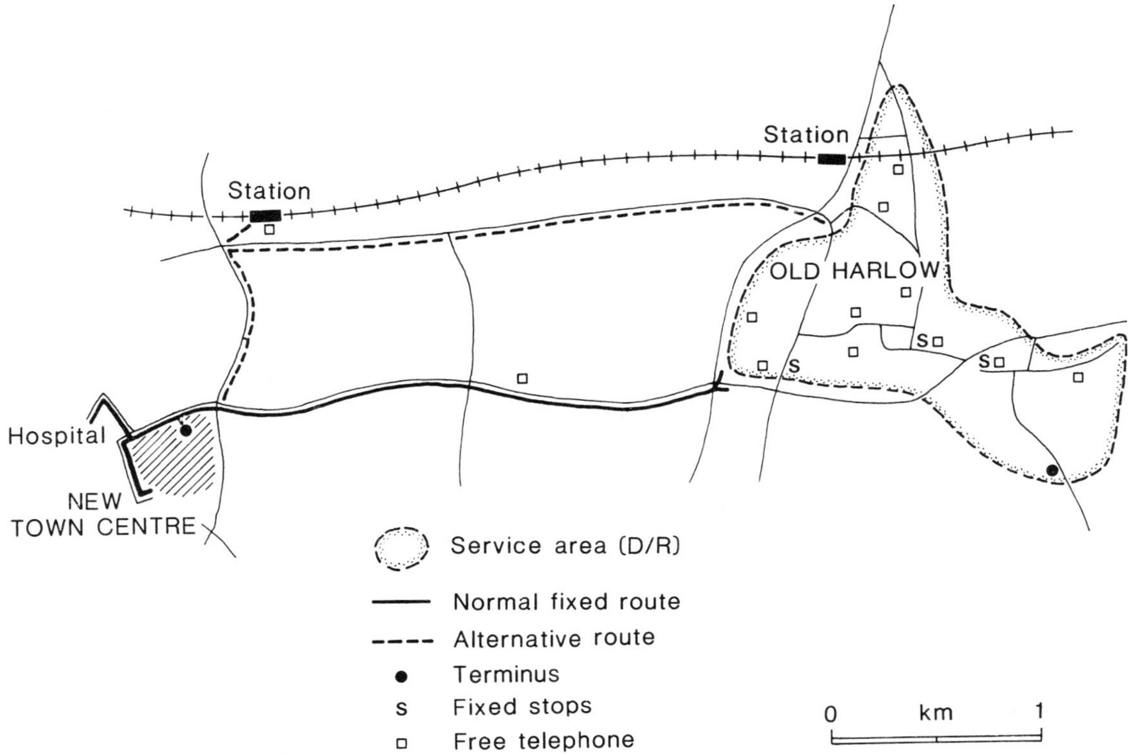

FIGURE 3.6 Dial-a-ride in Harlow 1974-6 (after: Watts 1976b).

with this kind of service. The variation was wide, from 38% of households in Carterton to 93% in Hampstead, but it was never intended that transport should be available only by this mode of contact. Standing order or subscription booking was possible for regular trips. Casual (unbooked) passengers could in most cases 'walk-on' at termini or fixed stops, or 'hail-stop' the bus anywhere along the route. Some systems, especially Harlow and Woughton, provided free direct-line public telephones. The smaller systems were generally able to manage with less specialised control centres, using existing bus company or local council offices, while the larger 'second generation' systems required dedicated control centres, usually with two dispatchers. Communication with the buses was by two-way radio, which allowed 'on demand' diversions to be directed from the control centre. In two cases however - Harrogate and Carterton - radio control was found to be unnecessary and was removed after a time, and drivers merely collected pick-up schedules before departure. Dependence on the telephone for contacting the service varied enormously, from only 4% at Carterton (where most boarded at the terminus), through 20-30% for larger systems (where walk-on/hail-stop was permitted), to 79% at Sale (where routes were most flexible).

A feature which seems most surprising in retrospect is that DaRs were seen as additional to, and not substitutes for, pre-existing conventional bus services in the operational zones, and moreover were not generally coordinated with them. It is not clear how much damage was caused to DaR economics by this unnecessary competition. The exceptions were railway station connections at Hampstead, Harlow and Knowle/Dorridge.

The rural DaR in Sawtry & Yaxley is a quite different case and not comparable with the others. This DaR service was a part of the 'St. Ives midibus' experiment (see chapter 4) in which a single 25-seat vehicle operated as a multiple service bus, and included a DaR mode on three weekdays. This involved a circuit of seven villages, with D/R pick-up arrangements, connecting with the small town of Sawtry for shopping, and a similar circuit of another seven villages connecting with Yaxley primarily to visit the health centre. There was no radio control. No data have appeared on the patronage or economics of this service.

3.3 TRIP GENERATION

Generalized data on passenger loadings are given in tables 3.2 and 3.3. First generation systems carried an average of 50-100 passengers per day and second generation systems 500-1000 per day. Standardized by BROH, patronage levels show wide variations, although the larger systems generally have higher loadings. Trip rates certainly suggest that only a minority of the population were using the services, but this has to be compared with other transport. Again, the larger systems achieve higher trip rates due to their higher BROH (except Sale, which serves an exceptionally large population).

Surveys of DaR users have been carried out on nearly all systems. One objective of the mode was to generate more off-peak travel and 'fill in' the trough between peaks. A wide range of journey purposes has been demonstrated by users: most DaRs were able to handle journeys to work (sometimes by acting as railway feeders), all provided shopping and social trips in the off-peak, and some allowed access to entertainment facilities until late at night. Fairly high off-peak frequencies did appear in most cases to smooth out the normal peaking effect (Oxley 1977).

Most attention should focus on the record of DaRs in meeting their 'social' objectives. First, the new services were intended to stimulate additional journeys which previously had not been possible, but this is complicated by the inevitable competitive element in that most trips would be abstracted from other transport in the same area. From surveys of mode foregone (table 3.4), it can be seen that 5-12% of trips are newly

TABLE 3.3 PSV dial-a-rides: costs and revenues (source: Oxley 1977).

	(1976 prices)				Busiest week	
	Av. no. passengers per BROH	Av. revenue per BROH £	Av. costs per BROH £	Operating ratio rev/cost	Passengers per BROH	Operating ratio
Abingdon (Tue)	5.5	0.42	1.50	0.28	n.a.	n.a.
(Sun)	1.2	0.09	>2.00(est)	<0.05(est)		
Maidstone	4.8	0.82	2.54	0.32	c.5.0	0.33
Harrogate	13.6	2.09	c.6.00	c.0.35	18.9	c.0.47
Eastbourne	7.5	1.69	2.89	0.59	10.4	0.74
Carterton	9.6	0.93	c.6.40	c.0.15	19.3	0.50
Harlow	13.2	1.73	7.41	0.23	17.7	0.38
Hampstead	9.6	1.95	7.29	0.27	14.3	c.0.26
Sale	9.1	1.41	c.3.15	c.0.45	11.3	c.0.64
Woughton	9.7	1.28	5.79	0.22	15.9	c.0.33
Knowle & Dorridge	20.4	0.72	9.42	0.08	24.2	0.09

Data refers to maximum period of opera tion up to 1976

TABLE 3.4 Mode foregone by DaR users (source: Oxley 1977).

Mode foregone	% of trips					
	Harlow	Hampstead	Sale	Knowle Dorridge	Carterton	Eastbourne
Car (driver)	2	3	1	6	} 5	} 2
Car (passenger)	4	4	0	10		
Bus (conventional)	62	38	61	12	10	84
Taxi	4	13	6	1	5	3
Walk	12	35	22	57	70	2
Other	5	2	4	4	0	0
Would not have made journey	13	5	8	11	10	9

generated. This is equivalent to between 40 extra trips per week at Carterton and Eastbourne, and almost 700 at Knowle & Dorridge. However, taking into account other transport in the area, the 'new' trips are probably only about 1% of total movement. Also, it is uncertain whether these results could have been bettered by conventional bus or any other mode.

It is also clear from table 3.4 that there is little realistic chance of effecting any significant diversion from the private car. The highest figure - 16% in Knowle & Dorridge - reflects the higher car ownership rate, but even this constitutes less than 1% of all car trips in the area. Thus even the door-to-door 'taxi-like' facility of DaR is unable to match the perceived advantages of privacy and independence obtainable from one's own vehicle. It is unlikely that any DaR development could improve this, and it was clear that the mode should aim at a different niche in the travel market.

Most DaR trips would previously have been made by conventional bus or on foot; variations between these are explicable in terms of area characteristics, but they add up to at least 70% in all cases. In Carterton, Knowle/Dorridge and Hampstead the small size of the service area and lack of direct bus competition meant most DaR trips were previously made on foot. In Eastbourne, Harlow and Sale there are clear signs of bus competition, and this must emphasize the need for closer coordination with other modes. It is also interesting that, given the nature of DaR, there was significant abstraction from local taxi services.

There are firm grounds for believing that the most successful aspect of DaR was the improved quality of service. Attitude surveys in Harlow showed high levels of appreciation of the doorstep collection/set-down (49-64% of sample) and of the "no waiting at stops" (24%). Most DaR systems attempted to run within a 10-minute time envelope around the quoted arrival time, and this wait could be spent in indoor comfort. Harlow statistics show that "as soon as possible" requests achieved a pick-up within 15 minutes in 60% of cases and within 30 minutes in 90% of cases. It is a feature of the mode that it ought to achieve shorter waiting times and avoid a walk to the bus stop, but on the other hand, diversions to pick up other passengers could lead to a longer journey time. In Harlow, an average 16-minute wait was followed by a 10-20 minute journey time, for distances of 3-5 km into the town centre. While the journey time by conventional bus was the same or slightly less, incorporating the D/R element and short wait time produced a quality of service more akin to a taxi. Only 8% of a passenger sample thought DaR journey times were too long. Further, the small buses were more successful in penetrating neighbourhood streets: the Harlow DaR covered 86% of the service area road network and the conventional bus only 26%. Results of this type have produced a consensus of opinion among observers that DaRs are at least a practical proposition under British conditions (Oxley 1977, Watts 1976b, Grimmer 1979).

It was of interest to DaR supporters to see if the services proved particularly beneficial to certain social groups. DaRs were very well patronised by women, the young and the elderly, but there seemed little significant difference from the national average for public transport. Usage of DaR by elderly persons accounted for a much higher proportion of their total trips than other social groups, but there is little evidence of the elderly making trips which were previously not possible. Surveys indicated that travelling difficulties reported by the elderly and handicapped could to a great extent be alleviated by DaR modes, and this idea was sustained into the 1980s when more specialized forms of DaR were developed.

3.4 ECONOMICS

DaR in Britain was killed by high costs. It is difficult to believe this problem could not have been anticipated by the operators, given the 'technological' and in many cases 'intensive' nature of the operation. Basic data are in table 3.3. Average costs vary widely, between about £2 and £9.42 per BROH. Deficits per BROH vary between £1.16 and £8.70. Costings are generally comparable, and made on a full average basis including all relevant overheads, except for Abingdon, Maidstone and Eastbourne, where the DaRs were provided as a marginal operation within the resources of a larger organization. Marginal costing produces lower costs in these three cases. All but two (Eastbourne and Sale) recovered less than 35% of their costs from revenue. Full cost data are in Oxley (1977).

Breakdown of the cost structure reveals that, typically, 25% of costs are due to the control and dispatch system and staffing; this has the effect of bringing the crew costs down to 40% - a much lower proportion than on a conventional bus. The problem therefore is that the high costs are due to the very feature that distinguishes the mode - its D/R operation - and that is most appreciated by passengers. The ability to dispense with an exclusive control office in those systems run as part of a larger operation was a great cost advantage. DaR must be seen as an intrinsically high-cost mode.

An exception is Sale, which has been fully costed and still seems relatively inexpensive, and achieves a higher operating ratio. As shown above, this was a fairly large and autonomous system, but was able to keep costs down by using part-time drivers, varying service levels according to driver availability, and contracting out vehicle maintenance. Despite a large service area the trip rate was low (table 3.2) and it has been calculated that demand stimulation might have raised the service to financial viability - the only British DaR where this was conceivable (Oxley 1977).

Various aspects of DaR operations have been examined to find ways of improving the very poor financial performance. The first possibility is whether higher passenger loadings could produce enough revenue to close the gap. Patronage levels predicted by a TRRL demand model for the Harlow service turned out to be 32% too high (Martin 1977). In four systems - Maidstone, Harrogate, Eastbourne and Woughton - the population catchment was extended during the operational period (data in the tables refer to the maximum size). If passenger loadings achieved in each system's busiest week can be taken as a measure of capacity, it can be seen (table 3.3) that even on this basis none covered more than 75% of its costs, and most considerably less.

Fares on DaRs were originally set at roughly the same level as on other contemporary buses, but the structure was simplified to either a flat fare or a crude zoning pattern. The financial position strongly suggested that the mode was under-charging for a 'superior' service, and that fare levels closer to those of taxis would be more appropriate. Most systems increased fares during their life, but met consumer resistance long before the break-even point was in sight. Theoretical work on DaRs indicates that revenue is maximised when fares are between one and two times the equivalent conventional bus fare; regardless of network system design and fare levels, these cannot break even (Grimmer 1979).

In practice, some DaR system managers realised that the only way to mitigate the high costs was to simplify, or downgrade, the nature of the operation. This inevitably meant withdrawing from the ideal D/R concept. It has already been noted that the Carterton and Harrogate systems after a time dispensed with their costly radio control equipment. This meant that passengers calling with "as soon as possible" requests after the bus had departed would have to wait for the next one; this might be acceptable on high-frequency systems. Also, if services worked to a basic timetable and on a "fixed route with diversions" basis (as did Carterton and Harrogate), then passengers would generally still be able to hail-stop the bus. At Carterton, the dispatcher was removed from the control office and telephone calls were recorded by a store-call system and replayed by drivers prior to departure. This would give considerable cost savings, but would only work where a small proportion of

UWCC LIBRARY

contacts were by telephone and where it was not essential to give passengers an estimated pick-up time. Minor operational changes occurred in most other DaRs, but the major case of service simplification was Maidstone which switched from a many-to-few taxi service to a fixed bus route with D/R diversions; this produced a trebling of patronage and revenue!

Considerable interest was generated in the late 1970s into the comparative economics of DaRs, minibuses (at a time when these too were thought rather *avant garde*), and conventional buses. Some studies were theoretical (Martin 1978a,b), while others followed up data from the Harlow DaR experiment (Tunbridge & Hale 1978, Tunbridge 1980). Comparing vehicle running costs only, Martin (1978a) calculated that each additional seat increased costs by £1 per week, and therefore a minibus was likely to be about £30 per week cheaper to run than a conventional single-decker. This advantage to the smaller vehicle does not apply under marginal costing. As vehicle running costs are only a small part of the total operation, overall costs for an equivalent service are likely to reverse this relationship. Thus a small DaR operation would cost between 25% and 65% more than an equivalent minibus service, and both would cost more than a conventional bus (1976/7 prices). At Harlow, the fixed-route minibus and conventional bus services were also monitored by TRRL to allow comparison with DaR. If frequencies and fares are standardised, the DaR and fixed-route minibus were equally attractive to passengers. The DaR is less efficient in that four vehicles are needed to maintain a 15-minute frequency compared with only three minibuses, and thus the former is more expensive for this reason in addition to its extra control costs. The conventional bus at a lower frequency was the only one to produce a net social benefit roughly covering its resource costs, and had a higher operating ratio (0.51) (Tunbridge & Hale 1978).

3.5 RETROSPECT

The DaR experiments which were tried out in 1970s Britain have been judged an almost complete failure. In retrospect, the early enthusiasm and faith in 'technology' seems to have been extremely naive. However, the 'learning process' was invaluable. It has at least been established that (a) DaR operations are practical under British conditions, and (b) they are popular with the public. Reasons for failure can be summarised as:

(i) The additional capital and labour costs are high, which is an inevitable feature of a mode which aims to be true to the D/R ideal. There is little or no realistic chance of covering costs from revenue; there is little scope for attracting more patronage by service improvements, and limited scope for raising fares to reflect the higher quality service.

(ii) The record on social objectives was also disappointing, in that the great majority of trips had been diverted from other modes and 'new' trip generation was unimpressive; there was very little switch from the private car. While disadvantaged groups certainly benefited, it is uncertain whether much latent demand was satisfied.

(iii) DaRs were inserted into areas already having a basic level of public transport, but were not coordinated with it. Deliberate coordination with pre-existing services would maximise the trip generation potential without any competition for the DaRs; this would also allow realistic fare levels to be properly tested.

(iv) There was no firm evidence on the ability to handle truly dispersed many-to-many trip demands - only the Sale DaR attempted this. Different types of D/R mode might be necessary to deal with this, but there was no willingness to experiment further: for example, in the USA, many DaRs are integrated with taxi services.

All DaRs had terminated by 1980. Most were replaced by fixed route bus services, some of which were minibuses with hail-stop facility (e.g. Abingdon, Hampstead, Harlow). Comparisons of the Hampstead DaR with its fixed-route replacement showed the latter to have reduced costs per BROH by 23% and increased passengers by 56%; the Harlow minibus replacement reduced costs by 40% compared with the DaR (Oxley 1977). In reality, the DaRs had proved the traffic potential for their replacements. Eventually, substitute minibuses tended to revert to conventional buses.

Towards the end of the 1970s it was becoming clear that, if DaR had any future, it was with the simplified versions such as D/R diversions from a fixed route, flexible or alternative routeing, and hail-stop. Small buses had proved their popularity, the value of local street penetration, and more contentiously, lower running costs. The simplified versions suggested they would only work in areas of low demand, such as outer suburbs or rural areas. Oxley (1980) saw the rural Sawtry & Yaxley DaR as a precursor to the kind of DRT which might succeed in low-density environments. Other forms of public transport with a D/R element were springing up in the later 1970s, and some were included in RUTEX. In retrospect, while the 'technological' approach to DRT has been abandoned, there could be said to be a convergence with conventional buses in the form of flexible routeing, such as the Ripon Flexibus and Strathclyde 'Gypsy' minibuses (see chapter 4). However, to regard these as 'third generation' DaRs, as does Oxley (1980), is incorrect and misleading. The best candidate is the Exe Valley Market Bus, which is run as a DaR by community volunteers (Devon Rutex 1978); this is regarded as a community bus and described in section 10.1.

There were suggestions in the late 1970s that DaRs were particularly suitable for people with personal mobility problems such as the elderly and disabled, who would appreciate a door-to-door service, and that the potential of this might be explored. This does nothing to reduce the cost problem, and might exacerbate it, unless another contemporary suggestion was taken up - that responsibility be offered to local communities to run such services on a voluntary basis. There was in fact very little continuity between the 1970s 'technological' public service DaRs and the community transport equivalents that flourished in the 1980s. The economic climate of financial stringency at the end of the decade discouraged further experimentation and development (Bailey 1984, Sutton 1987), and the origins of the CT versions of DaR for the elderly and handicapped are rather different (chapter 9).

Chapter 4

FLEXIBLE BUS OPERATION

4.1 DEMAND-RESPONSIVE DIVERSIONS

In this chapter it is intended to cover a number of types of more flexible bus operation, such as introducing more flexible routeing and timing, multi-purpose vehicles and schedules, buses financed from alternative sources, and large scale coordination of public transport resources over a wide area.

Since the failure of dial-a-rides in the 1970s (chapter 3), it was clear that there was still a strong case for a bus system that was not tied to rigid routes and times, but able to respond to dispersed demand patterns. Public transport which responded immediately to demand, any time, any place, could be provided only by taxi, as even the most ambitiously technological DaR system had not succeeded in doing this, indeed, had not really attempted it. Moreover, during the course of their short lives, several DaRs had downgraded the D/R element of their operation in order to cut costs. Some were operating a fixed route with D/R diversions; others had a D/R pick-up/drop-off zone at one end of a fixed route. Since the consensus of opinion at the demise of DaRs was that a D/R bus could only work at a simplified level, without any 'technological' component, then the possibility remained of adding D/R diversions to an otherwise conventional stage bus service. This would be feasible in areas of low demand, such as rural areas, whereas the earlier DaRs had needed the higher passenger numbers of the towns to justify their investment.

Typically, rural stage buses would divert from the main road to villages a short distance away, in order to collect every possible passenger in areas of sparse demand. With a fixed timetable, the problem of course is that most often such a diversion would set down or pick up no passengers at all, and thereby cause an unnecessary delay to other passengers on the bus, and increase costs. Conversely, omitting such diversions leaves village inhabitants without a service and loses revenue for the operator. The possibilities of instituting D/R diversions from rural fixed route services were examined in desk studies by Gallop (1975) and Dickinson (1977), in Dorset and Cumbria respectively. Both showed such a scheme could be practical, the first demonstrating the need that exists in off-route villages, and the latter concluding that time savings could be achieved by diverting only on demand, but cost savings would certainly be cancelled out by additional costs of managing the D/R operation.

(a) The Ripon Flexibus

A small number of D/R diversion arrangements had been tried, independently, in the mid-1970s. Gallop (1975) cites the United Counties 'Nightrider' service in Bedfordshire, Everton Coaches' Droitwich to Worcester service, and an attempt by the Alder Valley bus company in the Basingstoke area. The Sawtry & Yaxley dial-a-ride, mentioned in chapter 3, may be more closely related to these rural diversion schemes rather than to traditional DaRs. No data have been reported on any of these. However, probably the best known and best documented D/R diversion (as it was a specific experiment) was the Ripon Flexibus.

This was implemented in December 1977 as part of RUTEX in North Yorkshire; essential details are given in a TRRL report (North Yorkshire Rutex 1979). The area chosen was west of Ripon, characterized by several small villages, whose pre-existing bus service was demanding unacceptable levels of subsidy and facing review. Some kind of flexible bus experiment was determined by the desire to serve as many people as possible within the catchment area and to test the extent of latent demand for transport. Possibly also there was a wish to explore other forms of DRT suitable for rural areas, in context of contemporary experience with dial-a-rides of the formal urban type. A DaR was actually the first choice for Ripon, but this would have required a minibus which the local bus company did not have available.

D/R diversions from a regular route were an acceptable alternative, as it was found that a circular route out of Ripon with three possible diversions could reach all the villages in the area (figure 4.1). This novel mode could then be fully tested, questions at issue being the uncertainty of timetabling and reliability, and people's willingness to book diversions in advance. United Automobile Services (a NBC subsidiary) did not have minibuses but made available a 35-seat vehicle. Narrow roads were a constraint in planning. The Flexibus was the only experiment under RUTEX to attempt flexible routeing with a large bus operated by a public sector company.

One existing bus route was withdrawn and two others re-routed to make way for the new service. The most remarkable aspect of phase I of the experiment was its high frequency. There were three circular tours every day (Mon-Sat) plus two return trips from Ripon to Fountains Abbey - a major tourist site. This had been encouraged by a prior household survey which revealed an overwhelming preference for a frequent service, but also provided an opportunity to test the potential of demand stimulation in a rural area. Booking facilities were kept simple: people wishing to be picked up on a diversion could telephone the bus company office in Ripon up to 20 minutes before departure of the bus, or use booking cards which could be either posted or given to the driver on a previous trip; people wishing to be dropped off on a detour could simply inform the driver on the bus.

The timetabling problem was met by including a generous time allowance for possible diversions, but when no detours were made this led to buses running ahead of schedule, which caused a few complaints. Passenger numbers on phase I were almost identical to those on the previous bus service, as were trip patterns (mainly women going shopping). The high frequency was clearly

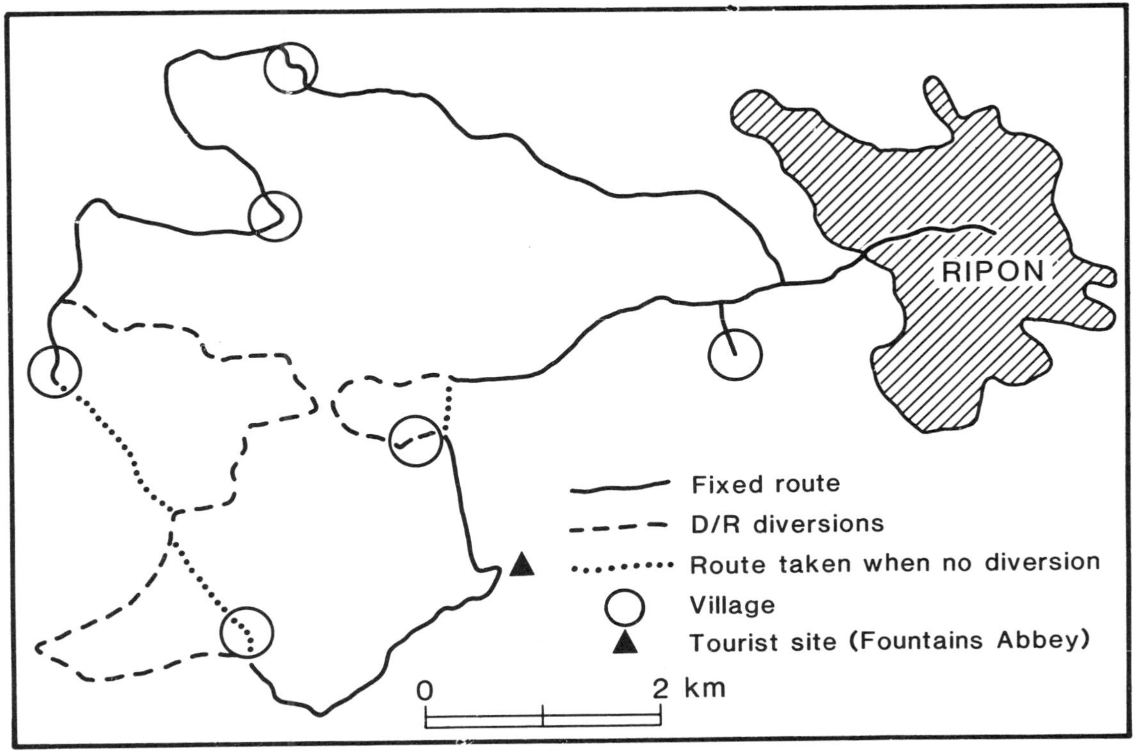

FIGURE 4.1 The Ripon Flexibus 1977/8
(after: North Yorkshire Rutex 1979).

unnecessary and not generating any extra traffic. The timetable was therefore revised for phase II of the experiment, in which the complete circular route was run only on the two most popular weekdays, but the Fountains Abbey workings were increased slightly in time for the 1978 tourist season. Excluding tourists, however, average passenger numbers were the same as before. Time allowed for diversions was cut back, and it was found that when diversions were made the bus was generally able to make up lost minutes.

It is the usefulness of diversions which determines the value of the Flexibus mode. They were, however, little used, with an average of about one per week. Nearly 80% of these were 'set-downs', which are much easier to arrange. It did appear that the detours were much appreciated by those who used them, mainly the elderly, who were able to avoid long walks with shopping bags.

Extrapolated over a whole year, the operating ratio under phase I would have been only 28%, but under phase II it would have been 53%. These compare with 45% for the original bus service (North Yorkshire Rutex 1979).

At first sight, this experiment was hardly a dramatic success, but this has to be qualified. It could be claimed that the D/R diversions and the circular route were at least practical (like dial-a-rides). Increasing frequencies in phase I proved fruitless, and confirmed the experience of other attempts to stimulate demand in rural areas (such as fares reduction experiments), that although one should try to meet latent demand it exists only at a basic level. The low usage of diversions partly reflects the fact that these served fewer habitations - all the villages were on the main route (figure 4.1). Balcombe (1980a) estimated that if the service was available throughout the area (as in a many-to-one DaR or social car), passenger numbers would have increased by only 2-3%. This suggests that D/R diversions should be tried in a different type of area, or in a different form, for example where the bus keeps to a main road and diverts to villages, instead of connecting villages in a 'rural tour'. As the financial results under phase II were slightly better than the previous bus, and a few people were benefiting from detours, then such a service might as well be maintained.

North Yorkshire County Council took over responsibility for Flexibus when the experiment ended in December 1978, but the service was eventually withdrawn in 1980. Three years later, a D/R shared taxi scheme commenced between Kirkby Malzeard and Ripon, which might be suitable for part of the area abandoned by Flexibus.

(b) Others

Two other schemes within the RUTEX programme had D/R diversion or variable route facilities. These were both in Strathclyde: the 'Medwyn Gypsy' in the Blackmount area (Lanark District), and the 'Ruralink' service in Cumnock & Doon Valley District.

The Medwyn Gypsy was a minibus designed to provide a service to numerous small settlements north of Biggar and east of Carnwath, and included a D/R element in that two hamlets were served only 'on demand' on certain journeys. Apart from the purely experimental motive, this mode was encouraged by the desire to reach as many people as possible throughout the area and be responsive to local needs, but at a lower cost than existing public transport. This experiment is also notable in involving the planned integration of two UCMs (three services), as well as connections with conventional transport. Two previous bus services in the area were withdrawn on economic grounds. A new postbus was inaugurated between Biggar and Lanark which took a devious route to tap several villages, and thereby allowed the Medwyn Gypsy to confine itself to a smaller area that it could cover more effectively. In addition, an existing postbus was extended to cover the Biggar-Carnwath area in parallel with the new minibus (see chapter 7). These two modes operated a fully integrated timetable, the postbus being constrained by mail schedules, but the 'Gypsy' free to run at other times of day. The D/R component in the latter comprised a note in the timetable to the effect that intending passengers from the places concerned should telephone the bus company within specified times on the day of travel (Scottish Rutex 1979, Lugton & Macniven 1979).

In June 1978, nine months after the original innovations, the Gypsy was extended to a new area north-east of Carnwath. This was to make better use of the minibus, and to compensate another locality with inadequate public transport. Because of the dispersed population, the new service was timed at two villages only and called elsewhere only if booked in advance. Frequencies in the original service area were reduced from five to three and a half days per week, to enable three days in the new area. Later again, two D/R locations were made part of a fixed route. Such changes reflect the inevitable learning process necessary in matching service types to demand: survey data emphasized that travel patterns could not have been predicted without local knowledge. This is of course a general conclusion valid for almost all UCMs.

As with the Ripon Flexibus, the D/R diversions were expected to make only a small contribution to passenger numbers. Over the life of the experiment (1977-9), the diversions resulted in an average of 2.4 trips per week, which constituted about 3.5% of overall patronage. Problems reported were the unavailability of telephones and, sometimes, passengers not waiting for the bus after having made a booking. Generally, integration with the postbuses and intensive use of the minibus (including some schools work and private hire) were successful. It is difficult to assess the D/R aspect, save that it did reach a few more people who welcomed the new service, without doing any damage to the finances of the operation. Compared with a subsidy of £30,000 demanded by the pre-existing buses, the Medwyn Gypsy cost under £6000 per annum while the associated postbuses made a small surplus (Lugton & Macniven 1979).

The other scheme under RUTEX in Scotland to incorporate D/R diversions was '<u>Ruralink</u>', which was simpler than the above in character and showed some differences. The basic rationale was to link two valleys which had lost a bus service some years previously, but also to provide access for rural dwellers to hospitals - a need that had been identified as deserving priority in the Scottish RUTEX programme. A 27-seat bus - itself an unusual size - was contracted to operate a basically linear route between Dalmellington, Cumnock and Ballochmyle Hospital. There were three D/R diversions: one an extension of the main route to the small town of Patna, and two others from Cumnock to the district council offices and to another hospital. Timetables were strongly influenced by the hours of hospital visiting and outpatient clinics (Scottish Rutex 1981a, Lugton & Macniven 1979).

Demand for the diversions was of the same small scale as found elsewhere - the longest diversion attracted only 0.9 trips per week, and the hospital detour 2.2 per week. Together, these represented an average of just 1.1 passengers per diversion, and accounted for 1.5% of total patronage in the experimental period. On the other hand, the cost of diversions was only 1.3% of total costs, but nevertheless the low usage still meant a subsidy of £2.20 per D/R passenger. The operating ratio was only 18%, and the annual net cost of £12,300 compared badly with the Medwyn Gypsy.

Apart from the implication that a service of the latter type is more successful, results from all three RUTEX cases appear to show D/R diversions having little economic justification. It is easy to demonstrate that incorporating the detours in a permanent schedule would be excessively costly with little return, and the D/R mode is an improvement on this while still meeting the few trip demands of the people involved. It is probably fair to say that results merely reflect the problems of running conventional buses in sparsely populated areas, rather than the merits of D/R operation itself. But the impression remains that the more sparsely populated an area, the greater the D/R component needed to serve it. A tentative analysis by Balcombe (1980a) suggested that areas with lower demand densities are more successfully served by D/R operations (as in the Medwyn Gypsy and Exe Valley Market Bus (chapter 10)) than those with higher (as in the Ripon Flexibus).

After RUTEX, Ruralink was replaced by a different form of service, but Medwyn Gypsy was continued together with two further schemes of the same type introduced by Strathclyde Regional Council. The 'Barrmill Gypsy' operates around Beith, and the 'Loudoun Gypsy' serves a wider area around the Irvine Valley east of Kilmarnock. Two more followed in 1980/81. Throughout the 1980s it has been impossible to assess the extent of adoption of D/R diversions, as there is no list of such operations, most are probably small scale, and unless any example acquires wider publicity its existence will not come to light. Some new schemes supported by the Rural Transport Innovation Grant have similar features. One example is the 'Ruralride' services run by Durham County Council, in which two minibuses operate routes in Weardale and Teesdale, mainly scheduled but allowing for variable routeing within a defined area according to passenger demand.

D/R sections of some of these services are often referred to as 'dial-a-ride', as if they were a continuation of the urban 'technological' types of the 1970s. Oxley (1980) cites the Ripon Flexibus and Medwyn Gypsy as 'third generation' DaRs. This is inappropriate, as they are quite different, but it could be concluded that the future of DaRs (apart from special services for the handicapped) lies in a downgrading to simpler operations, such as D/R diversions, and a convergence with the trend towards minibuses.

4.2 MULTIPLE SERVICE BUSES

A common approach to improving efficiency of transport in sparsely populated areas is to combine two or more different types of traffic on the same vehicle. It was explained in the classification in chapter 2 that there are two ways of doing this. The vehicle might carry two types of traffic or client groups simultaneously, such as passengers/mail (chapter 7), adults/schoolchildren (chapter 6), or passengers/freight (chapter 8). Alternatively, a single bus might cater for different traffics consecutively, by making individual trips for specific purposes which vary from day to day. True multi-purpose operation can be achieved, as more than two business sectors can be provided for. Dead mileage and layover time could be used for different purposes, and it is possible to achieve intense vehicle utilization and cover a wide area, although this might be limited if it is desired not to exceed one driver's working week. A few schemes of this sort have become widely known in Britain, and they are difficult to classify using accepted terminology, so for want of a better term they have been called 'multiple service buses'. The best known cases are the 'St Ives midibus' and the so-called 'Bassetlaw community bus'.

(a) The St Ives Midibus

This ambitious project was set up in 1976 to solve a multitude of rural accessibility problems over a wide area by the use of a single bus. Great concern had been expressed in the early 1970s that a period of bus service withdrawals had had a severe impact on many parts of rural East Anglia. Contemporary surveys exposed a large tract of western Cambridgeshire and north-east Northamptonshire as having only a vestigial degree of local public transport, although a major railway line and long distance coach routes traversed it. Joint studies were done by the two county councils and the Eastern Counties and United Counties bus companies to produce proposals for the area.

It is interesting that the first proposal, made in conjunction with the Ford Motor Company in 1973, was for a dial-a-ride service in a large area around Sawtry encompassing 20 villages. This approach was symptomatic of the technological ethos of the time, in which the urban DaRs were born (chapter 3), and reflected a desire to try out DaRs in a rural environment. For various reasons, especially the realization that local demands necessitated a much more diverse pattern of services, this plan was still-born, but a fragment remained in limited DaRs based on Sawtry and Yaxley.

The area of interest was overwhelmingly agricultural with about 50 villages, focused on the towns of St Ives, Huntingdon, Sawtry, Yaxley and Oundle. Expressed needs for transport were largely for shopping and market day access to St Ives, Oundle and Sawtry, but also, following a rationalization of local surgeries there was demand for village access to the health centre at Yaxley. Further surveys revealed a need for local connections with Huntingdon railway station for work and school journeys. The emergent scheme was nothing less than the incorporation of all these trip demands on to a single bus. The original proposals are summarised in Eastern Counties (1975) and early progress reported in Cambridgeshire (1977). At the start it was known simply as the 'Rural Midibus Project', but later became known as the 'St Ives Midibus' (the vehicle being based here), or alternatively as the 'Cambridgeshire Pick Me Up'.

The diversity of services can only be appreciated by listing them in full. Their geographical spread is illustrated in figure 4.2.

(i) *Monday to Friday early morning and evening peaks*. A commuter link facility between St Ives and Huntingdon station to connect with London trains. Passengers booked in advance could be picked up and set down at their homes, while others boarded from specified bus stops.
(ii) *Monday to Friday morning and afternoon peaks*. A regular work and school service from the Alconbury area into Huntingdon. A similar advance booking and doorstep pick-up facility was available.
(iii) *Mondays*. A 'shop or ride' service into St Ives from numerous villages to the north, and in the afternoons from Holywell. An entirely novel feature here was to give villagers the alternative of travelling to do their own shopping or placing orders with the bus driver to do it on their behalf. Thus the driver could take orders for groceries, magazines, prescriptions or even library books, and collect them from town during

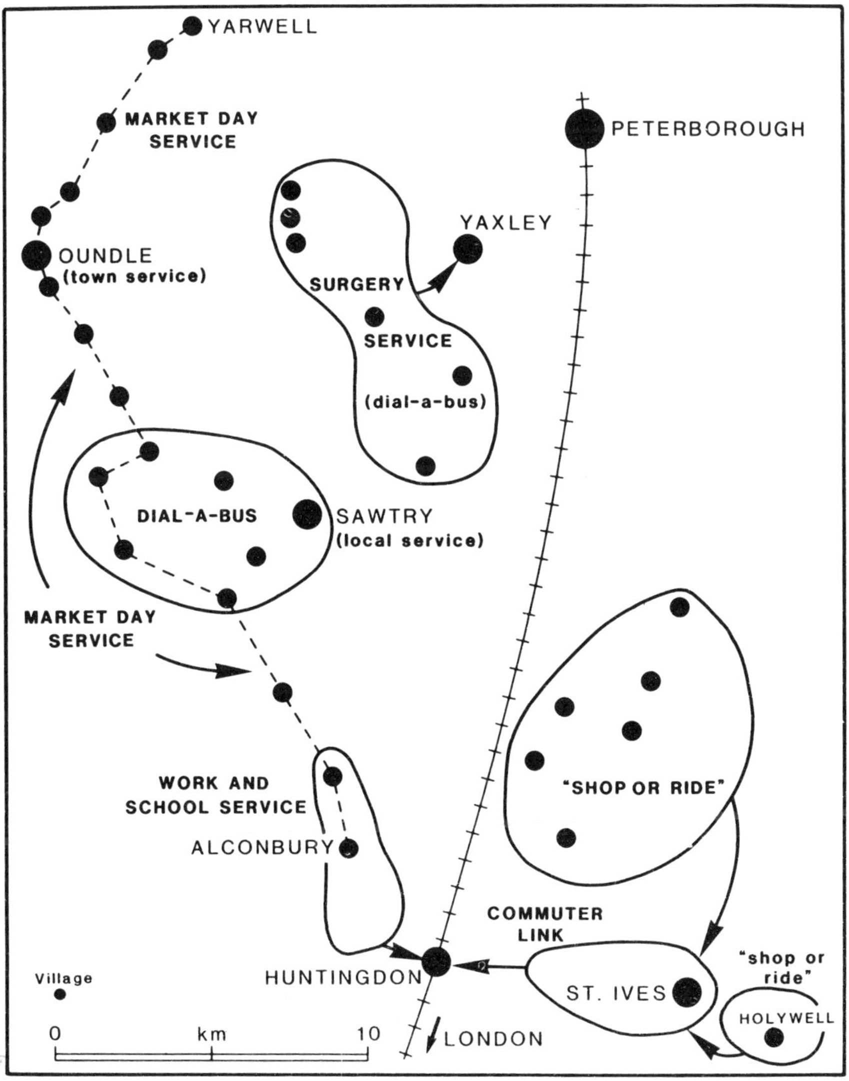

FIGURE 4.2 The St Ives Midibus 1976 (after: Eastern Counties Omnibus Co. 1977).

the bus layover time; there was a charge by weight for carriage of goods.

(iv) *Tuesdays and Fridays*. A 'dial-a-bus' service available in an area to the west of Sawtry containing seven villages, for shopping or bus connections in Sawtry. Passengers were required to book by telephone up to 0900 hrs on the day of travel and arrange a pick-up point, or could hail the bus en route.

(v) On the same days, the bus ran a 'local service' in Sawtry before returning to the villages on the DaR circuit.

(vi) *Wednesdays*. A 'surgery service' operated in dial-a-bus mode to connect an area west and south of Yaxley to the health centre in the town. Two circuits were run, each connecting a group of three or four villages. A convenient arrangement was the ability to book transport at the same time as making a doctor's appointment. (See also Oxley (1977) for the DaR contributions).

(vii) *Thursdays*. Two 'market day services' into Oundle, one from Alconbury to the south-east connecting ten villages, and a shorter route from the north-east.

(viii) On the same day, the bus ran a 'town service' in Oundle during the layover time of the above workings.

This network was originally run with a 25-seater midibus in Eastern Counties livery, using two drivers. Peak loadings were not therefore anticipated despite the commuter services and a high level of publicity. With such a diverse operation it would be expected that some services would be more successful than others, and thus it proved, with the rail commuter link quickly building up substantial patronage, the market day and surgery services having much less but steady custom, and the Sawtry DaR being very disappointing (Cambridgeshire 1977). The overall operating ratio in the first few months was only 30%. As might also be expected, there were frequent revisions to schedules, with the Sawtry DaR being withdrawn and new shopping connections being arranged for Huntingdon and Peterborough. However, "almost nobody" took up the option of asking the driver to do their shopping on the St Ives 'shop or ride' service. Since 1977 the project was a great success, with its high profile and publicity and responsiveness to local needs producing an impressive build up of traffic. The midibus gave way to a full size bus, and at times a double decker was needed for the commuter link services. Higher than anticipated usage eventually led to a reorganization of the scheme, whereby a large bus was devoted exclusively to the peak hours rail commuter link and school contracts, the Yaxley DaR was abandoned, and the large rural area west of Huntingdon served by smaller vehicles operating multi-purpose, partly D/R, networks.

The St Ives Midibus could be said to have been the victim of its own success. It had in effect acted as a catalyst, in identifying local trip demands and consolidating them into viable traffic flows, aided by community responsiveness and strong marketing, to such an extent as to restore public transport use to 'conventional bus' levels. This sometimes happens with other UCMs.

(b) The Bassetlaw Community Bus

While it shares the diversity of the St Ives scheme, the Bassetlaw Community Bus is quite different in many respects, notably its orientation towards health services, its facilities for the disabled, and its organization by a management committee comprising representatives of local community groups. It is this last point which gives it the title 'community bus', but it clearly does not qualify under the normal understanding of a community bus, nor under the definition given in chapter 2. A true community bus has to be managed and operated by local volunteers (see chapter 10), whereas the Bassetlaw scheme hires the bus from a commercial company and uses professional drivers. It is certainly 'unconventional', but primarily in that it operates a highly diverse pattern of services, aiming to satisfy many different trip purposes and client groups, with the same vehicle, and is therefore best classified as a multiple service bus.

Origins of the project lay in a series of travel surveys undertaken in Bassetlaw District (north Nottinghamshire) in response to a number of minor cutbacks in conventional bus services in 1975/6. The area is large with mixed environments ranging from rural through small towns to the larger centres of Worksop and East Retford. At the time, the area was also experiencing changes in its settlement pattern with some villages expanding, and a rationalization of some service facilities, especially health centres. Travel patterns were diffuse and complex, not adequately matched by existing public transport. Surveys uncovered the poor accessibility of various smaller service centres, but particularly, widespread dissatisfaction with the transport available for hospitals and doctors' surgeries. Centralization of the latter was causing great problems (Childs 1978).

An informal consortium of local organizations set up to consider the transport options effectively became the management committee of the new bus service. It is generally acknowledged that the Bassetlaw Council for Voluntary Service was most influential, but the planning group also included officers from various local authority departments, health authorities and the community health council. It was soon clear that the scale and responsibility of the operation was too great for a normal community bus. The catchment area covered 250 sq km with a population of 90,000 including the two main towns, and it had been decided at the outset that special attention was needed for elderly and handicapped passengers. Hence it was essential to contract a professional bus company with full-time drivers. An independent operator won the contract for the first experimental year (1978/9). A new vehicle was purchased, with a side-loading wheelchair lift (this was such an innovation at the time that it caused problems in licensing the bus).

Services offered when operations commenced in May 1978 were a combination of the following: five weekly trips to doctors (from five locations to four surgeries), a fortnightly trip to each of three different clinics, two weekly services for hospital visiting in Worksop, and on alternate Sundays to two other hospitals, four weekly shopping trips to the small town of Bawtry, plus a variety of weekly or fortnightly excursions on behalf of half a dozen local associations (mainly for youth and the elderly). Some of these were partly D/R. The complexity of such a schedule puts great weight on carefully targeted publicity and the vital role of the local coordinator. It led also to the problem of potential competition with other bus operators, and agreements had to be made for the purpose of obtaining road service licences.

As would be expected, the service pattern underwent frequent revisions in the light of experience. Although an initial heavy deficit was expected, the first few months were disappointing, not even meeting the 30% target operating ratio. Doctors and hospital trip needs had been over-estimated, but shopping journeys had been running to capacity and welfare group work was very popular. A high proportion of dead mileage also suggested a need for concentration of the route pattern. After revision, the schedule was as follows (1978/9).

Monday am: villages-surgery; pm: villages-clinics; evening: villages-surgery.
Tuesday: Bawtry shopping; evening: local welfare associations.
Wednesday: for Age Concern and other local associations.
Thursday am: villages-surgery; pm: villages-clinic; evening: youth club.
Friday am: shopping in Bawtry and Doncaster; pm: playgroup; evening: villages-Retford.
Saturday: various group hires.
Sunday: from Worksop and Retford to two psychiatric hospitals.

(Nottinghamshire PTP 1979/80)

The main change was for local hospital visiting services to be withdrawn in favour of greater availability for community groups. Although much more successful, this schedule was never static and changed constantly through the 1980s. It does however give a reasonable impression of the long term balance of service types.

Social benefits due to the innovation were immediately apparent and appreciated. Access provided for voluntary groups to clubs and day centres relieved them of the burden of hiring their own transport, as well as integrating them into the community through use of a public bus. There were indirect benefits in savings permitted to various public bodies, for example, it allowed the rationalization of rural child health clinics to go ahead without loss of access, with the health authority meeting any increased travel costs. Centralization of other health facilities was permitted in the same way, together with a reduction in practitioners' house calls, and avoidance of subsidized taxi fares for visitors to psychiatric day centres (Childs 1978). Cooperation amongst the relevant public and voluntary bodies was also enhanced by joint discussion of problems through the management committee.

From 1979 the system was operated by the Lincolnshire Road Car Company with a 20-seater bus, a tail-lift and places for two wheelchairs. A school contract was added to the already diverse mix of services, and the bus continued to run for 80 hours per week. The clearest statistical indicator of success is the increasing patronage: from 8300 in 1978/9 to 23,900 in 1983/4. Operating costs have remained high and deficits were running at about £9-10,000 p.a., but the subsidy per passenger has declined from £1.26 in 1978/9 to £0.52 in 1982/3. In its more recent form, the Bassetlaw project has been continually praised by the County Council, with its high patronage and undoubted popularity apparently justifying its costs (Nottinghamshire PTPs 1980/81-1985/6).

Unfortunately, the Bassetlaw Community Bus was brought to an abrupt end by deregulation. After the Lincolnshire Road Car Co. was privatized, the new owners demanded a contract price which was too high for the County Council and the scheme collapsed in 1986. The importance of maintaining continuity for welfare organizations meant hasty adjustments were made, and the former routes were split up among social services transport and other operators.

(c) Others
No other scheme resembling a multiple service bus has become as well known as the two described above. It is probable that there are many other bus services which might qualify, through catering for different trip purposes and client groups consecutively, but perhaps less ambitiously diverse. An 'extended school contract' might qualify, through running school trips in the peak and stage carriage in the off-peak (chapter 6). Similarly, the school minibuses made over by Grampian Regional Council to local groups for use in the off-peak as community buses (chapter 10) might be regarded as 'consecutive' dual-purpose transport. The Durham 'Ruralride' service, mentioned in the last section, might also qualify through its partial use for education and social services clients. Some aspects of the ESCORT scheme (see below) also resemble this mode of operation. Finally, the Rural Transport Innovation Grant has assisted various new bus enterprises aiming to serve diverse passenger markets in a similar way.

4.3 ALTERNATIVE FUNDING - BUS CLUBS AND OTHERS

It could be argued that the motivation behind any UCM is ultimately economic, and new modes are developed to do the same job with less money, but the main interest remains the modus operandi which is the prime criterion for classification. There are however a few cases where the *modus operandi* and vehicle types appear quite conventional, but which are financed in a different way. Of course, the voluntary sector has often relied on local fund-raising events to supplement their income; many CT schemes have been able to claim from various local authority budgets other than transport, and have been able to obtain volunteer workers through government employment measures such as the Manpower Services Commission and the Community Programme. Apart from these, otherwise conventional bus services that are needed but not provided by the free market nor supported by local authorities might be funded by other organizations not normally having transport responsibilities.

(a) **Subscription Bus (The Horncastle Bus Club)**
The rationale for the subscription bus (or cooperative bus) is to guarantee adequate financial backing for a local bus service by forming a club of potential users from amongst the community and paying for the bus by subscriptions rather than fares. Club members then travel 'free', but the advantage is that bus revenues are not directly dependent on patronage but on the number of travellers who think it worthwhile to join the club. Non-members may travel on payment of a normal fare. Subscription buses have been popular in the United States, but always in urban areas, usually involving a group of commuters clubbing together to hire a bus in order to avoid the cost and aggravation of peak-hour motoring. No interest has appeared in the UK for this mode of travel, and the only example of a subscription bus is located in a rural area - the Horncastle Bus Club in Lincolnshire.

This scheme has sometimes been called a community bus (Kaye & Kite 1983, Kaye 1986), as the services are determined by the Bus Club committee, composed of local people. However, like the so-called community bus in Bassetlaw, the vehicle is contracted from a commercial bus company and driven by their employees, and therefore it should not be described as a community bus. Moreover, the nature of the services are fairly conventional, and the scheme is distinguished primarily by the club organization which is merely an alternative form of financing.

The origins of the Horncastle Bus Club have been described by Mawford (1976) and Kaye & Kite (1983). Concern over the steady contraction of rural services run by the Lincolnshire Road Car Co. led to the initiative being taken by local organizations for some kind of replacement where most needed. In this case, it was the Horncastle Divisional Liberal Association who saw an opportunity to promote 'community politics' through running their own bus service for the benefit of local people. Accordingly, in June 1974 a bus contracted from an independent operator commenced an experimental service, free to passengers, between Donington-on-Bain and Louth via Horncastle. These free runs were very successful and continued for almost a year. In mid-1975, 'conventional' LRCC services in the area finally ceased and for six months another independent ran these on lower frequencies, but in the meantime the Bus Club concept was being established and succeeded in gaining a road service licence for the Horncastle-Louth route. This commenced in January 1976.

Originally this was operated by a 29-seater bus hired from a local company, with a frequency of three return journeys on Wednesdays and Saturdays. Thus, seven villages on the route were provided with a level of service higher than for many years. Early surveys endorsed the value of the innovation for non-car owners (Mawford 1976). In its early years the Club just managed to break even, with 40% of its revenue coming from members' subscriptions, 50% from fares from casual passengers (charged at a higher rate), and 10% from the proceeds of other Club activities such as excursions and social events. Subscriptions have proved a successful basis for guaranteeing income and inducing stable travel patterns; this is combined with some positive attributes of voluntary transport such as the unpaid work of the committee and its sensitivity to local needs.

However, inflation soon upset this careful balance, and despite frequency reductions and more active fund-raising measures, subsidy had to be obtained from Lincolnshire County Council. No economies could be found in using smaller vehicles, as passenger numbers in the early 1980s were demanding a 45-seater. By the Club's tenth anniversary, the climate was more pessimistic (Kaye 1986). Changes in travel patterns had weakened the Saturday services and the county council encouraged the Club to withdraw them, leaving only a single journey on Wednesdays. Finally, like the Bassetlaw scheme, the Horncastle Bus Club fell foul of deregulation. Despite the previous subsidy, the operator registered the route as a commercial service, but it was impossible for the Club to meet the full costs and it folded in early 1987. A conventional bus is still running on the route.

(b) **Free Shoppers' Bus**
There have been examples of buses being financed by traders or retailers expressly for the purpose of bringing customers from outlying areas, where other public transport was thought inadequate, to shopping centres or individual stores. These are usually 'free' to passengers.

The best known early example, and the most exclusively rural case, was the Cromer Market Bus in Norfolk. Started in 1972 as a publicity stunt by a Cromer market trader, it achieved rapid popularity (Moseley et al 1977, vol 2). The bus was contracted from a local coach firm, and was designed to pick up shoppers from six parishes inland of Cromer which previously had few or no other bus services into the town. Another motive was to divert consumers from the competing market town of North Walsham. Absence of fares had the intended effect of filling the bus (although it was impossible to guarantee that all passengers would patronise the retailer backing the service), and by the mid-1970s two 50-seat coaches were needed. The operator was able to cut costs by working the market service into school contract schedules. With the success of the original scheme, contributions were sought from other retailers in Cromer, and by 1976 about 40 shops and some market stallholders were participating. At the same time patronage was about 100 people per week, representing 5% of the total population of the source villages, and considerable social value was reported to be derived from the facility. This novel service ended some years later, and most of the rural area involved has since been provided with a basic shopping service run by commercial bus operators.

In the 1980s shoppers' buses provided by retailers have become more common, although they are nearly always associated with large stores and hypermarkets in edge-of-town locations. The Asda and Tesco chains have been particularly active in providing free buses. As purely commercial propositions, free buses do not necessarily seek out areas of particular need or disadvantage but may compete directly with conventional (perhaps subsidized) bus services, and deregulation would facilitate this. Rather than exploring a sparse rural hinterland, free buses are more likely to seek custom in suburban housing estates. An interesting scheme to resolve conflicts was agreed between Cornwall County Council and retailers, to the effect that free buses would not run where there were conventional stage services, but passengers could travel free on the latter with tickets issued by the stores to customers spending a certain amount. Retailers reimbursed the council for bus tickets, and hence private money was being used to subsidize local bus services for mutual benefit (Cornwall PTP 1985/6).

(c) Contract Bus

Buses or coaches hired by organisations or individuals for their exclusive use for specific journeys would not normally qualify as UCMs, as this practice has been widespread and normal since the origins of the bus industry. Intermittent publicity given to such arrangements as 'works buses' or 'commuter coaches', even if they appear novel at any point in time, are no more than regular contractual arrangements. As a UCM, classified in chapter 2, 'contract bus' refers to an otherwise conventional bus contracted for normal public service purposes by agencies without statutory transport responsibilities. Thus as a last resort, district, town or parish councils might 'buy in' a service out of their own meagre resources; this would now have to be put out to tender. A few cases are mentioned in county PTPs, but it is unclear how common they are.

One example which deserves to be cited is the participation of a health authority in supporting bus services giving access to hospital. Normally only responsible for 'institutional' travel (i.e. the ambulance service), the North West Surrey Health Authority agreed to collaborate with Surrey County Council in setting up an alternative bus service to provide general public access to St Peter's Hospital, Chertsey. Following the withdrawal of a conventional bus, communities in the Byfleet area lost direct access to the hospital. A converted 16-seat minibus was provided by the Health Authority, managed and operated by them under a Restricted Operator's Licence, in order to meet this demand. Financial support was agreed by the County Council. Named 'Peterbus' after the hospital, this is the only such example in the UK (Surrey PTP 1984/5). Since its commencement in 1984 the new bus carried 10,000 passengers a year, and a second was introduced two years later to connect the Chobham area.

4.4 COMPREHENSIVE MANAGEMENT - THE 'ESCORT' PROJECT

(a) Concept

Planners in the public services have often proclaimed the ideal of 'coordination' or 'integration' as means of making the best use of resources, but have found this extremely difficult to implement in practice or only possible in a very restricted form. Successive Acts of Parliament have urged 'transport coordination', while in the changed conditions of the 1980s 'efficiency' in the public sector means 'rationalization' of existing resources. The latter need not necessarily mean cutbacks, but if institutional barriers can be surmounted there is an opportunity to harness total existing resources towards meeting a wide range of travel needs experienced by different social groups. Widening the legal definition of 'public transport' has pointed in this direction, but nevertheless coordination is usually confined to services of the same mode, to the integration of schools traffic, or to local dual-purpose schemes such as postbuses. What might have been *envisaged* was the further integration of conventional public transport with taxis, school, health and social services transport, and facilities offered by voluntary organizations, over a large area. This had always been deterred by traditional thinking, institutional barriers, and the ambitious scope of such a reorganization.

The nearest approach to coordination in the broader sense is a project which commenced in East Sussex in 1982, known as East Sussex Coordinated Rural Transport (ESCORT). The conception of this is described in Charnock et al (1983) and Brown & Barnes (1985). Its origin lay partly in a transport study conducted in the Lewes area by the County Council and the Department of Transport, which exposed a high level of available resources other than conventional services. This highlighted expenditure by the County Council and Health Authority on social services buses, ambulances, schools transport, group hire minibuses, community buses, contract taxis and expenses paid to voluntary groups and private car drivers for carriage of clients. Although this is true of all local authorities, East Sussex expenditure suggests the potential of these 'hidden' or more specialised facilities - education transport (£2.1 million), social services (£0.4 million), health authority transport (£3.5 million), when compared with subsidy to public transport of £1.8 million (1982/3 prices).

The question therefore arose whether the physical and financial resources deployed were really being used as efficiently as possible and providing the maximum public benefit. It is possible that each service may be reasonably efficient in isolation, in catering for its own client group, but it was felt that there was much duplication and a frustrating inability to satisfy a proportion of local needs because the persons concerned did not qualify to use certain types of transport. Apart from cases where 'specialized' transport was essential, institutional segregation of both suppliers and consumers was a severe barrier to a rational coordinated service. The overriding need was to remove such barriers, so that all users could be considered collectively and transport resources deployed in a more flexible manner, implying a more comprehensive management or corporate approach.

A particular problem was that public transport had been declining while the specialist services - contracted schools transport, social services transport and health authority transport - had been growing over the medium term (this is a nationwide characteristic). Thus the scope for meeting *general* needs had fallen while, ironically, there was perhaps more transport available in the wider sense. Transport need was to be regarded as a total market, not segmented by client group, and served by the best appropriate means regardless of which institutional sector was formerly responsible for any group. Some considerable reorganizaton of transport is required, but the potential benefits are great as figure 4.3 graphically demonstrates.

Based on the previous study of accessibility and needs in the Lewes area, the strategy was to establish a commercial 'trunk' network, identify all needs remaining and then to coordinate transport resources of all types in order to satisfy them. Coordination was dependent on two pivotal roles - the banker and the broker. It was found not difficult to reclaim the cost of a multi-purpose ESCORT service against a number of different local authority budgets, via the banker. Day-to-day management of ESCORT required the key role of a broker. This is a concept developed in the United States, and operates in Britain mainly at a small scale as 'local transport advisers' (e.g. Oxfordshire), but otherwise Hampshire is usually credited with pioneering a brokerage system. The East Sussex broker system was developed with the assistance of TRRL (Davies 1987). Requests for transport are received from public sector agencies (the education and social services departments and the health authority) on behalf of their clients. A computerized transport directory is used to locate appropriate services already available, or in some cases to arrange new services if nothing suitable is running. The broker is a communications focus able to respond to changes in demand patterns and to adjust service supply, and hence gradually to refine or widen the scope of transport provision.

(b) Operations

The experimental area for ESCORT is in parts of Lewes and Wealden Districts of East Sussex, from the Newhaven/Seaford area on the coast inland to

FIGURE 4.3 The ESCORT concept (Brown & Barnes 1985, (c) Transport Design & Development Ltd.).

beyond Lewes and Uckfield (figure 4.4). ESCORT occupies a large gap in the transport hierarchy between, on the one hand, the trunk or commercial bus network and rail services, and on the other, various specialized or informal demands such as urgent medical cases or lift-giving schemes. To fill the gap, four main types of service have been developed.

(i) Conventional size buses integrating public and schools traffic.
(ii) Multi-purpose midibuses catering for the public, schoolchildren, social services and health clients, in an integrated manner.
(iii) Minibuses operating in a similar multi-purpose mode, at a more local level, and involving the 'community transport' sector.
(iv) Shared hire-cars, with a more complementary role.

ESCORT does not fit neatly into the classification scheme drawn up in chapter 2, as it is best thought of as a comprehensive management strategy embodying several different UCM types. Most vehicles can be described as 'dual-purpose' or 'multi-purpose' in terms of social groups and trip demands catered for, and according to the list of UCMs in figure 2.2, they contain elements of the school bus, the D/R diversion and multiple service bus. There is also a shared hire-car and a link with a community bus.

Specific new services developed between 1983 and 1985 are briefly outlined below, and mapped in figure 4.4. More detailed descriptions are given in Watts (1985) and Grigg (1987).

(i) Public/school buses
This is by itself not a new idea, but school/stage integration is obviously compatible with the ESCORT principle. Bulk school demands are still carried on 'bespoke' contract buses, but where numbers are lower integration was considered. A previous stage bus and a school contract in the Lewes-Haywards Heath corridor were reorganized as a new public service with improved frequencies. Passenger numbers were unchanged, and a saving of 14% was recouped.

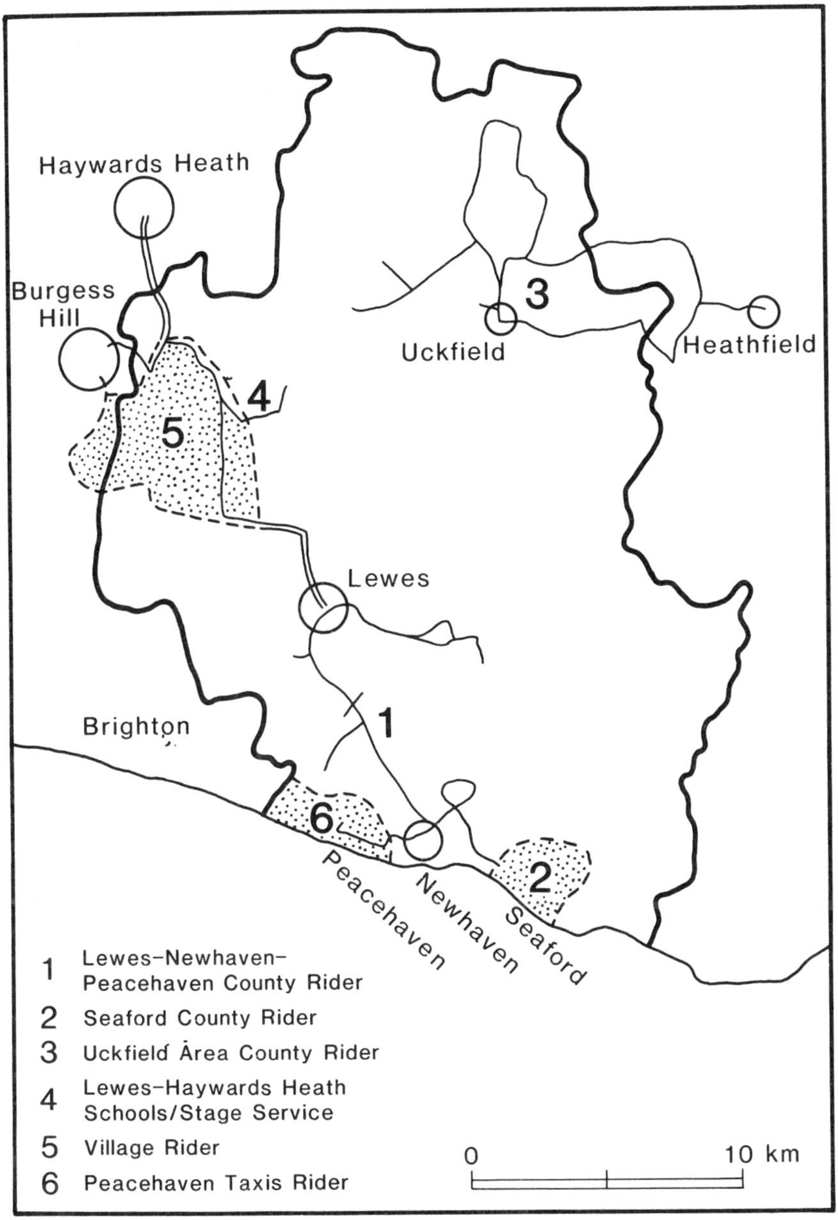

FIGURE 4.4 New services under the ESCORT project (after: Watts 1985; Grigg 1987).

(ii) Multi-purpose midibuses

Three of these were implemented in the first phase of the project, each marketed under the name 'County Rider'. Vehicles used were 24- or 30-seater buses with rear wheelchair lifts, the first of which was built to order. They were operated by Southdown Motor Services under contract to the County Council. It is central to the ethos of the project that such services should be available to the elderly, schoolchildren, social services clients and the disabled, any of whom might require a D/R doorstep facility. An advance booking system was therefore set up, under the control of the broker. On some journeys an attendant was felt to be necessary to assist passengers with difficulties, and this is an extra cost item.

The new services were:

$1. *Lewes-Newhaven-Peacehaven County Rider.* This replaced two conventional buses, one social services bus and some voluntary car trips, and took some schoolchildren (allowing a rationalization of other school contracts), mainly focusing on facilities in Newhaven and Lewes. Many passengers were disabled and D/R diversions were common.

$2. *Seaford County Rider.* Replacing two conventional buses, this operates a town service within Seaford and a connection to Newhaven. It also connects with the first County Rider to provide access to a special school and day centre in Lewes.

$3. *Uckfield Area County Rider.* This replaced three low frequency bus routes around Uckfield, part of another, and also the loss-making postbus at Heathfield. Some social services and health trips are incorporated but not education.

The general conclusion on County Rider buses was that they had been operationally successful and popular, with users receiving a higher quality service and patronage increasing. A survey was done by TRRL on the first experiment, Lewes-Newhaven - Peacehaven (Watts 1985). On complementary services by the County Rider and a conventional bus the weekly loading in late 1983 was 1478, of whom 888 were general public, 570 were schoolchildren and 20 social services clients. Pre-ESCORT equivalents carried 1230 passengers, and hence the increase in 'general public' patronage was 17%. This level of custom remained steady during the second year (1984/5). The Seaford County Rider was conveying 3040 'general public' trips per week in 1984, an increase of 5% over pre-ESCORT equivalents. Weekly loadings for the Uckfield County Rider showed a similar increase, carrying 430 general public, 60 social services and 45 health authority clients. Use of the wheelchair lift was no more than 2% on any of the new services, with an additional small proportion making use of the rear doors and lower steps (Grigg 1987). Surveys on the Lewes and Seaford buses concluded that 75-80% of passengers were female, 36-40% were aged 65+ and over 90% were able bodied; about half of trips were for shopping. General attitudes towards the buses were overwhelmingly favourable, and there seemed few problems over the mixing of client groups.

Cost savings were expected from the rationalization of vehicle use, although there would be cost implications of the higher levels of service offered. Setting-up costs were significant as new vehicles or adaptations were required plus much planning and implementation work. Features of the operating account in each case are the extra costs of the new services, outweighed by savings of under-used capacity on conventional buses and in displaced education and social services transport, plus enhanced revenue due to higher patronage. Overheads such as the on-bus attendant and the booking system meant that overall public expenditure was little changed in the case of Seaford and Lewes (in its first year only). The latter in its second year (largely due to increased revenues) and the Uckfield service showed overall savings of up to 10% (Watts 1985, Grigg 1987, Brown & Barnes 1985).

(iii) Village minibus

Serving smaller settlements over a more dispersed area, a minibus known as 'Village Rider' was inaugurated to fulfil a flexible multi-purpose role. Its territory is around the Plumpton/ Ditchling area, with connections to Lewes, Haywards Heath and Burgess Hill (figure 4.4). A further innovation is its joint operation by the Southdown bus company and the Downland Community Bus committee. The former is responsible for peak hour school and social services transport and midday stage carriage, while in the evenings and weekends it is taken over by volunteer drivers for local services as a community bus. A wheelchair lift is carried, and D/R pick-up is available via the ESCORT broker. More journeys are offered than before, some to previously transport-less places, and weekly loadings were about 500 (late 1985). This improved service was costing about 2% more in revenue support.

(iv) **Shared hire-car**

In the Peacehaven area, a shared hire-car was introduced to provide a small scale local facility, complementing other transport. Known as the 'Peacehaven Taxi Rider', it is a 8-seater minibus licensed under the 1981 Public Passenger Vehicles Act, with a basic schedule of departure times but D/R routeing.

The various new transport services instituted under the ESCORT umbrella could all claim some success, in terms of public expenditure savings, higher service levels, increased patronage, or more subjectively through the public acceptability of integrated traffic types. A summary of results is given in table 4.1. It appears consistent with these early figures for Brown & Barnes (1985) to claim that overall expenditure savings of 10% on a total budget of £200,000 should be attainable. These are contradicted somewhat by financial data given in Grigg (1987), taking into account differences in service levels, which produce an overall saving of 6%. Attributing the costs of the booking system and a proportion of setting-up costs would cancel out this. An overall 6% increase in general public patronage vindicates a policy of not only rationalizing, but improving, transport supply. The brokerage system seems to have been satisfactory (Davies 1987) but with scope for further development.

TABLE 4.1 Performance of ESCORT services 1984/5 (source: Brown & Barnes 1985).

	Start date	Improvement in service level	Passenger increase %	Public Cost Band £'000	Change in Public Expenditure %
Lewes-Newhaven-Peacehaven County Rider	7/83	Good	17	20-25	-9
Seaford County Rider	9/84	Good	5	20-25	0
Uckfield County Rider	11/84	Marginal	5	45-50	-10
Peacehaven Taxi Rider	12/84	Very Good	n/a	10-15	n/a
Village Rider - Plumpton/Ditchling	4/85	Very Good	9	25-30	+20
Schools/Stage Service Lewes-Haywards Heath	4/85	Good	8	20-25	-40

It was the intention in East Sussex to develop other integrated transport schemes within the ESCORT area, and also to extend the concept to other parts of the county. In 1986-8, further multi-purpose County Rider buses were launched in Rye, Battle and Bexhill, and another shared taxi in Polegate. Bus deregulation could have been a disruptive influence but commercial services have not interfered. The big question that remains of course is why the concept has not been adopted by other local authorities. There seems to be nothing in the pre-conditions for ESCORT which is peculiar to East Sussex, and it could in principle be implemented anywhere. Other authorities are presumably waiting for the newly deregulated system to settle down before considering another major upheaval.

Chapter 5

URBAN PSV MINIBUSES

5.1 THE 'MINIBUS REVOLUTION'

Other UCMs in this book have been based on small capacity public service vehicles, e.g. dial-a-rides, postbuses and community buses, but in these cases their 'unconventional' status depends upon their mode of operation with vehicle size being a subordinate factor. As explained in chapter 2, vehicle type is, by itself, only a valid UCM indicator if it incorporates new technology. The case for including urban PSV minibuses in a review of UCMs is therefore arguable, and rests overwhelmingly on the phenomenal growth of such services in British towns and cities since 1983, often referred to as the 'minibus revolution'. It is the sheer novelty and impact of this transformation that has led to the unquestioning assumption that these constitute "innovations" on a par with other new modes. Although some minibus services have a demand-responsive element, they do not normally come within the UCM definition in chapter 2 and were therefore omitted from the list of modes in figure 2.2. Neither can they claim to be catering for minority needs or unprofitable environments. The novelty is that a common mode (the minibus) is used in an unaccustomed manner (as a high frequency public service) in unfamiliar environments (city centres). It is included here largely in deference to the widespread appraisal of its 'new' image.

The irony about minibuses is that they had long been considered candidates for rural areas but had not succeeded in making any significant impact, and then suddenly made a dramatic and successful appearance in urban areas where it was thought they had few obvious advantages. Minibuses were often the popular 'solution' for rural routes in the belief that low loadings made conventional size buses superfluous and smaller vehicles would be cheaper. This was usually countered by the conventional wisdom on minibuses that (a) labour costs were 70% of the total and therefore small vehicles would make little difference to costs, and (b) most rural routes had some occasional peak loads, particularly schoolchildren, and reserve capacity was therefore essential. In towns, it was generally assumed that loadings were too high for anything other than a full size bus.

Historical aspects of small buses can be obtained from the transport press (Anon 1986), and are briefly summarised in Hall (1988). Small vehicles were normal in the early years of the bus industry but had disappeared by the time of peak demand in the 1950s. It was not until new PSV Conditions of Fitness Regulations were passed in 1958 that small buses could legally be reintroduced by converting delivery vans for passenger use. Apart from the economic question, design problems, particularly lack of space, made these 'converted bread vans' unattractive. Later conversions achieved better standards of space and physical access, with up to 16 seats, such that a few urban authorities experimented with minibus shopping services between 1967 and 1972 (Hall 1988). In the 1970s, the advent of dial-a-rides, postbuses and community buses popularised small vehicles to some extent, mainly in rural areas. On 'conventional' bus routes in the countryside, minibuses were introduced only where narrow lanes and steep hills forbade anything larger. However, it was becoming appreciated that they could play a useful role in suburban housing estates, where a smaller vehicle could achieve greater penetration of residential streets and would be more acceptable to the inhabitants. A few distinctive mini- (8-16 seats) and midi- (17-35 seats) bus services were operated by NBC subsidiary companies in the 1970s, for example at Beaumaris (Anglesey), the Gower and Vale of Glamorgan (South Wales), Blandford (Dorset), Hereford and Evesham/Pershore (Worcestershire), Truro (Cornwall), Alton (Hampshire) and the High Wycombe/Maidenhead area. Most of these were small scale operations with rarely more than one vehicle deployed (Select Committee 1977-78). These were contemporary with the 'Cambridgeshire midibus' and various RUTEX schemes (see chapter 4).

The modern phase of commercial minibus deployment, in which numerous vehicles run high frequency urban routes, is usually credited to the NBC company, Devon General, which inaugurated such services in Exeter in February 1984. There were however predecessors worthy of mention. In 1982, the Potteries Motor Traction company started running small vehicles in the Stoke-on-Trent area, under the name 'Flexibus'. Initially, these were 7-seater converted taxis of the London type (Austin FX4); in addition a trial was conducted on behalf of the Department of Transport of the 'CR6' Range Rover conversion, designed to be suitable for the disabled. These were later supplemented by 12-seater minibuses. In 1983, Strathclyde PTE started experiments with different vehicle types in areas of Glasgow. By the following year there were nine 'Microbus' (11-17 seats) services in various parts of the city, some of which were partly demand-responsive. Devon General's 'Exeter experiment' started with 12 minibuses on one route, and by the end of the year 1984 this had expanded rapidly to four routes with 38 buses. Replacing conventional buses, the 16-seater Ford Transits ran at higher frequencies in central locations but also penetrated more deeply into residential estates and allowed a 'hail-stop' facility (NBC Annual Report 1984, Turner & White 1987).

Wishing to try out minibuses in smaller towns, NBC implemented them in St. Austell (Cornwall) in late 1984. In the following year, urban minibus services started to proliferate. The largest fleet introductions were in Leicester (68 vehicles), Worcester (61) and Weston-Super-Mare (46), while Exeter expanded further. Other towns seeing the introduction of minibus fleets of ten or more vehicles were: Ryde (Is. of Wight), Coalville and Hinckley (Leics.), Oxford, Bridlington, Norwich, Taunton, Shrewsbury and Gloucester (Turner & White 1987, Hall 1988). The situation at the end of 1985 is shown on figure 5.1. Simultaneously, great efforts were made to present minibuses as a 'new product' with bright marketing packages, unusual names and eye-catching colours. The permanent nature of the phenomenon

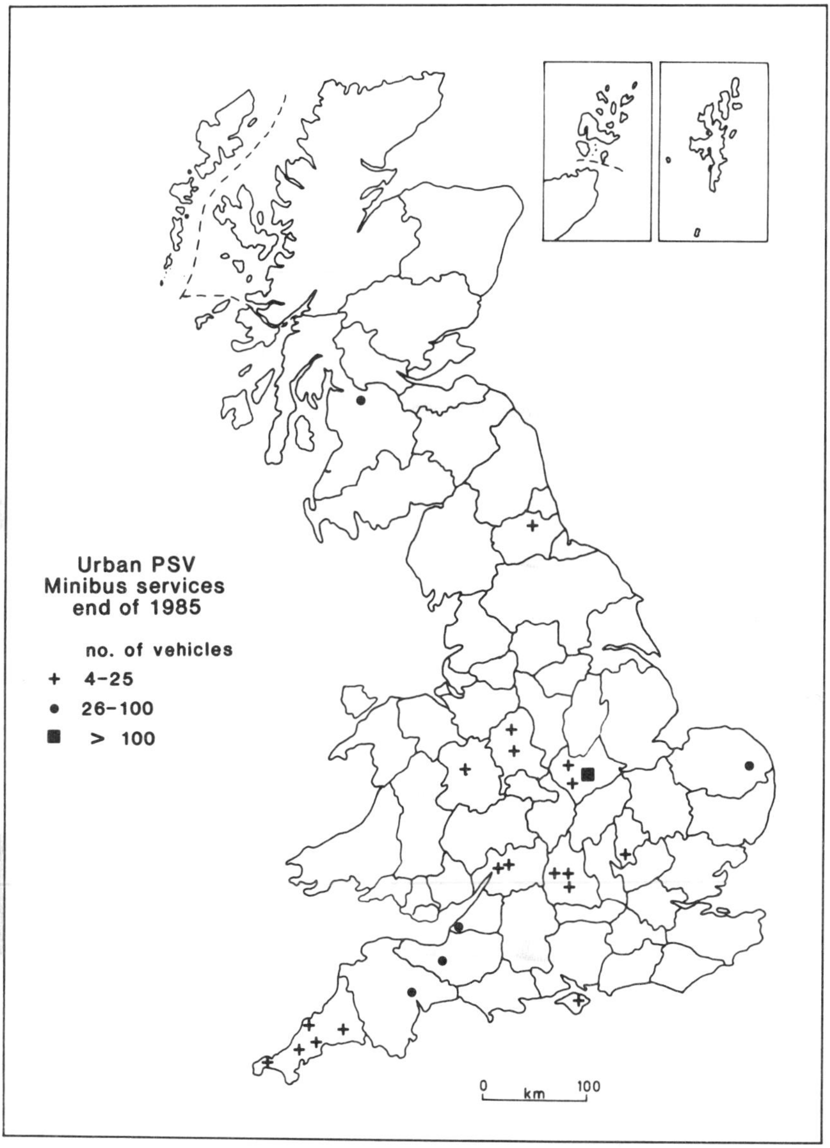

FIGURE 5.1 Urban areas with minibus services 1985 (after: Hall 1988)

was attested by NBC ordering 1200 new minibuses in 1985.

Clearly, a dramatic change in attitudes had occurred in a very short time. In urban environments, the small capacity of minibuses can only be compensated by higher frequencies, and this would normally mean a larger vehicle stock and higher labour costs. If the latter could be overcome, the other aspects could be turned to advantage. High frequencies, shorter waiting times and the 'user friendly' characteristics of small vehicles all have the potential to generate patronage. So does the greater penetration of housing estates and the 'hail-stop' facility - arguably an 'unconventional' feature - where it can be allowed on traffic and safety grounds. Improved speed and manoevrability in traffic were further selling points. Marketing was designed to take advantage of all these.

Second, improvements in vehicle design had led to higher standards of comfort, durability and reliability, although these were still conversions of standard vans such as the Ford Transit and Mercedes-Benz.

Third, changes in the labour market had played into the hands of managers and allowed cost-cutting. High unemployment rates together with the Conservative government's "union-bashing" legislation had facilitated the erosion of national pay bargaining and the imposition of inferior wage levels and employment conditions.

No doubt the main reason for the climate of change was anticipation of the 1985 Transport Act. The Exeter experiment had begun just a few months before publication of the *Buses* White Paper (Department of Transport 1984). Removal of quantity licensing and the allowing of direct competition on commercial routes spurred both established operators and hopeful rivals to look for an attractive new product to give them a competitve advantage. New entrants to the industry saw no reason to risk the high capital cost of conventional size buses, and for many only minibuses were affordable. Also, the anticipated privatization of the NBC subsidiary companies and of the PTE bus fleets in the conurbations encouraged the public sector to forestall future opposition by adopting similar tactics. The low cost but limited life of minibuses was an appropriate response to short term uncertainty (Murrell & Newport 1985, Stonham 1985).

5.2 MINIBUSES AFTER 1985

Rather than causing any dramatic switch in bus company policies and operations, the advent of deregulation merely intensified the trend towards minibuses, as the new regime had been realistically anticipated. A further factor was the privatization of NBC subsidiaries, completed between 1986 and 1988. The immediate rise in numbers was even more dramatic than before: between October and December 1986 the number of minibuses being operated by current or former NBC subsidiaries jumped from about 1800 to about 3000 (Hall 1988). Adoption by former municipal and PTE companies and independents was slightly later than NBC but, including these, there were about 4000 minibuses in service by mid-1987. Estimates for the end of 1987 vary between 5000 and 6000. This constitutes about 10-15% of all passenger vehicles on local services (Hall 1988, Balcombe et al 1988). In addition, many more minis are in use for private hire and schools work. While under 50 locations were served by minibuses at the end of 1985, this rose to 250 by mid-1987 and nearly 400 by the end of the year. These are mapped on figure 5.2, which shows an even distribution of urban centres with few obvious gaps. Lower minibus penetration in Scotland is in contrast with the extent of bus competition after deregulation; possibly the decision not to privatise the Scottish Bus Group may have had an effect.

A few examples could be presented to draw attention to relevant aspects. Service details can be found in Turner & White (1987) and Knowles et al (1988), while Hall (1988) gives a case study of the Tyne-Wear area. Three years after it began, Devon General's Exeter fleet had taken over

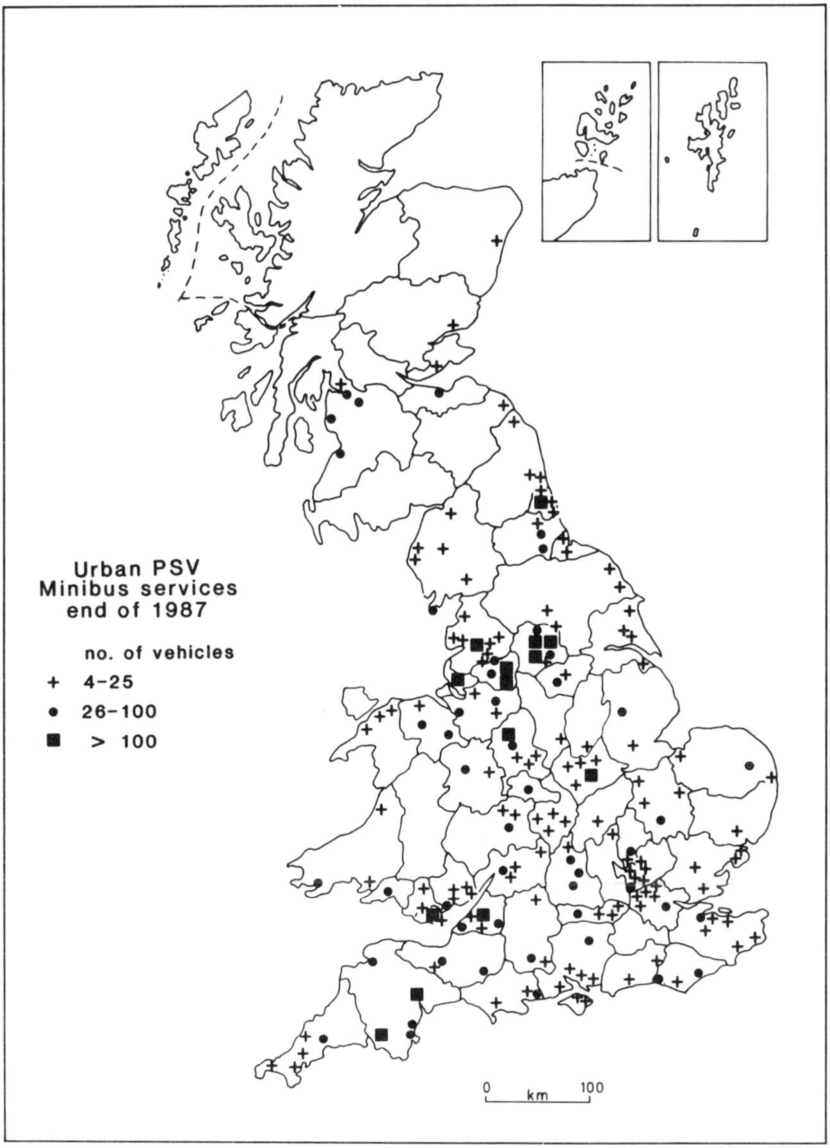

FIGURE 5.2 Urban areas with minibus services 1987 (after: Hall 1988, Balcombe et al 1988).

all routes in the city, with 100 buses. It was making a profit, receiving no subsidy, carrying nearly twice as many passengers as before, and employing more drivers, albeit at lower wage rates. The company expanded to the other main towns in the county, and in 1987 tested deregulation to the full by starting a minibus service in distant Oxford in competition with the local company. Other examples of direct competition among minibus operators include Manchester, Preston and Blackpool. This is primarily an attribute of the deregulation of route licensing and not due specifically to the use of small vehicles. An analysis of the results of direct competition between minibuses and conventional size buses would be of great interest. In Leicester, Midland Fox (ex-NBC) put a fleet of over 100 minis into competition with the municipal operator. The wholesale conversion of complete networks from 'conventional' to 'minis' is becoming a common objective, especially by ex-NBC companies. Apart from Exeter, this has occurred in Worcester, Taunton, Weston-Super-Mare, Milton Keynes, Clacton (Essex), Washington (Tyne-Wear) and other places. Otherwise, more selective introductions have occurred for a variety of reasons. PTE bus operators began to order minibuses in 1985/6, before privatization, the most energetic of which was Greater Manchester Transport with deployment of over 100. Despite being immune from deregulation, London is experiencing a similar situation with some routes being put out to competitive tender under the 1984 London Regional Transport Act. A scattered set of suburban routes have gone to minibuses run by London Buses, London Country or independents; the most coherent 'mini' network is in Orpington.

Minibus deployment strategies may be classified as follows (Hall 1988).

(i) *Low frequency, low capacity.* Either a pre-deregulation rural mode, or on tendered routes without competition.
(ii) *Substitution of conventional networks.* To cover a whole town, usually introduced in phases; examples given above. Pre-existing schedules would be altered to allow greater penetration of

residential areas, faster running times and hail-stop.

(iii) *New networks as a competitive device.* To encroach upon another operator's territory where existing or potential demand is not yet fully satisfied.

(iv) *Selective introduction.* To complement conventional bus services by putting 'minis' on selected routes, such as:

(a) identifying gaps in the network and/or responding to expressed demand;

(b) 'creaming off' profitable routes run by a competitor;

(c) on off-peak workings of routes normally served by full size buses;

(d) complementing, or competing with, other modes, such as urban railways.

(v) *New networks as a transitional device.* Either:

(a) to test the market or to stimulate travel, or

(b) to attract a new workforce and justify new employment conditions.

Despite constant adjustments through experimentation or competition, some general trends can be discerned in operational characteristics (Turner & White 1987). In terms of route configuration, it is common for cross-town routes to be replaced by minibus services terminating in the central area. At the outer ends, there is much greater use of 'loop termination' where minibuses complete a circuit through residential estates before commencing their next journey into town. These sections are commonly hail-and-ride with no fixed bus stops. In many towns, residents demanded such services be brought into their estate. A variety of routes originating in different suburban areas very often merge in the central zone to give a high frequency service on the main commercial streets. In Oxford, to cope with the peak, a 5-minute frequency service was provided in the centre, splitting at the outer ends to serve separate estates. In Exeter, the policy is to build up high frequency corridors, with few diversions off the main roads. Access is therefore improved both to residential areas and also to congested parts of the centre. In some towns, narrow streets in the core area may now be penetrated by minibuses where they were formerly out of bounds to larger vehicles (e.g. in Cornwall). Minis might be allowed in some semi-pedestrianised zones. On the other hand, their high frequencies have in some cases led to traffic flows high enough to cause the local authority to close streets to buses. The 'hail-stop' facility also varies among towns, generally being easier to implement in smaller towns where it might be possible anywhere outside the central zone. In larger cities (e.g. Norwich, although Exeter is an exception), it might be confined to certain parts of housing estates. Also, there is some evidence of a switch from cars in favour of informal park-and-ride. In Norwich, some commuters leave their vehicles in suburban car parks and take a minibus to the centre.

Higher service quality is also apparent in terms of operating hours. This varies widely by area and route, and sometimes depends on subsidy; there may be some implicit cross-subsidy in sustaining evening and Sunday services. Devon General for example has a policy of maintaining dense coverage throughout the day for seven days a week. Departures up to 2330 hrs all week have reportedly stimulated evening entertainment activities. This has been made possible partly by the imposition of flexible working practices with no overtime rates for unsocial hours. Improvements in frequencies upon the conversion of previous networks to minibuses, although general, are difficult to quantify. Routes might not be comparable, and frequencies vary by time of day. An increase of about three-fold is a reasonable guideline, although this may vary in specific cases between factors of 1.5 and 6. Headways are correspondingly reduced, most often 10 minutes in the peak but sometimes 5 minutes. Less time is needed for passenger boarding, and this plus the greater manoeuvrability of the vehicles improves average speeds. However, bunching in traffic does occur and timetables may deliberately be kept flexible (i.e. not quoting specific departure times). Observed headways may get down to one or two minutes (survey by Turner & White 1987). Flexible working or radio control can alleviate this, with drivers given discretion to alter schedules as appropriate.

Since deregulation, minibus services have been marketed more vigorously than ever, and imaginations have run riot with respect to trendy brand names and psychedelic colour schemes. Development of vehicle types has also been given a stimulus. Three stages have been recognised by Hall (1988). Until mid-1986, the 'first generation conversions' were virtually limited to two models - the 16-seat Ford Transit and the 20-seat Mercedes-Benz 608D. Without the market incentive for better designs, lack of space caused complaints about "converted bread vans". 'Second generation conversions' appeared in 1986, primarily in the form of restyled Transits and Mercedes plus 16-20 seat versions of the Freight Rover Sherpa. In 1987, Dodge/Renault and Iveco-Ford produced larger purpose-built bus bodies, which heralded the third phase of innovation; particularly attractive were the Optare City Pacer and StarRider models; these all had capacities of 25-33 seats. Design standards were improving in these later models, providing more space and headroom, better passenger access and greater durability. The trend was towards larger vehicles, which perhaps should be described as 'midibuses', with some operators upgrading from older 'minis'. Currently, the market is strong enough to support the development of purpose-built integral (chassis-body) models, such as the Metro-Cammell Weymann Metrorider and the Peugeot-Talbot Express Pullman. It should be noted that the small bus market has also been enhanced by demand from 'community transport' organizations (see chapter 9).

A combination of high frequencies, better origin and destination access, improvements in vehicle design and wide publicity have generally succeeded in raising levels of patronage over the conventional bus predecessors. As 'before' and 'after' networks may differ substantially, the level of growth is difficult to calculate on a realistic basis. Simple estimates can produce extraordinary figures. In towns surveyed by Turner & White (1987), the increase in traffic was estimated at 25% in Leicester, about 40% in Taunton and Norwich, about 50% in Exeter and Weston-Super-Mare, and 75% in Stafford. These seem to relate more to frequency increases of the levels cited above than to total seat capacity (which takes account of different vehicle sizes). Where previous services were weak, extremely large increases may be apparent: Cornish towns averaged 100% while St. Austell claimed 400%. These of course vary widely among routes within towns, and by time of day.

The composition of this increased traffic has several points of interest. It is derived from a combination of previous users making more trips, and new bus passengers either transferring from other modes or making new trips. In Exeter for example, 64% of minibus passengers formerly travelled by bus, 18% walked, 9% used a car, and 3% used a taxi and/or train. In many places it was thought that the increase was mainly due to bus users travelling more often, but this varies: new users were largely credited in Norwich, and in Cornwall where 80% of the increase was from former walkers or car users. Average trip lengths are often shorter, reflecting transfers from walking; but this may be encouraged by 'flat fares' policies. Large increases in leisure traffic were widely reported, reflecting the new evening and Sunday services. Greater usage by economically active males was recorded, implying more work journeys and transfer from other modes. Some informal park-and-ride has already been noted. In larger towns, additional traffic has been generated from higher income areas, as minibuses are perceived as an acceptable alternative to road congestion and parking charges. Taxis have also suffered from minibus competition in many areas, leading in some cases to vehement protests from the taxi trade. In Exeter however, taxi use has increased, as minibuses have attracted more people into town at night, who then take a taxi home.

Traffic increases on minibuses also reflect a generally positive attitude by the travelling public. Surveys have shown favourable responses to frequencies, route access, speed, and value for money. More subjective factors, deliberately intended as part of the 'package', also seem to be successful, as attested by wide approval for 'friendliness' and staff courtesy. The distinctive fleetnames and colours are generally liked, and the marketing ploy of presenting minibuses as a 'new product' - neither a bus nor a taxi - has struck a responsive chord. Approval of interior comfort has been more variable, and is lower on the older van conversion models where the crucial factor seems to be lack of space for people and especially for luggage. Passenger access in terms of step height and door design has provoked complaints, especially by the elderly. For these reasons, many operators are converting to larger vehicles. Overcrowding often sparks complaints, but this is obviously variable by route and time of day. Some company policies are able to respond flexibly to this, such as by radio-paging additional vehicles. Hail-stop could easily be abused, causing the bus to halt too often, but there are relatively few reports of this.

Finally, it should be pointed out that there has been renewed interest in the use of minibuses in *rural* areas. The 'conventional wisdom' referred to in the previous section probably still applies, but in the new situation there may be some scope for reducing staff costs. Rural PSV minibuses seem now to be associated with new entrants to the industry who are small firms only able to afford small vehicles, and possibly owning just one. Also, they are likely to be supported by the Rural Transport Innovation Grant, for which use of a minibus *per se* seems to be regarded as "innovative". Almost half of the schemes approved by the end of 1987 are PSV minibuses, apparently without any other 'unconventional' features (see section 14.4(b) and figure 14.2).

5.3 THE ECONOMIC ARGUMENT

In the run-up to deregulation, much of the intense argument over the policy was devoted to the economic performance of minibuses vis-a-vis conventional ('big') buses. Little empirical data were available and the argument was almost entirely theoretical. Earlier studies (e.g. Martin 1978b, Tunbridge 1980) and limited experience concluded that minibuses were very expensive means of public transport, only appropriate where physical restrictions prevent access by larger vehicles or where traffic is consistently light, and where there are no peaks. This view may have reflected the lack of any determined effort prior to 1983 to test the mode, as even before deregulation minibus innovation seemed strong enough to contradict this. Many factors had of course changed.

Theoretical work comparing minibuses with conventional buses can be technical and complex, and only a very brief review is offered here. Glaister (1985) attempted to model competition between big and small buses on a heavily used corridor route, using data from London. Correctly predicting the introduction of smaller buses on deregulation, results were based on the belief that fares would be much higher. Minibuses operating at low load factors and high fares would attract passengers who place a high value on their time, benefiting from the faster speeds. Hence, product differentiation would segment the market, but there would be fewer cheap big buses and so poorer people might be worse off. However, the overall improvement in service quality would yield an expansion of the total market, within which there would be room for everyone.

Similar conclusions were reached by Oldfield & Bly (1985) and Bly & Oldfield (1986). Also working on London, they concluded that the optimum size of bus was 60 seats, but also that minibuses of 10-25 seats were likely to be profitable and to give a small net social benefit if they operated along the same routes as existing LRT services. This was based on the alternative premise that, operating on the same routes, passengers would take the first bus to arrive regardless of its size. Minis were therefore likely to achieve high load factors and to cover costs at competitive fares, so long as their numbers were not so great as to cause congestion and undermine big bus load factors too seriously.

A more negative view was advanced by Gwilliam et al (1985). It was estimated that based on current differences in cost/vehicle mile and seat capacity, the same subsidy rate and load factor would mean minibus fares nearly four times as high as on conventional buses. Time savings on minis would be insufficient to offset this, and they could only be commercially attractive if they could achieve much higher load factors than their bigger competitors. This could only work if they operated only the denser route sections at the more popular times. This in turn would force big bus operators to a combination of reduced services and higher fares, and hence the prospect of inequitable social consequences. It was therefore essential to restrict minibus numbers, i.e. to retain a degree of regulation.

In practice, fare levels changed very little upon the introduction of minibuses and are generally equivalent to those on conventional buses. This seems to invalidate much of the foregoing argument.

Despite a survey of NBC minibus operations, Turner & White (1987) were unable to use empirical data for economic comparisons due to the new commercial confidentiality. Instead, a complex simulation model was developed to compare the likely performance of different bus types and sizes over a hypothetical urban corridor route. It took a whole-life view, and allowed for variations in essential parameters such as vehicle lifespans, wage levels, vehicle capital costs, availability, maintenance costs, route lengths and round trip times. As is normal, simple replacement of big buses by minis was assumed to be at a 1:3 ratio to give the same capacity; this 200% increase in service mileage raised total discounted costs by 70%. The additional patronage needed for viability implies a service elasticity of +0.4, a level which is achievable, but not everywhere. Results were also shown to be sensitive to changes in most of the parameters listed above. Peak (normal) and 'flat' profiles gave similar results, but excessive peaks (as for schools traffic) could be handled more cheaply by conventional bus.

It is to be emphasized, however, that real cost comparisons are impossible until long term operating experience has been acquired. As with any aspect of deregulation, definite trends are obscured by short term uncertainties and fluctuations such that the situation is in constant change. A few minibus services were withdrawn in the post-deregulation turmoil, yet the new mode will probably survive into the medium term even if its degree of market penetration cannot be predicted. The advantage of low wage rates may not be sustained if the economy improves with a real fall in unemployment. Worsening city centre congestion might enhance the minibus' advantage, or it might deter bus services of all types.

5.4 PSV CARS

The logical extension to the small bus concept is to use a car for public services. At present the only example of a car being used in the same way as a bus is the 'Countrycar' service in rural parts of Essex. Other uses of cars for public transport are virtually confined to 'community transport' modes (chapter 11); otherwise the nearest approaches are cars used by the Post Office as postbuses (chapter 7) and some shared hire-car or taxibus schemes (chapter 12). In Essex, the car was operated by the Eastern National Omnibus Company on regularly timetabled routes, using a professional driver. The vehicle is a Peugeot 504 estate car converted to carry seven passengers, and embellished with distinctive bus company livery.

Innovation was stimulated by the bus company's withdrawal of two routes in Braintree and Halstead in November 1981. Countrycar was then deployed on a diverse weekly schedule, allowing the redrafting of other conventional bus routes. The original schedule was: Tuesday- into Halstead from villages to west; Wednesday- villages to Braintree; Thursday- villages to Sudbury (Suffolk); Friday- into Halstead from villages to north; Saturday- one route into Braintree, and one route into Witham. There was a significant cut in fares upon conversion from bus to car, and the service was subsidized by Braintree District Council (Kaye & Kite 1983). This schedule has changed over the years, but since deregulation the privatized Eastern National bus company has continued to operate the service in a similar manner. Again, a single car runs different routes on different days, connecting outlying villages with Colchester, Braintree, Halstead and Sudbury. The new routes have replaced conventional buses. All are subsidized by Essex CC and the district councils.

Chapter 6

SCHOOL BUSES

6.1 THE INTEGRATION OF SCHOOL TRANSPORT

Transport of pupils to and from school may be 'unconventional' through use of a dual-purpose mode, that is, where schools traffic is combined with adult fare-paying passengers. In most cases this is normal anyway, and UCM status depends on the existence of transport provided exclusively for the journey to school, which may have potential to accept the general public. The principle of school transport provided or organized by the local education authority, and free to pupils, was laid down by the 1944 Education Act on the reasonable grounds that, as the state made it a legal requirement to go to school, then the state must provide the means of getting there. Qualification for free travel was based on minimum home-to-school distances of two miles for pupils under eight years of age and three miles for pupils over eight. These arbitrary limits have been a major bone of contention ever since, implying that children living within the distance bounds are able to walk to school in possibly unpleasant or dangerous conditions, or be taken by their parents or else pay a fare.

Hence, there are three ways by which LEAs may meet their statutory obligations to provide school transport.

(i) By paying a local bus operator to carry them on existing stage services, where the routeing and timing are convenient. This should be regarded as the 'normal' mode. Although buses are by far the most common, cars, taxis and trains are used where appropriate, and in the Highlands of Scotland ferries and in extreme cases aircraft may also be used.

(ii) By contracting an operator to provide a bus exclusively for school pupils on routes and at times determined by the locations of schools and pupils' homes.

(iii) By exclusive use of the LEA's own vehicles.

The first of these could be interpreted as dual-purpose transport in that pupils share the bus with ordinary passengers, but this is never regarded as 'unconventional' because it is the normal means of provision established for over forty years. School buses as a UCM are those under (ii) and (iii) which by various means have been made available to the general public. Minibuses owned by schools for the purpose of serving extra-mural activities are best considered as 'community transport' (chapter 9).

The school bus is very much a minority mode. The 1978/9 National Travel Survey reported 61% of education trips to be made on foot, 15% by car and 18% by bus or coach. These are of course nationwide aggregate figures dominated by urban conditions, and includes tertiary education. More appropriate data have been assembled from various sources by Rigby (1979). Primary school journeys by bus or coach comprise 6-18% of all school trips, and secondary school journeys 19-48%; the latter involve greater distances. For both primary and secondary, there is a clear correlation between journey length and the proportion of school journeys made by bus or coach. In urban areas distances are shorter, but more importantly, there is a much greater likelihood of existing stage bus services being adequate for school access. Contracted, exclusive school buses are more likely to be found in rural areas where distances are greater but where conventional stage bus services are less likely to be suitable.

Wider use of exclusive school buses in rural areas soon became the main issue. As conventional rural public transport declined through the 1960s, it was seen that school buses provided on contract hire, and LEA vehicles, were a grossly under-exploited resource. They may have spare capacity which could be used by adult passengers, and the vehicles may be idle in the off-peak when they could be used for public service at marginal cost. In many rural areas by the 1970s an extraordinary contrast had arisen, between the withering sparsity of conventional bus services and the dense intricate network of statutory school transport, picking up children from many remote settlements. The former may have declined to one or two workings per week, while the latter ran five days per week. There were often significantly more vehicles dedicated to schools work than to public service. Trips made on school buses could vastly outnumber those made on stage buses in rural districts (x4 in parts of Dyfed in 1977). Local authority expenditure on education transport typically exceeds subsidies to public stage services by a considerable degree (x2 in Dyfed in 1978). Transport cost per pupil is much higher than the capitation fee for educational resources (x3 in Oxfordshire in 1976). Many rural bus operators owed their survival to the 'school contract' with the LEA, which was relatively profitable and allowed much implicit cross-subsidy to their other operations.

Integration of schools and public transport was therefore seen as a vitally important task, to rationalize use of resources and avoid duplication of expenditure from different local authority budgets (transport and education). Declining demand in rural areas, the mounting subsidy bill and government pressure on spending, all made this urgent. One means of doing this was to make use of section 30 of the 1968 Transport Act. This allows the public to be carried and fares charged on school contract buses (but not LEA vehicles) if the operator obtains a permit from the traffic commissioners; PSV standards would not apply but fuel tax rebate may be claimed. It should be mentioned that informal carriage of adults on school buses has always occurred in rural areas, which would have been illegal if fares were charged. 'Section 30 permits' have proved a useful mechanism for schools/public transport integration, but is only one of various methods, and its application has been variable. Problems with this method are as follows.

(i) There is frequently insufficient spare capacity on the school bus for adult passengers. Also, three pupils under fourteen may share a

double seat to fill the bus, in which case adults could not be carried.
(ii) Routes which deviate to pick up and set down pupils at remote places may be too inconvenient or slow for adult passengers.
(iii) Many adults dislike sharing the bus with noisy and boisterous children.
(iv) Bus timings suited to the school day may leave too long a stay time (seven hours) in town for shopping or other purposes.
(v) On the other hand, while the morning inbound trip may be suitable for the journey to work, a return at 1530 or 1600 is too early for employees.
(vi) Although a five day a week frequency is high by rural standards, the school bus will not be available on weekends or, more seriously, in school holidays.

It is uncertain exactly how many section 30 permits were in force at any particular time, as such data are not readily available. A survey by the National Consumer Council (1978) found there were at least 80. At this time, permit operation was dominated by four authorities: Northumberland, Somerset, Devon and Borders. A scan of PTPs/TPPs gives incomplete information but confirms that take-up of permits was highly variable. Northumberland had 45 in 1978/9, more than any other county, rising to 52 in 1984/5. Apart from the authorities mentioned, Cornwall and Hereford & Worcester had a dozen or more, but all others mentioning the subject claimed to have just a few (up to five). A number of counties claimed to find school permit buses not convenient or unsuccessful, through some combination of the reasons given above.

After 1974, local authorities appointed 'transport coordinating officers' who were able to achieve more cooperation between county surveyor's (transport) and education departments. Various means of schools/public bus service integration were possible.

(i) *Direct conversion of school contract buses for public use* under section 30 permits.

(ii) *Re-routeing and/or re-timing public stage buses* to cater more appropriately for schools traffic.
(iii) *Converting school contract buses into a 'normal' public service* by obtaining PSV and road service licences. These might be expanded by adding mid-day trips between the peak school runs.
(iv) *The 'extended school contract'*, by which school buses and drivers are deployed on public service duties in the off-peak at marginal cost.

In practice the most common policy was to adopt (ii), (iii) and (iv) as much as possible, leaving (i) only to fill in gaps where there was no other solution. This removed some, but not all, of the problems enumerated above. The necessity for integrated planning of schools and public services has been emphasized by Awdas (1978), in the interests of both the viability of operators and economy to local authorities. Placing school contracts with operators also running stage carriage, other education trips and private hire allows costs to be shared and is in the interests of both parties. Allocating more children to existing stage services, with season tickets, would be cheaper for the LEA and enhance the viability of the bus service. Even within the catchment area of a single school, planning the most efficient system can be extremely complex. In more urbanized areas, the coincidence of school and work peak demands in adjacent localities produces a high demand for vehicles, and the staggering of school hours might need to be considered.

A case study exemplifying the complexity involved is given by McCracken & Millar (1974). The benefits of developing unavoidable school bus provision by using the same vehicles for off-peak shopping trips are shown by Winfield (1978). This was thought superior to the section 30 permit method, as the bus company could provide a suitable large vehicle, not be constrained by LEA contracts, and could run both types of service on the same road service licence. This became a widespread policy in Dyfed. There is no correlation therefore between a local authority's level of activity in school/public bus integration and the number of section 30 permits in its area. Many active authorities have few or none: Hertfordshire finds permits "not convenient"; Dyfed and Gwynedd have very few. In the Scottish Highlands, coordination of school buses and postbuses works well. Such is the dominance of school demands within the road public transport market in rural areas that a logical strategy might be to design all public bus services around the schools network. The main proponent of this idea was Oxfordshire. In a radical plan the network of school services was first established, and any public service route desired by local communities would be included in a 'package' together with a school route, and the operator would then tender for both on the 'extended school contract' basis (Ennor 1977). While successful in cutting costs, this policy was highly controversial at the time due to its lack of commitment to providing a complete public service network.

One school bus experiment was inserted into the RUTEX programme in 1977-9. As one of the Welsh schemes, four school contract bus routes linking villages with schools in Llandovery were opened for public use by section 30 permits (Dinefwr Rutex 1979). In the case of one route without spare capacity, a larger vehicle was provided by the operator. Over a 40-week school year 930 fare-paying passengers were carried, an average of 5.8 per route per week; this was only 5.5% of the spare seats available. Theoretically the additional costs, over and above the costs of carrying schoolchildren, are zero, unless one includes the cost of permits and additional running costs for the larger vehicle. The little extra revenue was no doubt useful, but was considerably less than the fuel tax rebate!

Escalating costs also put pressure on the principle of free travel. It was thought this might at least remove the iniquity of pupils living within the statutory minimum distances having to pay their own costs while others travelled free, perhaps by levying a flat rate for all. However, a Bill to authorize charging on school buses was rejected by the House of Lords in

1980. Nevertheless, existing rules were applied rigidly. Where before, rural LEAs had generously allowed many pupils living within the distance limits to travel on buses free, government pressure on budgets no longer permitted it. A few legal test cases have been brought, but the courts have interpreted the two- and three-mile limits strictly.

Relatively few LEAs operate their own bus fleet for school transport, as they would have to pay the full costs and vehicles would probably be under-utilized. Specific local circumstances might justify them. Amongst authorities which run their own buses are: Lincolnshire, Isle of Wight, Dorset, Devon, the Inner London Education Authority, and the Education & Library Boards of Northern Ireland. These could not be used for non-school passengers until the 1980 Transport Act, which contained a general provision for school contract buses *and* LEA buses to carry fare-paying adult passengers. Permits were no longer needed. In addition, LEA vehicles were allowed to be used for normal public service when not being used for school purposes. LEA vehicles were exempted from PSV standards, even when carrying the public, although this was partly retracted by the 1985 Act. Some educational vehicles operate under minibus permits (see chapter 9), which are not regarded as PSVs. The current legal position is summarised by Atherton & Taylor (1988).

These provisions afford better opportunities for making full use of available resources. A prominent example is Hampshire's 'Plus Bus', in which a LEA vehicle is deployed on off-peak rural services on three days per week (Hampshire PTP 1985/6). Another example is the school minibuses owned by Grampian Regional Council being made available for off-peak use as community buses (Grampian 1985). Integration is further encouraged in various sections of the 1985 Transport Act, but this of course is only continuing existing trends. A survey of local authority responses in the immediate aftermath of deregulation was made by Masey & Hopkin (1986). This found that almost half had integrated schools and public services to some extent before the Act, but almost a fifth had no integration at all, and the rest were moving towards it. If correct, these results must be disturbing, as the impression gained from a reading of PTPs/TPPs is that most or all authorities were going to great lengths in rationalizing these two transport sectors from the mid-1970s. The need for integration had been imperative for at least ten years prior to the 1985 Act, and it seems extraordinary that some authorities had neglected the opportunity.

While in some rural areas deregulation caused little change (see chapter 14), in others school services underwent sharp changes as they were reorganized in conjunction with tendered public 'local services'. Another problem has been lack of provision for scholars not eligible for free school transport, as their regular bus might not have been registered and a subsidized alternative might be difficult to justify. Some children's concessionary fares have been withdrawn.

Meanwhile, long term concerns over prosaic matters such as safety and supervision on school buses still need attention (Atherton & Taylor 1988). Consistent PSV standards might be a desirable improvement, but care should be taken to exclude school minibuses used for extra-mural activities as these are usually driven by teachers who could not be expected to get a PSV licence. The importance of schools traffic to the viability of other UCMs might be mentioned here, for example, the ESCORT scheme, the demountable vehicle, the Border Courier, while postbuses and community buses may carry small loads. There must now be a case for ceasing to regard school buses as UCMs, as integration proceeds and becomes more 'normal', no special legal dispensation such as permits is needed, and the remainder are conventional exclusive contract hires.

Chapter 7

POSTBUSES

7.1 POSTBUS ORIGINS IN BRITAIN

Postbuses are a dual-purpose mode in which two different types of traffic - mail and passengers - are conveyed simultaneously in the same vehicle. It is based on the simple premise that as large numbers of vehicles are needed every day to collect and deliver the mail in all parts of the country, then they might as well be carrying passengers also. The costs of this would be marginal to the unavoidable duty of running the mail service, and passenger revenues would contribute usefully to Post Office income. Such arrangements would be ideally suited to areas of low passenger demand, at the local scale, where conventional buses are unsuitable, uneconomic or in decline. Hence the 'community service' role of the Post Office is augmented by providing fragile local communities with vital physical linkages to the outside world.

This is by no means a new idea - antecedents may be found in the Royal Mail Coach services of the eighteenth century. More recently, trains and buses contracted by the Post Office to carry mail have also been open to passengers. Apart from dial-a-rides, postbuses are the major example of British UCMs being influenced by overseas comparisons. In continental Europe, integration of mail and passenger carriage is normal practice and widespread, notably in Sweden, Norway, Finland, West Germany, Austria and Switzerland. Subsequent to British experience, a single postbus has started in the Republic of Ireland, and discussion has commenced in the USA.

While the European example had always been recognized in Britain, fundamental differences mean that it is impossible to emulate the continental type of service. A detailed comparison of practice in Britain and Switzerland is given by Holding (1983). Historically, the national post offices or PTTs (Post, Telephone, Telegraph) of the European countries cited were the main providers of road passenger transport in rural areas, typically commencing about 1910-30. Fully integrated networks evolved so that there is little competition from private carriers or the railways, and the latter play a supportive or coordinated role. Other important differences concern the lower emphasis on door-to-door mail delivery in other parts of Europe: this includes the provision of personal mail boxes at 'road ends' instead of remote houses and farms, the requirement in some remote areas for households to collect their own mail from the local post office, and the practice of sorting mail at more local levels which means a greater proportion of mail transport is for bulk movement only. This last point means that most continental postbuses carry mail in bulk between sorting offices, and can therefore operate 'trunk' routes at higher speeds. Vehicles are otherwise conventional buses which carry mailbags and sometimes other freight. PTT buses can also handle school contracts, which is not normally possible on British postbuses. Finally, a vital difference is that in Britain the demands of the mail come first and determine postbus schedules, whereas on the continent there is much more freedom to schedule PTT buses according to the demands of the passenger market (e.g. there is no insistence that mail be delivered early in the morning).

Useful reviews of postbus origins in Britain are Turnock (1977) and Watts et al (1978a). The Jack Report (Ministry of Transport 1961) recognized that the Swiss or German system could not work, and instead preferred that the Post Office placed more mail contracts with bus companies. Nevertheless, considerable representations had been made, especially from Scotland, that the integration of passenger and mail services be considered. The Post Office was incurring losses in rural areas due to its labour-intensive operation and small volume of business, and it was thought evidently wasteful for both a mailvan and a bus to run below capacity on the same route. Eventually, the Post Office was prevailed upon to try out postbuses, and in

1967-8 mounted four experimental services: Llanidloes-Llangurig (Powys), Dunbar (Lothian), Honiton (Devon) and Penrith (Cumbria). Initially, these were no great success and withdrawal was considered, but a major and timely constitutional change transformed the picture. Under the Post Office Act of 1969, the Post Office changed from a civil service department to an autonomous public corporation. The statute specifically empowered it to carry passengers for hire or reward in vehicles used for postal business. The new status made post office vehicles used as buses eligible, under the 1968 Transport Act, for fuel tax rebate, new bus grant (then at 50%), and local authority subsidy.

These subventions certainly rescued the viability of postbuses. The four experiments continued (and survive to the present), and a fifth, from Broadford to Elgol on the Isle of Skye, was started in early 1972. It was soon realised that the new concessions reduced the marginal cost of postbus operation in that the cost of the minibus was little in excess of that of the mailvan which has to run anyway, and raised the possibility of small but worthwhile profits to set against the cost of the mail business. Subsequent growth was phenomenal. At the end of 1972 there were 15 services; in 1973 this more than doubled to 36 (mapped on figure 7.1); the following year it nearly doubled again; by 1978 there were 157 postbus services - a tenfold increase in six years; since then the rate of increase has tailed off and numbers have stabilised (see graph, figure 7.2); in 1985 there were 173 (mapped on figure 7.3). The main growth phase of postbuses occurred distinctly earlier than that of other UCMs.

The great majority of this development was in Scotland (figures 7.1 and 7.3). Explanations for this depend mainly on the strong initiative of the Scottish Postal Board and the personal enthusiasm of its chairman at the time, Mr T. Carpenter, in recognizing the suitability of the postbus for rural Scottish conditions. This was largely an extension of the accentuated social role of the postal service in Scotland, where a

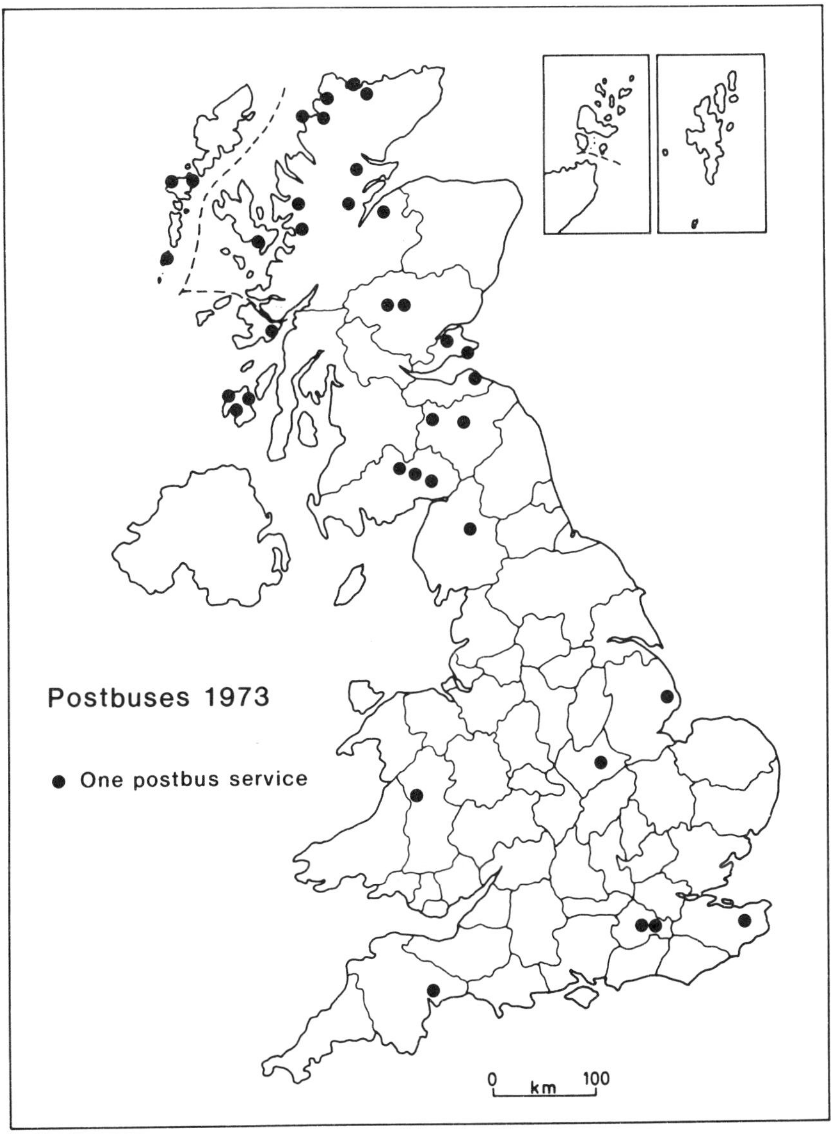

FIGURE 7.1 Postbuses 1973.

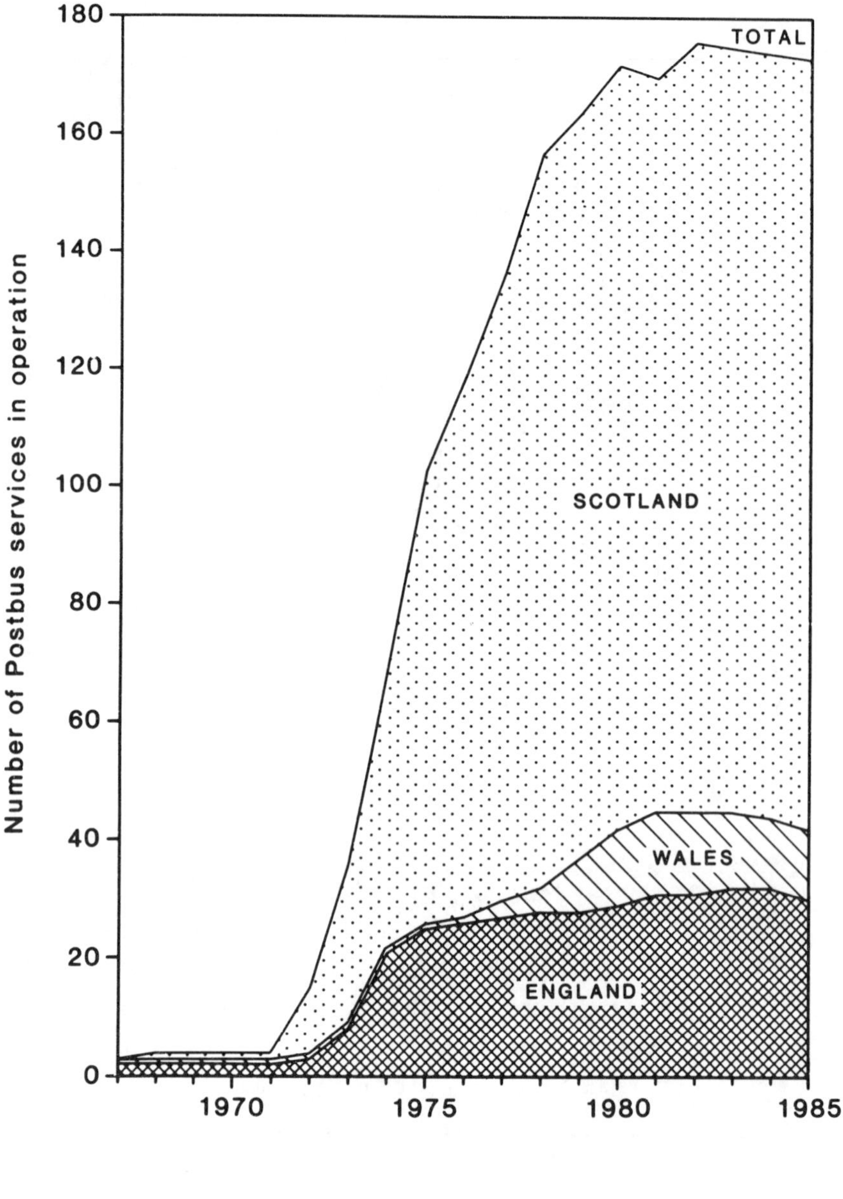

FIGURE 7.2 Growth in postbus services 1967-85.

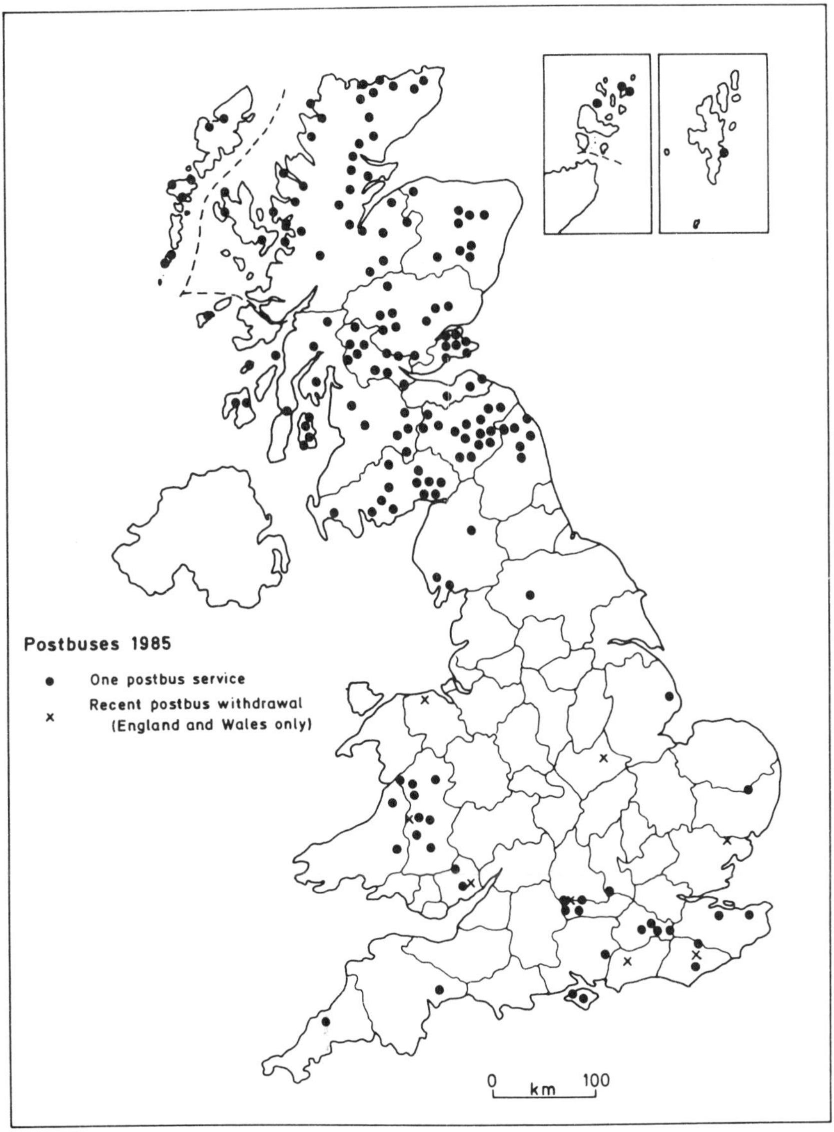

FIGURE 7.3 Postbuses 1985.

greater proportion of communities are in remote locations with a weak service structure and fragile public transport, yet the Post Office has a statutory obligation to reach every household. The mailvan, sometimes together with the school bus, provided often the only permanent and reliable link with the outside world. In these circumstances, postbus expansion plans got the wholehearted support of local communities, Post Office staff, Traffic Commissioners, local authorities and other relevant agencies. Particularly notable was the enthusiasm of Post Office drivers and their unions who, despite the need to undertake PSV training for a nominal extra payment, welcomed the opportunity to supplement the service they could offer to their own communities (Carpenter 1977).

The other explanation commonly given for the success in Scotland is the suitability of the topography and settlement pattern. While population density is very low on average, settlement is often confined to linear patterns by valleys and the coastal strip, and thus transport services may cover relatively long distances without diverting very much from the direct routes. By contrast, a lowland area with a truly scattered settlement pattern induces very devious mailvan/postbus routes which are inconvenient for passengers. While there is some truth in this, it is not a convincing explanation as topographic and settlement factors did not seem to help conventional buses in the Highlands nor did they encourage postbuses in upland areas of England and Wales. The standard of pre-existing transport, and the policies and commitment of the regional postal boards lie behind the extraordinary discrepancies on the postbus maps (figures 7.1 and 7.3). A further point is that the Scottish Postal Board took a more liberal attitude than others to the general policy that mail requirements have first priority - alterations to mail services could be considered if they were acceptable to the community.

That Scottish postbuses were successful was not in doubt, and there was some euphoria in the mid-1970s, when for a time it seemed that a new service was being inaugurated every week, almost to the extent that a rural transport panacea had been found. The innovations were widely known and popular; there had been relatively little demand on local authority subsidies. There was talk of linking up the individual services to form a whole network; there were hopes of expanding the 100 services to 500, raising profits of £250,000; it was even hinted that any of Scotland's 1500 mailvans could be converted to postbuses (Turnock 1977, Carpenter 1977).

It can be seen from the maps that postbuses are fairly evenly distributed throughout Scotland. They are by no means confined to the remotest areas - some are remarkably close to the major urban areas of the central belt, for example the cluster of seven services in eastern Fife. Most were introduced in areas previously without bus services, but there were several cases of conventional bus operators surrendering difficult routes in favour of postbuses where the latter might prove more adaptable and require less subsidy. Postbuses also proved successful on many islands, in some cases taking over the entire road passenger transport system on larger islands such as North Uist and Islay, and providing the only internal transport on many small ones such as Barra, Tiree, Colonsay, Luing, Rousay and Sanday (Orkney), and Bressay (Shetland). One problem occasionally mentioned in Scotland was the inability of the postbus to carry the larger freight items which could be handled by a full-size bus or a boat: Turnock (1977) cites the case of Lochaline, which declined a postbus in favour of maintaining its sea link with Mull and Fort William. Out of the 1985 national total of 173 postbuses, 131 were in Scotland. A crude indication of their penetration is given by the fact that, of about 220 local bus services listed in the Highlands and Islands Development Board area, 58 (26%) were run by the Post Office.

Outside Scotland, the pace of innovation was much slower and more cautious, and the pattern very sparse and geographically selective (figures 7.1 - 7.3). The subsidiary concentration in south-east England has provoked comment, with unexpected clusters in Surrey and Berkshire, indicating pockets of transport deprivation away from the main traffic routes so close to London. The other significant English grouping is Northumberland, where conditions similar to the Scottish Borders might be hypothesized. In Cumbria the three postbuses hold a place within a diverse provision of UCMs - social cars, a community bus and shared taxis - within the county. Other cases stand in isolation, having survived for a decade or more without inspiring the postal or county authorities to introduce any others in the area - Honiton (Devon), Wadebridge (Cornwall), Diss (Suffolk), Louth (Lincs.) and Masham (N. Yorks.). There were 30 postbuses in England by 1985. Many English counties claim to have searched thoroughly for possible routes but have found few or none which offered a realistic public service, was not too devious or time-consuming, and did not compete with other transport. There were also a few cases of trade union opposition.

Wales hosted Britain's first postbus - Llanidloes-Llangurig - in 1967, but the irony is that the principality had to wait another ten years for its second. This was the Llandovery services introduced under RUTEX (see section 1.4). During this ten-year gap eighty routes had been investigated, but strict application of official guidelines prohibited all of them (Stark 1979); conventional bus penetration in rural areas was surprisingly high. After 1977, with the RUTEX demonstration and the gradual disappearance of some rural buses, the earlier caution was relaxed. Postbus conversions were then effected more readily, although with great variation among counties (figure 7.3). There were twelve in Wales by 1985, of which eight were in Powys (Powys 1981).

According to Turnock (1977), studies of the potential for postbuses were also made in Northern Ireland, but had been unable to find any route not operated by Ulsterbus where postal requirements could be married to likely passenger trip demands. This would not now be re-considered due to the security situation.

Regional variations in the adoption of postbuses depended to a large extent on the postal boards' different interpretations of the criteria laid down by Postal Headquarters. These were originally that:

(i) a need for the service should have been demonstrated;
(ii) there should be no competition with currently operating bus services;
(iii) the passenger service should be compatible with existing mail handling arrangements.

A few years later (by 1976) there was more concern about financial guarantees. More services were seeking local authority subsidy, and some new routes were specifically requested by regional/county councils. The 1972 Local Government Act had required councils to coordinate and, where necessary, financially support public transport. Experience was showing postbuses to have highly variable economic results, with average profits not matching earlier expectations and possibly not justifying the staff time and effort involved. A fourth criterion was therefore:
(iv) necessary financial support should be guaranteed.

There is clearly scope for different interpretations of all these criteria - how great a 'need' should be proved under (i); whether the existence of a low level bus service should be allowed to prevent, under (ii), the initiation of a postbus with a higher frequency; whether financial guarantees under (iv) should allow a definite profit level. Criterion (iii) was the most problematic, causing a notable difference of opinion among the regional postal boards. Generally, boards in England and (at first) Wales inclined to the view that the mail was the foremost priority, and no adjustment to schedules could be countenanced for the convenience of passengers. This was often accompanied by a strict view of 'competition', that the Post Office has no right to threaten the livelihood of existing bus operators. The alternative view, which prevailed in Scotland, is that the Post Office should serve the community in a more general way, and hence should take any opportunity to offer additional services provided these did not actually harm the mail business (Watts et al 1978a). A good example of this flexibility is on Islay, where the local authority and public sector bus company jointly requested the Scottish Postal Board to take over the total bus network on the island; this included certain stretches of route not strictly required for postal purposes, which had to be fully costed and subsidized. A few services are also known to convey groceries, prescriptions and other goods, and there are some school contracts where pupil numbers are small.

Also, the relative weight of problems and pressures associated with other, more vital, aspects of postal business might have deterred serious consideration of postbuses in some regions. The influence of key decision-makers was crucial, and it is probably the varying enthusiasm of postal board directors coupled with the readiness to adopt a flexible attitude towards operations which explain the inter-regional disparities in postbus innovation.

In the late 1970s the initial optimism was severely diluted, partly due to very low patronage on some services and partly due to the high rate of cost inflation typical of the period. Even in Scotland, where few services originally required local authority subsidy (as opposed to central government subsidy such as new bus grant), councils were now more often expected to make a contribution in order to retain them. Some services began to appear self-indulgent and over-optimistic. Alarms were sounded and the Post Office was impelled to take various economy measures. A number of postbuses were withdrawn, although overall numbers continued to increase (Nickson 1981). New projects were not deterred, but were more rigorously assessed in terms of passenger demand and subsidy requirement.

Inevitably, the pace of postbus expansion slowed considerably after 1980, and the idealistic objectives once hoped for in Scotland were forgotten. There were fewer introductions and rather more closures, so that the graph (figure 7.2) has tailed off, although total numbers are relatively stable. After 1970s inflation and the oil price rise of 1979, two further financial shocks put a severe dampener on postbus development. New accounting procedures introduced by the Post Office had the effect of passing more of the costs over to local authorities, and second, the phasing out of new bus grant by 1984 meant that income could rarely cover vehicle replacement costs. There was also the tendency, especially in England, for councils to compare postbuses with other UCMs or low frequency PSV minibuses and find the latter a cheaper option.

7.2 POSTBUS OPERATIONS

(a) Vehicles

Conversion of mailvans to postbuses obviously requires a completely new vehicle type, expressly designed for the purpose. The staple postbus is a PSV standard Commer (later, Dodge) 'spacevan' with seating for 11 passengers. Minimum adaptations concerned the need for security of the mails (and drivers had to handle money for the first time); a grille is placed behind the driver, and the space otherwise occupied by the front passenger seat is made available for letters and packages. The Post Office's ability to purchase vehicles centrally is a great advantage, but there has been no excessive standardisation and other vehicle designs are used where the terrain and passenger demand requires.

In Scotland mail services are often provided over unmade roads and farm tracks, and it was normal to use Land Rovers for these. A direct conversion was possible in these cases, with the installation of four passenger seats. Elsewhere, the Commer van may be excessively big for narrow roads or likely passenger demand in remote areas, but the terrain may not require a specialist vehicle. Routes in Scotland and Wales where these conditions obtain are covered by estate cars, again with four seats, the Morris Marina and Hillman Avenger being the most common types originally used.

There are a few workings which concentrate on bulk mail transport rather than collection/ delivery, and larger vehicles might be appropriate. A Leyland midibus was employed on the Achnasheen-Laide route in the West Highlands, which has more space for mailbags and 12 seats. More recently, new vehicle types have been introduced: the postal estate cars are now likely to be Talbots, and there are a few cases of larger minibuses being deployed such as Ford Transits, and the Freight Rover Sherpa which has 16 seats and more headroom.

Reserve vehicles are kept to cover breakdowns on any of a group of services, and this is another cost element. The Post Office does its own vehicle maintenance and is normally able to achieve economies of scale in this respect. For most of the period of postbus development drivers were required to obtain a PSV licence, which also adds to service establishment costs and puts more demands on staff, although this seems to have produced few problems. Licensing of public services was normally done by taking advantage of Section 30 of the 1968 Transport Act, which had initiated a simplified permit system for minibuses.

(b) Service Configurations

In detail, service patterns vary enormously, as would be expected from the particular arrangements that have to be made to collect and distribute mail in a great variety of local circumstances, and any further adjustments that may have been conceded for passengers' benefit. Perhaps the most important generalisation that can be made at the outset is to stress the advantages of the *high frequency* service that postbuses provide. Most operate Monday to Saturday every week, and this is far more generous than remote communities could normally expect, or could reasonably be provided in most cases by a conventional bus operator. This indicates the problems of the 'no competition' rule in that, if a low frequency bus already exists on a route, then an inferior service could frustrate the establishment of a superior one. Saturday services may often be restricted, and certain parts of routes may be run on fewer weekdays, but generally the high frequencies greatly enhance the potential accessibility of local people to service centres or to connecting transport, and there is no necessity to organise one's weekly activities around the one or two days when the bus used to arrive.

To set against this are the constraints due to the responsibilities for mail. On any one day, there could conceivably be a letter for delivery to every single house in the area, and this would mean extremely slow and tedious journeys with very devious routeings. Collection rounds and backhauls would of course be faster than delivery rounds. A postbus is not strictly demand-responsive - one cannot telephone in advance to be picked up at the doorstep - but a door-to-door service is possible if both origin and destination are on the route. The route itself has to be flexible, as on any one day, a different set of households will have mail for delivery, and therefore routes will vary in detail from day to day (especially on delivery rounds); this also implies timing will vary in a similar manner. Timetables are therefore fairly loose.

Apart from the likelihood of slow journeys, adherence to original mail schedules might not be convenient for passengers in that: arrival in town is unnecessarily early, or the stay time before return is too short or long, or the postbus arrives after the doctor's has closed, or connections with other transport are poor. On the last point, mail schedules accept no responsibility for postbus passengers achieving connections with onward transport, and this is a problem in many areas. In the Scottish Highlands however, carriage of mail has long been part of the tradition of transport integration, so that postbuses were able to retain connections with trains, ferries and even aircraft (less commonly with other buses), and mail and passengers benefit from the same chain of distribution.

Classifications of operational types have been attempted by Turnock (1977) and Watts et al (1978b), although it is very difficult to overcome the complexity involved. First, there is some confusion between numbers of vehicles, numbers of separate operations and numbers of routes. A 'service' usually means a separate operation with one vehicle, which might involve more than one route. Different routes might be run in mornings and afternoons, or on alternate days. Due to this difficulty, authors might not agree on numbers of services. The average is 1.5 routes per operation. The average route length is about 37 km (a circular route or outward plus return distance on a straight run); as might be expected, Scottish mainland routes are above average length and English routes below. That Scotland has some routes of 80-135 km is remarkable given the essentially local nature of the mode, and indicates the urgency of the need if such gaps previously existed in the transport network. Rural populations served by postbuses might be as low as 200-500 (excluding the towns where they are based), but in areas of evenly dispersed settlement over 1000 might be within walking distance.

Typical postbus route configurations may be either out and back along the same roads, from a base in a small town or large village with a sorting office, or else circular from a similar base, in each case with the necessary diversions for collection/delivery. They may also be classified by the number of rounds per day (Watts et al 1978b). Most commonly (about 50% of services) there are two round trips from the town sorting office into the rural hinterland: the first would be a slow early morning delivery circuit, returning direct making collections from sub-post offices and letter boxes; the second would be a later delivery/collection tour about mid-day, or a collection run in the afternoon. From the point of view of rural residents, however, these two round trips provide only *one* opportunity to get into town and back; this is normally adequate, although the stay time allowed may be anything between one and six hours. On Saturdays there is normally only one collection/delivery tour in the morning, and thus rural residents are offered only a one-way journey.

According to the survey by Watts et al (1978b), about 5% of services offered *three* return journeys daily. This would be extremely useful to passengers in giving a choice of a short morning stay in town, a longer afternoon stay, or an all-day visit. On the other hand, about 10% of services, all in Scotland, make only one journey out and back from a town, usually later in the day. This is not convenient for passengers, who must use another mode of transport for one leg of a return trip. Another 10% of services, again confined to Scotland, centre on the postbus making a connection with bulk mail transport (e.g. meeting a train), sorting the local mail and delivering it on the return leg. Convenience for passengers in this case depends on the usefulness of the connection point.

Quite different are the 5% of services devoted largely to the bulk carriage of mail, making few or no local deliveries or collections. Such 'mailbuses' are likely to operate only on main roads at higher speeds, and are similar to the continental type of postbus. The most striking example centres on Achnasheen in the West Highlands, where mailbags for onward distribution are off-loaded from the train out of Inverness. These are loaded on to the postbus - uniquely, a Leyland midibus with more space - which then departs on its 'trunk' route, delivering the appropriate bags to sub-post offices along the way. At Kinlochewe and Poolewe local mail is sorted and transferred to Commer and Land Rover postbuses for delivery rounds; at Gairloch the mailbus also connects with two public/school buses (figure 7.4).

This classification is over-simplified, and there are many cases of combinations of the above types and more idiosyncratic systems. On the Scottish islands, both mail and passenger services must connect with the arrival of ferries and aircraft, and all modes may in turn be affected by sea and weather conditions. To exemplify the enormous variations that exist, figure 7.4 depicts three postbus systems which differ greatly in scale, route configurations, schedules, and relationships with other transport.

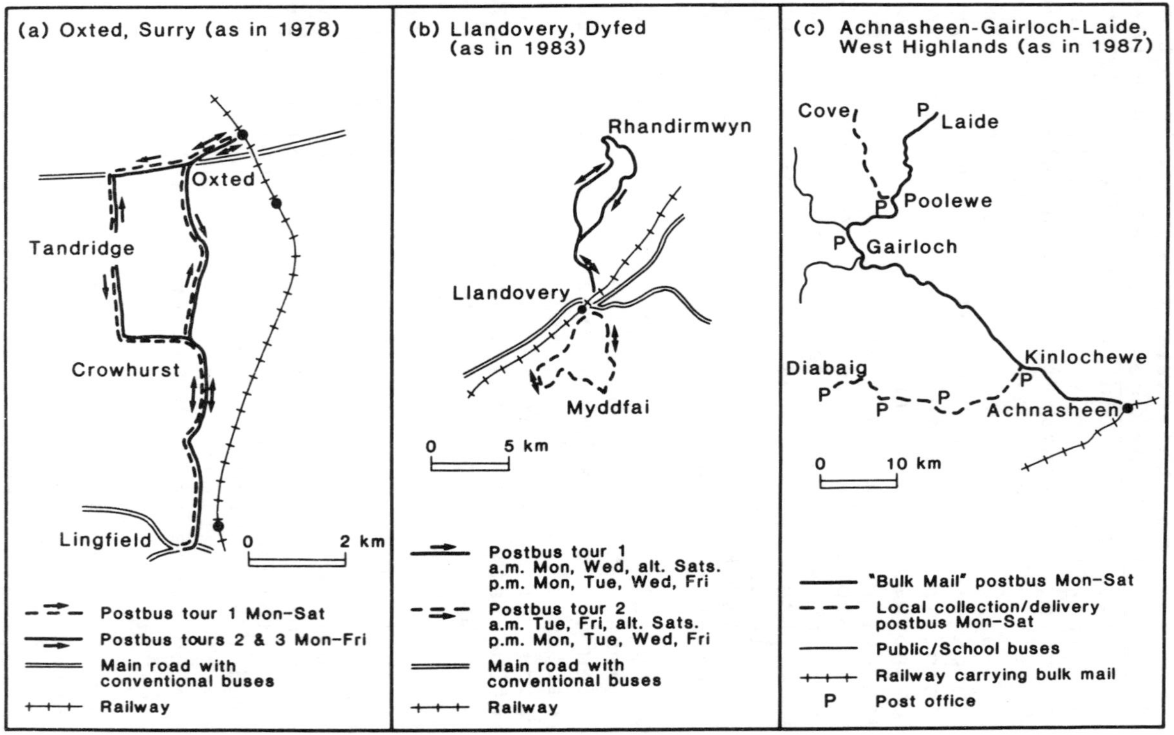

FIGURE 7.4 Examples of postbus operations.

Two postbuses were incorporated into the RUTEX programme in 1977-9, in Llandovery (Wales) and Blackmount (Scotland). While the Welsh project may have been motivated partly to rejuvenate the flagging interest in Wales, both were good examples of the prospects for coordinating postbuses with other public transport. In Llandovery, one of the two postbus route circuits was previously covered by an independent bus company's market day service, and both postbus routes (figure 7.4) were also followed by school contract buses opened up to adult passengers. Such duplication would not previously have been allowed under the 'no competition' rule, but in practice it gave consumers much flexibility in that timetables often complemented each other, and different modes could be used for each leg of a return trip into town (Dinefwr Rutex 1979, Nutley 1983). In the Blackmount RUTEX area the main innovation was a flexibly routed PSV minibus, but the necessary geographical range of this was reduced to manageable proportions by reorganising the area's postbuses in a complementary manner. An existing postbus between Biggar and Newbigging was extended to Carnwath, and re-scheduled to connect with conventional buses at both termini, without harming the mail service. A second (new) postbus service was inaugurated on a roundabout route between Biggar and Lanark, conveniently using a 'spare' vehicle and thereby incurring only marginal costs. The Biggar-Carnwath postbus and the partly D/R minibus covered the same territory with an integrated timetable (Scottish Rutex 1979). Again, this would not have been possible under a strict 'no competition' rule.

To conclude this section it would be useful to give an example of the difficulties of planning new postbus services. The possibilities of converting existing mailvan routes to postbuses were investigated in a part of Radnor District, Powys, an area with very little traditional public transport. This was part of a project to evaluate all UCMs as policy solutions for the area, and is described in more detail in section 14.2 below and in Nutley (1985). Figure 7.5 shows the mailvan routes serving the area. The overall view is one

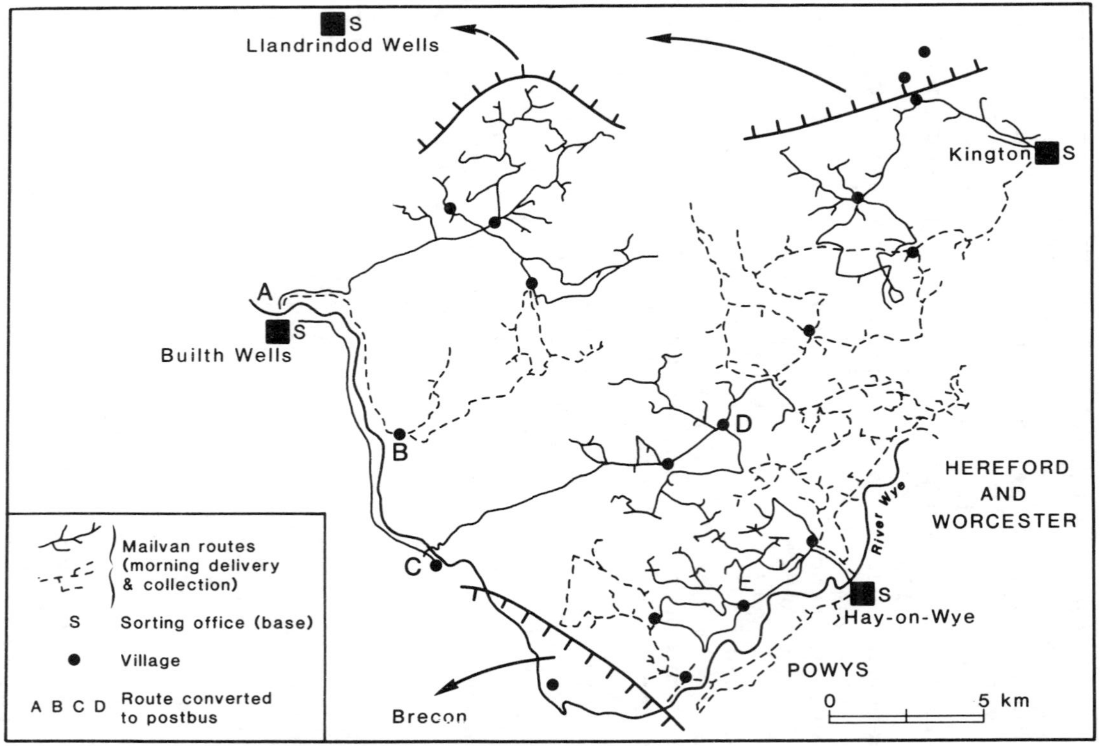

FIGURE 7.5 Mailvan routes in southern Radnor District 1982.

of extreme complexity, with most of the area covered by eight mailvan circuits based in three towns, and with operating fields not corresponding to any natural or local authority boundary. Population is sparse but fairly evenly spread, the 19 villages shown on the map being generally small. Mailvan routes are therefore very intricate, to reach every habitation, and cover small pieces of territory. No single route covers more than four of the villages. Therefore several postbus conversions would really be necessary to make a significant impact on accessibility levels in the area. Most mailvan routes are too limited in scope to offer useful conversions; others follow minor roads too narrow for a Commer postbus, although a postal car might be feasible. But there are other problems: afternoon collection rounds in the area do not correspond to the morning tours shown on the map - the former are much simpler, so that one afternoon round may replace two or three morning rounds, and this would make a postbus impossible or useless. One part of the area receives morning services and afternoon services from two different towns.

The only practical postbus conversion was the mailvan route marked ACD on the map, which is longer and has fewer intricate branches and serves three villages. This was eventually converted to a postbus (Builth Wells - Painscastle) in 1984, enhancing its viability by altering the 'trunk' section of its route to include also village B (figure 7.5).

7.3 TRIP GENERATION

The review of British postbuses by Watts et al (1978a) found that patronage levels varied widely, between about 5 and 150 passenger trips per week, implying annual totals between roughly 250 and 7500. Publicly available evidence on the patronage levels of specific services certainly endorses the extraordinary variations in popularity, but also suggests that most postbuses generate trip levels that congregate around the lower end of the range claimed above. For example, a commonly quoted figure is that all postbuses in Scotland had carried a total of 100,000 passengers by 1976: dividing this by the number of services then existing gives an average patronage level of 1300 trips/year or 25 per week.

Some relevant data on postbus usage can be obtained from a trawl of county PTPs and regional TPPs. Not all authorities with postbuses on their territory give such data, and that which exists is usually incomplete and limited to passengers carried, mileage, and sometimes the subsidy costs. Specific evidence on postbus patronage obtained from this source is presented in table 7.1. The most common annual patronage seems to be between 1000 and 3000, with few higher than this but quite a number below. Only two cases could be found which exceed 4000 trips/year (80 per week), and nothing to justify the upper limit claimed by Watts et al (1978a). On the other hand, there are several cases whose passenger loadings are disturbingly low - even less than one per week. One in Central Region carried only 66 in the year 1983, one in Orkney only 39 and one in Dumfries & Galloway only 34 in 1983/4, and one in Tayside carried only 17 in the year 1981/2. There is even a report (Nickson 1981) of a postbus on the Isle of Skye which had only *two* passengers in the year 1980!

Loadings as low as these could conceivably have been handled by a social car scheme, and the postbuses in question have presumably been reviewed by the authorities. Reasons for such wide variations in patronage would have to be sought in detailed local investigations, and there is insufficient evidence to generalize on the type of postbus service most likely to succeed. The nature of some remote glens and small islands make the low usage of certain Scottish services perhaps understandable. At the other extreme, it is known that passenger figures for some Scottish and Welsh services have been inflated by summer tourist use. The novelty value of a postbus ride is even 'sold' in some local tourist offices; the Broadford-Elgol route on the Isle of Skye has acquired particular fame for its spectacular views of the Cuillin Hills. It is not clear whether the Post Office has tried sufficient demand stimulation. Publicity

TABLE 7.1 Trip generation on postbuses (source: County PTPs, Regional TPPs, Dyfed CC, Powys CC).

	No. of services in sample	Trips per year range	Trips per year mean	Date of information
Central Region	6	66-2557	1403	1983
	6	95-2545	1250	1978
Dumfries & Galloway	12	34-3007	1250	1983/4
	12	95-2922	1448	1978/9
Tayside	11	17-2441	647	1981/2
Scottish Rutex (Strathclyde)	2	2392-3692	3042	1977-9
Orkney Is.	2	39-248	144	1983/4
Western Isles	6	n.a.	987	1977
Dyfed & Powys	9	385-4022	1657	1981/2
Gwent	2	408-1116	762	1980
Berkshire	4	904-3798	2607	1983/4
	4	1309-2897	2137	1979/80
East Sussex	3	886-2944	1803	1978/9
Isle of Wight	2	2115-2711	2413	1981
	2	2037-3677	2857	1975
Hampshire	1	-	1573	1984/5
North Yorkshire	1	-	>4000	1983/4
West Sussex	1	-	3120	1979/80

leaflets are normally available at local post offices and elsewhere, and a comprehensive timetable is produced for all services in Scotland, but the mode is not marketed as forcefully as are commercial buses.

A survey of users is also reported in Watts et al (1978a,b) although this covered only five services for one week. Travel patterns are unremarkable, in the context of rural dwellers without access to cars. About three-quarters of trips were made by women and half by pensioners. Most trips in the survey were made by regular users, but the low patronage on other services suggests this might not be universal. Most passengers were from non-car households. The most common journey purposes were shopping and social visits, mainly at local destinations. Personal business, including health, trips were fewer but widely spread. Other purposes such as work and education tend to be possible on some services but inapplicable on most. It is interesting that, in this survey, only a third of postbus passengers used this mode on both legs of the journey. As explained above, it is not always possible to make a round trip, but even when it is, complementary modes are still used, e.g. for some social trips the stay time may be too short. In these cases, the most common complementary mode (another third of all trips) is a lift in someone else's car; sometimes a school bus is available for one leg of the journey. It would appear that postbuses making only one daily trip are still useful to passengers. About one quarter of respondents would not have made the journey in the absence of the postbus, and these are the newly generated trips if the survey is representative; in many other cases walking or cycling were the only alternatives. Opinions of the services showed a distinct preference for two-way journeys giving a stay time of one to two hours in a centre with shopping and health facilities.

Few other surveys of postbus use have emerged. Monitoring of the RUTEX postbuses (three services, over a longer period than the above) revealed similar basic patterns (Lugton 1980). One point of interest is that the intensive publicity associated with RUTEX produced, in the Welsh scheme, a 100% awareness rate. This is an aspect which other schemes might consider more closely.

As most postbus services have been in existence for 10-15 years, some longer term passenger trends might be apparent. Available evidence is fragmentary. A few local authorities provide data allowing a time comparison, but this is insufficient to discern trends with any confidence. In four of the local authority areas on table 7.1 available data allows a comparison over periods of five years or so, but these are inconclusive with two showing a passenger increase and two a decrease. However, a reading of PTPs/TPPs shows local authorities becoming gradually more pessimistic, although this is expressed primarily in terms of cost rather than passenger numbers alone. In Scotland, where each region has a number of services, there is more concern about the wide variations in patronage. There are a few cases in England of carefully nurtured postbuses reporting steady usage, but these are outnumbered by those attracting negative comments (apart from those already withdrawn) due to passenger levels that are low or insufficient to justify subsidy.

7.4 ECONOMICS

Financially, the advantage of the postbus has always been its ability to adopt marginal costing, over and above the cost of the mailvan which has to run anyway, and second, central government grants such as new bus grant and rebate of fuel tax and VAT. These leave operating costs of a scale which could often be covered by fares revenue, or otherwise require subsidy much less than would be needed by a conventional bus. Local authority subsidy is the third item of cost support, but it has been the increasing demands on this that has brought problems for postbuses in the mid-1980s.

In the different conditions of the 1970s the economic equations seemed very attractive.

Watts et al (1978a) calculated the marginal costs of operating Commer (minibus) postbuses, postal cars and Land Rovers with typical mileages of 10,000-30,000 miles per annum (figure 7.6). Costs reflect Scottish conditions where the number of postbuses give certain economies of scale in respect of reserve vehicle cover and maintenance. A typical cost of a basic Commer service in the mid-1970s might be no more than £20 per week. Marginal costing also means a relatively slow cost increase with mileage, and a widening advantage over other modes as distance increases. Running costs of postal cars and Land Rovers were taken to be equal to those of a mail-only service, and thus their marginal costs are independent of distance.

Figure 7.6 also shows the effect of subtracting government grants. Fuel duty rebate roughly cancels out the Commer's greater fuel consumption when compared to the mailvan, and thus the operating cost of the Commer becomes virtually independent of mileage, at a mere £7 per week. For postal cars and Land Rovers the fuel tax rebate results in decreasing costs with increasing mileage, so that theoretically, a Land Rover would always show a profit as would a postal car on the longer routes. Although the methodology makes assumptions generous to postbuses, the advantages of the latter over minibuses run by conventional PSV operators appear dramatic.

Dependence on marginal costing and state concessions is clearly shown when looking at the costs of specific postbus services and their mailvan counterparts. The 'setting-up costs' are based primarily on the difference between the capital cost of a Commer minibus and that of a mailvan: in 1977 this was £2280 against £1050. To add to the former was the costs attributable to passenger conveyance - modifications to the vehicle, PSV certification, administrative, planning and publicity costs, and driver training - which raised the figure to £4050. Subtracting new bus grant (then at 50%) and the cost of the equivalent mailvan then reduced the overall costs of establishing a postbus service to £1950. This example comes from the Llandovery RUTEX scheme (Dinefwr Rutex 1979).

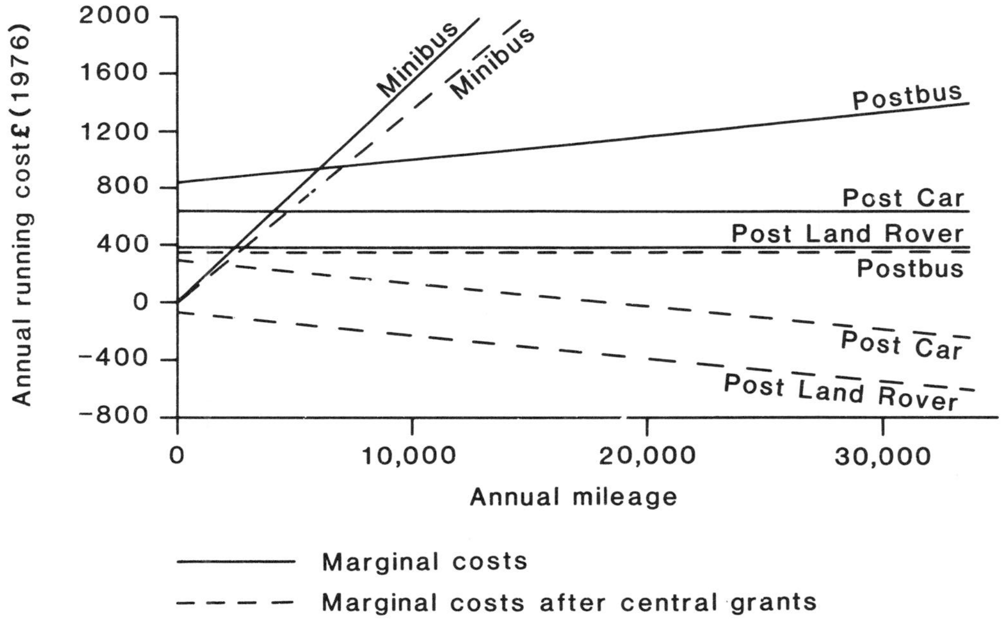

FIGURE 7.6 Costs of mail/passenger vehicles (1976 prices) (after: Watts et al 1978a)

TABLE 7.2 Postbus operating costs at the time of RUTEX (source: Scottish Rutex 1979, Welsh Office).

	Biggar-Carnwath FY 1977/8		Llandovery July 1977-April 1978	
	Postbus	mailvan	postbus	mailvan
	£		£	
Depreciation & Interest	806	485		
Recurring annual charges (insurance, licences,etc.)	629	352	505	313
Administration	609	-	70	-
Running costs	1473	1713	1695	1581
Less Fuel Tax, VAT rebate	-270		-163	-6
	3247	2550	2107	1888
Less Mailvan costs	-2550		-1888	
	697		219	
Post Office return	28		26	
	725		245	
Less Fares received	-552		-190	
Deficit	173		55	

Operating costs show the same effect. Two examples from the time of RUTEX are given in table 7.2. The importance of fuel tax rebate and the offsetting cost of the equivalent mailvan are clearly seen. Fares revenues are surprisingly low and might perhaps have covered remaining costs, but nevertheless the deficits are minimal.

Such costings seem to be typical of the 1970s and go a long way towards explaining the popularity of the mode. At this time, only about 15% of Scottish postbus services were in receipt of local authority subsidy, as were about 60% of those in England and Wales (Watts et al 1978b). Crucially, however, it was the tendency for gradual increase in demands on local authorities which led to a brake being applied to postbus development in the 1980s, and selective withdrawals. High inflation rates, the oil price rise of 1979, and disappointing passenger loadings in some cases all had their effect. Further, the Post Office introduced new accounting procedures which demanded financial guarantees from local authorities for a 5-year period (the assumed life of the vehicle), including the introductory costs. This made local councils much more critical of the economic performance of postbuses. Also, new bus grant was phased out in 1984, and this left considerable doubt as to the ability of many existing services to cover vehicle replacement costs. This has exposed the grant-dependent nature of the mode. Local councils had been keen on postbuses in the 1970s when most of these grants were reclaimable from central government, but now the boot is on the other foot.

To exemplify the new conditions, Watts (1984) re-calculated postbus 'setting-up costs' at 1983 prices. The 1977 costs of £2280 and £1050 for the Commer and mailvan respectively had risen to £6500 against £2100. With new bus grant then at only 10%, the excess of the former over the latter, after administration and training and part of the cost of reserve vehicles, had become £8100. Covering the introductory costs, and with no bus grant, it is estimated that each passenger/mail vehicle would need about £3000-3500 from fares or subsidy to break even.

7.5 THE CURRENT SITUATION

Scottish regional TPPs provide examples of the increase in numbers of subsidized services, and costs, after 1980. In Central Region the cost of postbus subsidy rose from £600 in 1977 to £10,850 in 1983 (six services, plus another five shared with neighbouring authorities). In Dumfries & Galloway none was subsidized in 1980/81 but 7 out of 12 were by 1985/6. In Fife, at least 4 out of 7 were subsidized by 1984/5. In Highland Region only 3 out of 37 postbuses received financial support in 1979/80, but 11 did so in 1985/6 at a cost of £13,230. In Strathclyde, 7 out of 19 services shared a £30,000 subsidy in 1984/5.

In this climate, the weakest services (low passenger loadings were noted in section 7.3) were vulnerable to closure. There were some cases of Commers being replaced by Land Rovers, and some complete withdrawals, even in Scotland where postbuses generally seemed permanent fixtures and enjoyed official confidence. Successive editions of the *Scottish Postbus Timetable* now show a greater turnover from year to year but overall numbers appear stable. Route closures have been more conspicuous in England and Wales where postbus numbers were lower. These are marked on figure 7.3: Billesdon (Leicestershire), which averaged one passenger per journey and fares covered less than 10% of costs; East Bergholt (Essex); Rhyl (Clwyd); Elan Valley (Powys), which carried only 25 passengers in two years... In other cases the postbus was replaced by other UCMs: in Petworth (West Sussex) the demand was covered by the Bignor & Barlavington community bus; at Heathfield (East Sussex) it was subsumed by the reorganization of public transport under the ESCORT scheme (see chapter 4).

This raises the question of the comparative value of postbuses against other UCMs. Now that the financial burden has been shifted from central to local government, councils are bound to consider whether other modes, especially in the voluntary sector, might be cheaper. If a postbus is uneconomic, the expressed demand not too high, and sufficient local volunteers available, then such a conversion would probably be beneficial. Even a 'conventional' PSV minibus might be cheaper, if it operates at a low frequency. Northumberland County Council (PTPs 1984/5 & 1985/6), concerned that only two of its six postbuses were well used and financially sound, considered that the less successful ones might be replaced by a minibus service covering similar ground but only once or twice per week. Limited demand for postbuses suggested that their high frequencies were unnecessary and, at a similar cost, this lower service level might be adequate. Also, timings could be faster and more convenient than on the postbuses. Alternatively, if a less comprehensive route coverage was acceptable then a cheaper or more frequent minibus service could be run. In this particular case, the situation was resolved by operating the four routes in question with postal estate cars.

The postbus era appears to be over. There does not at present seem to be any scope for further developments. The growth curve (figure 7.2) has flattened out, circumstances have changed, and despite steadily increasing political support for UCMs in general, postbuses are in danger of entering a downturn. While numbers are generally steady in Scotland, recent losses in England may presage a declining trend, so that the national total (figure 7.2) was dipping slightly by 1985. Ironically, there is still potential for growth if economic conditions are right. Watts (1984) conducted a theoretical study at various sites in south-east England, and found that in 70% of cases postbuses could be operated with little or no change to existing mail distribution arrangements, at costs which were still reasonable. Holding (1983) even concluded that a fully integrated passenger/mail system on the Swiss model could theoretically be introduced in Britain. Both authors accepted that financial and institutional barriers would prohibit any such development. Under deregulation, postbuses have been granted exemption from the tendering procedure until 1991, and they are presumably eligible for Rural Transport Innovation Grant although no awards have been made. It remains to be seen whether this is a stay of execution or an opportunity to consolidate.

Chapter 8

PASSENGER/FREIGHT SERVICES

8.1 THE 'COMMON CARRIER'

It sometimes appears that today's problems of public transport in rural areas resemble those of an earlier historical era. In the face of bus decline, centralization of consumer facilities and an economy which assumes widespread car-borne mobility, persons without cars may be experiencing a new isolation equivalent to that existing before motorized transport. If so, one might search for historical parallels. Before the motor age, village dwellers had little opportunity for regular journeys, the village may have been more self-sufficient than now, but otherwise, everyday goods not obtainable locally were provided by peripatetic traders. Goods came to the people instead of vice versa.

At one level there was the individual specialist trader, passing from village to village selling his own produce or offering his services from a small cart pulled by a horse or donkey, or even on foot. In addition, the 'country carrier' or 'village carrier' had four functions: to take orders from villagers for the purchase of goods in town; to forward bulky goods and parcels from the nearest town or railhead to their rural destination; to convey country produce into town for sale to merchants; and to act as a primitive form of *public passenger transport*. These, again, took the form of a horse-drawn cart. All these were made redundant in the early twentieth century with the coming of the motor bus and lorry, and the longer term restructuring of the rural economy and local travel patterns. A descendant of this form of service provision is what are today known as 'mobile services' - mobile shops, libraries, etc - the further promotion of which might offer a partial solution to rural accessibility problems (Moseley & Packman 1983).

The concept of interest is the 'common carrier', where passenger conveyance is shared with the transport of a wide range of goods items. By this means a single vehicle replaces numerous others which would otherwise have been needed, costs are saved all round, and the 'carrier' vehicle has a good chance of economic viability. Various authorities have from time to time speculated on the possibility of a modern version of the common carrier, in the certain knowledge that there are many public and private small goods vehicles circulating in rural areas, serving their own exclusive business, and probably all under-capacity and uneconomic, in common with the bus service. Rationalization is an obvious response, especially when recalling the historical precedents.

A parallel from more recent memory is the informal tradition in the Scottish Highlands and Islands of conveying an extraordinary variety of goods on stage buses. A report from the Outer Hebrides in the 1930s told of "mails, goods, animals, cloth, hay and timber" being regularly carried in the body of the vehicle with the passengers. This was a sensible and widespread arrangement in the peculiar circumstances of that remote region, where a sparse population, even in a time of lower car ownership, meant that a specialist service had no chance of covering costs. The more significant bus operators, such as David MacBrayne (later, Highland Omnibuses) and the Sutherland Transport & Trading Co. maintained this tradition for many years, although it was at its height in the 1950s. The latter company at this time obtained only 30% of revenue from passengers, depending mainly upon mailbag contracts from the Post Office, school contracts and private parcels. A compartment at the rear of the bus was partitioned off for the purpose. Personal orders for groceries, prescriptions, etc., were also taken, by collecting messages from boxes outside remote houses and then depositing loads on the return trip. Dependence on freight and mails meant that these controlled the bus schedules, so that timetables related to the arrival of trains and ferries. As an example, the historical flavour of this can be gleaned from Skewis (1963). More recently, improvements to road transport and the innovation of postbuses have reduced this activity, but it still continues in a less extreme form.

8.2 THE BORDER COURIER

(a) Operations

It may be due to this heritage that all current passenger/freight services are in Scotland. The Border Courier scheme, set up in 1979, has achieved particular fame due to its distinctiveness, its large scale of operation, and its continuing uniqueness within the UK. Borders Region is a large sparsely populated area with all the predictable problems of rural transport, with a fragile bus service already, by the late 1970s, having been boosted by several postbuses and one social car scheme. An urgently felt need for an overhaul of public transport coincided with the realization that numerous delivery vans were circulating in the rural areas on behalf of various public authorities. These were regarded as an untapped resource which could be combined with each other and with passenger services to satisfy a range of transport needs more efficiently.

The idea therefore was to combine the various goods carriage requirements of Borders Regional Council, its four constituent district councils, and the Borders Health Board, on to a system of buses which would operate on scheduled routes and also be convenient for the public. A feasibility study was commenced in 1976, grant aided up to 75% by the Scottish Development Department in recognition of possible wider applications. The scope for rationalization of sundry goods services was soon appreciated: the health board was running a van service for laundry and food supplies, and another for laboratory samples and pharmaceuticals; the environmental services and water and drainage departments of the five councils sent vans containing milk, food and water samples to laboratories in Edinburgh several times a week - sometimes these were shared between authorities, sometimes not. Internal mail, parcels and files were often transported between organizations in their own stores vehicles. It was believed that all these could be coordinated. Minibus schedules operated by the education and social work departments were also looked at, but could not be integrated with the above (Borders 1981).

Operation of the whole system by the Post Office was first considered, which would have had the advantage of integrating the letter post with the other freight. This could not be agreed, however, and the contract went to Scottish Omnibuses Ltd (currently run by its subsidiary, Lowland Scottish). Agreement on the contributions payable by the health board and the councils and the continuation of the SDD grant allowed the new service to get under way in May 1979 for an experimental period. As would be expected, there were frequent changes to operations and route patterns from year to year, and these need not be detailed here; any experimental scheme must be a learning process. There were various improvements for the second year (1980-81), at the end of which a report was issued on the results of monitoring (Borders 1981). The service was made permanent at this stage.

In practice, freight carriage was the main determinant of the route pattern. Routes were designed primarily to link two major hospitals - at Peel and Melrose, fourteen small hospitals, over twenty surgeries, health centres or clinics, a couple of health administration offices, and five local government offices, scattered over most of the region. Figure 8.1 shows the route pattern as it was during phase 2 of the experimental period (1980-81). It can be seen that while all the above establishments are connected, the network shows many deviations for public transport purposes, intermediate routeing being deliberately taken through areas with little or no pre-existing transport. Hence the goods function is the backbone of the service, and the passenger facility no more than a spin-off from it.

Vehicles used might be described as midibuses or short coaches, specially adapted to include a segregated goods compartment at the rear, and with 13 seats in front. Specific

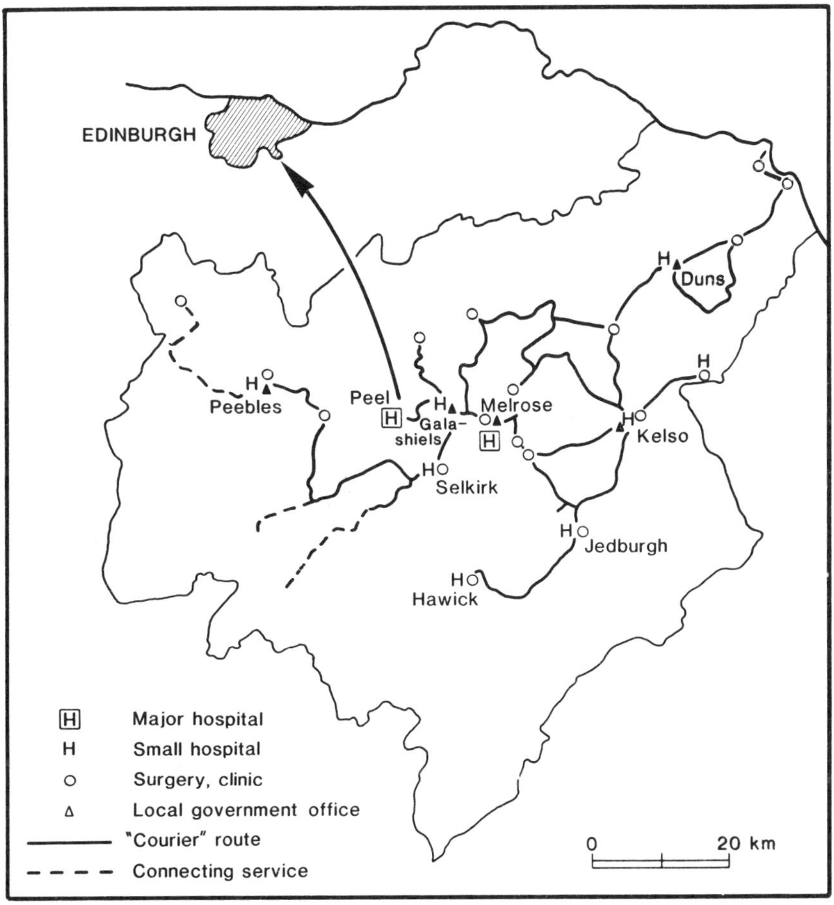

FIGURE 8.1 The Border Courier service 1980-81 (after: Borders 1981).

precautions were built into the design of the goods compartment to allow the safe conveyance of items like blood samples and gas cylinders. Four buses plus a spare were used, each bus being housed at a separate outstation - at Peebles, Hawick, Kelso and Galashiels. Routes in each case were all focused on the expanding town of Galashiels and the nearby Peel District General Hospital. The basic schedule for all buses was to operate an inbound collection run for medical samples, connecting all medical centres on the route after morning surgery hours, passing through Galashiels to deliver the samples to Peel Hospital laboratory, and then doing an outbound delivery run with supplies. Hence a one and a half hour shopping stay in Galashiels was made possible for people from many areas without other transport. On the other hand, surgery hours and a mid-day deadline for samples at Peel laboratory made timings often tight and imposed some problems in providing a passenger service for the rural areas. Long distances in the Berwickshire part of the network also imposed restrictions. The Courier based in Galashiels did a shorter return trip in the morning, before occupying the rest of the day taking samples between Peel and laboratories in Edinburgh. Each of the buses (marketed as Courier 1,2,3,4) took slightly different routes on different days to maximize rural coverage.

A few subsequent changes are of interest. Originally, a 'dial-a-ride' facility was included in the Coldstream-Duns area (Berwickshire) to mop up dispersed demand, but this was very little used and was dropped after the first year. From 1981, a new role for the Courier was to distribute tourist information literature to local offices. Later, a Courier route in the Ettrick valley was taken over by the newly founded Ettrick & Yarrow community bus. There has been an increasing degree of integration with conventional buses, some to facilitate school contracts and others to permit long distance connections with Edinburgh. One Courier service was integrated with the 'Border Harrier' tourist bus, which runs in summer to Peebles and the Yarrow valley, with the Courier running on schooldays only. More recently, Peel Hospital closed to be replaced by the Borders

Bute Library

Items on loan 17/11/2010 16:50

XXXXX60207

Item Details	Due Date
1605508931 * Unconventional and comm	01/12/2010
1610970965 * Carfree cities / J.H. Crawfc	01/12/2010

* Indicates items borrowed today

Overdue items will incur charges.

Renew items at: http://library.cf.ac.uk

General Hospital between Galashiels and Melrose; the Courier buses now converge on the new site in the same way as before.

Deregulation has had little effect, with the services going out to tender, but the existing operator - Lowland Scottish - winning a five-year contract. Unlike in England and Wales, public sector bus companies in Scotland were not privatized by the 1985 Act, so this has prevented any further uncertainty. There has been little competition from commercial operators, except for a single 'taxibus' on the Berwickshire coast.

(b) **Performance**
Most available information on the Courier's performance is derived from the final report of the monitoring period, during which various user surveys were done (Borders 1981). At this early stage, much effort went into publicity to raise awareness of the new service, and household surveys of potential users. Although sample sizes were very small, 52% of respondents in the area reported difficulties in making journeys, and 42% of potential users thought the Courier would contribute "a good deal" to meeting their transport needs. Ninety-four percent had heard of the service within the first year of operation.

The most impressive sign of success was the rapid build-up in passenger numbers, such that the feasibility study prediction of 200 per week was soon surpassed, and numbers averaged 369 per week in 1980/81. Initial service revisions helped increase numbers despite a fares raise, so that the predicted weekly revenue was overtaken. About three quarters of passengers were female and one quarter were over 60, over half from non-car households but about 85% lacking a driving licence. This is an unremarkable user profile, save that the Courier appeared not well suited to the younger age groups, apart from school contracts. The breakdown by journey purpose is also unremarkable - shopping 34%, work 25%, social and recreational 23%. It was for shopping trips that respondents had most often reported difficulties. A significant relationship was found between those with travel problems and those using the Courier, and also between degree of difficulty and frequency of use. Work trips were inhibited in many cases by the Courier taking different routes on different days, although it was useful for Peel Hospital employees. The proportion of health trips was perhaps low, at 5%, and it is ironic that connections with surgeries and hospitals were timed for the convenience of medical supplies collection and delivery, not necessarily for the convenience of patients. In summer, 16% of passengers were tourists, and some crowding occurred.

Similar monitoring was applied to the (arguably more important) goods service. Compared with the previous van distribution system, the level of service provided by the Courier was in almost all cases at least as good, while almost a third of the medical centres had more frequent deliveries. Speed and reliability of service was as good as previously, the main problems being the timing of sample collections, which was not always convenient for medical centre staff, and a shortage of space on some occasions, requiring some re-allocation of goods flows. After the second year new vehicles were deployed, with additional goods space and 17 seats.

The unexpectedly high patronage has already been noted, and the high mileage reflects intensive use of the vehicles (1980/81 data in table 8.1). Depreciation costs were high, as the original vehicles had a short life. Simple cost/revenue evaluations are hampered by the twin passenger/freight role. Fares revenue was only 11.5% of total costs, but if the proportion of costs due to passenger service is isolated, then the operating ratio becomes 20%. In comparison with other modes in Borders Region, the Courier's cost per mile at £0.31 was much higher than for postbuses (£0.03) but significantly lower than for conventional buses (£0.43-£0.57). Taking into account its far higher passenger loadings, the Courier's deficit per passenger at £1.23 was generally in line with that of other modes (Borders 1981). Costs were split among the public agencies consigning goods to the new service (table 8.1), but unfortunately the savings made by

TABLE 8.1 Border Courier results in second experimental year 1980/81 (source: Borders 1981).

	1980/1 (4 services, 5 vehicles)
Passengers	20757
Mileage	142901
Total Cost	£78349
Revenue from fares	£9024
Crude operating ratio	11.5%
Cost due to passenger service	£44714
Passenger service operating ratio	20.2%
Cost per mile on passenger service	£0.31
Cost to Borders R.C.	£25551
Deficit per trip on passenger service	£1.23

Contribution to Costs	%
Borders Regional Council	
Policy & Resources	36.1
Water & Drainage	1.1
Education	4.3
Social Work	5.1
Four District Councils (each)	0.4-0.9
Borders Health Board	18.4
Scottish Development Dept.	32.4
	100.0

these bodies in comparison with the costs of the previous van system were not disclosed in the report.

The Border Courier continues to be very popular. In 1985 it handled about 28,250 passenger journeys, at a subsidy of about £34,000. The deficit per trip is still, at £1.2, roughly the same as in 1980/81. 1987 passenger loadings were even higher, at about 38,950 (excluding school contracts and private hires); this is almost double the patronage of 1980/81. Revenue was £20,700. Substantial contributions from the health board and other goods users make the overall service appear cheap in cost per mile terms.

(c) **Postscript**
Ten years after its introduction the Border Courier remains unique in the UK. There is no immediately apparent reason why a similar scheme has not been introduced elsewhere. It has been considered and discussed by many local authorities. Perhaps in more continuously settled rural areas goods flows and health networks are more complex, or these cannot be married to passenger requirements, or a new service would clash with existing bus routes. Certainly, the large scale of organization and coordination would be a big job, rivalled only by the ESCORT project (section 4.4). Deregulation would make this more difficult, as one would need to plan for a large rural area free of commercial competitors.

8.3 THE DEMOUNTABLE VEHICLE

Combining passengers and freight in the same vehicle is only possible if the volume of goods or numbers of passengers are not excessive. For the carriage of bulk freight and full loads of passengers there is apparently no alternative to running two separate specialist vehicles. However, a 'technological' solution is available. As long as the passenger and freight services are not required at the same time, then there is still scope for the rationalization of operating costs by providing both services from a single unit.

There is not much history of the bus passenger business drawing lessons from the road haulage sector, but a solution to the problem could be found in that direction. What is needed is a single cab/chassis unit capable of accepting interchangeable bodies designed for different purposes, in this case a van body and a bus cabin. Although unprecedented in the public transport sector, such technology has been available since the mid-1970s in the road haulage and construction industries. In these cases, the vehicle might be converted into trucks, tippers, etc., but there have been examples of construction companies using temporary bus bodies to ferry workers in and out of sites, and other bodies for freight applications at other times (Rowlands 1984).

Extension of this principle to PSV operation does not seem to have been seriously considered by bus operators or local authorities until Strathclyde Regional Council included it in their discussions on passenger/freight modes in the early 1980s. Continuing the common carrier tradition in Scotland, it was felt there was a need for something better than a postbus which was recognizable as a medium capacity bus, suitable for occasional peaks such as schoolchildren and tourists, but also able to take bulk loads. One alternative was to coordinate public agency goods movement as in the Border Courier; otherwise, most goods haulage is in the private sector, but there were a few cases where the same firm operated buses and lorries. It is in operations like these that there is most potential for the demountable vehicle (popularly known as the 'swap-body bus'). Strathclyde PTE thought it important that such a solution should at least be tried out in passenger service. An apparently ideal testing ground was found in the Isle of Arran, where the island's passenger and freight services were provided by the same operator - the Arran Transport & Trading Co. The first problem was the technical one of designing and constructing the vehicle.

Four contractors were commissioned to build components. The basis of the vehicle was a Dodge midibus chassis, fitted with a special body underframe by a Peterborough firm who also provided the demounting equipment. A Cambridge firm provided the cab and the demountable 20-seat bus body, while the van body was built by a Linlithgow company. Several intriguing engineering problems were encountered, not surprisingly for such a novel concept: notably how to produce a compressible weather seal between cab and body, locking the bodies to the chassis, walk-through access from the cab, and heating for the bus saloon. The procedure for swapping bodies at the depot could have been a weak point, but is achieved by a simple hydraulic system integrated into the vehicle itself and powered from its batteries, and can be done by the driver alone in five minutes. The demountable has had to be licensed both as a PSV and as a goods vehicle. It was thought (Strathclyde PTE press release) "to be the first fully certified public service vehicle of its kind ever built".

Since September 1984 the vehicle has been operated by Arran Transport & Trading Co. on behalf of Strathclyde PTE. Replacing two existing vehicles - a midibus and a parcels van - the initial plan was to use the bus for school contracts in the morning and evening peaks, and the van for goods deliveries and collections during the day. The detached van body could be pre-loaded at the depot while the bus is out on its morning school run, and the bodies swapped when it returns. It is also used in season for transporting tourists, connecting with ferries at the island's main pier.

No indicators of performance have yet appeared, although the operator is reported to be "delighted" with the vehicle (Rowlands 1984). Like other such schemes, the crucial test is the savings incurred by replacing two vehicles with one, but these have not been disclosed. Strathclyde PTE were confident that, despite an initial cost of £30,000, the demountable would be financially viable over its write-off period.

Impressive as it is, the technology employed on the demountable vehicle is not the primary unconventional feature; it is a dual-purpose mode, with the technology being applied in order to facilitate this. A comparison with the Border Courier offers some points of interest, as both are primarily passenger/freight dual-purpose modes. The latter is based on buses with relatively simple design adaptations, conveying passengers and freight simultaneously, while the demountable takes both traffic types consecutively with each having exclusive use of the vehicle. Unless passenger numbers or bulk freight are excessive, one might ask whether the demountable technology is really necessary and whether a Courier-type solution might be better. Both have institutional constraints: the Courier is best able to coordinate the goods shipment demands of public agency customers, while the demountable is best suited where its operator has both passenger and freight businesses. This would be a major impediment to use of the demountable elsewhere - even in Scotland. The challenge is therefore not a technical one. The design is tried and tested, and could be expanded, conceivably, with more than two alternative bodies - a 'welfare' passenger cabin and mobile library have been suggested. As with any other multi-purpose mode, the real challenge is to overcome institutional barriers so that a wider range of organizations - public, private and voluntary - agree to combine their transport functions.

8.4 OTHERS

The demountable vehicle was part of a number of experiments carried out by Strathclyde Regional Council into different types of freight-carrying buses. There have been two other, rather similar, innovations along these lines, in both of which the smaller scale of the freight element allows it to be conveyed on the bus together with passengers (Strathclyde TPPs 1984-89 & 1985-90).

First, in the Cumnock-Ochiltree area (east of Ayr) a need was exposed after a network rationalization exercise on the part of the public sector bus company caused the removal of a stage carriage service. The opportunity was taken to replace the passenger service at the same time as the van which had previously distributed meals-on-wheels on behalf of the Social Work Department. Hence, a 16-seater minibus was provided by the Education Authority in order to carry fare-paying passengers, schoolchildren, meals-on-wheels and some mail. Since deregulation this has been a registered commercial service.

The second case stemmed from the same network rationalization exercise, in which a regular bus service was withdrawn from the village of Colmonell, south Ayrshire. Again, a replacement bus was designed to incorporate a wider range of traffic demands, such as for education and social work, and also some goods carriage by means of a compartment at the rear of the vehicle. Introduced in November 1983, the bus now provides a commuter and schools service, conveys handicapped trainees to an adult training entre and, most originally, distributes meal containers from a central kitchen to village primary schools. It is available for hire in evenings and weekends (described as a 'community bus'). Known as "Honeybee", the new service has produced significant savings over previous arrangements. Originally set up by the PTE, this is now run by an independent with the same vehicle.

Chapter 9

COMMUNITY TRANSPORT FOR SPECIAL NEEDS

9.1 THE VOLUNTARY SECTOR IN TRANSPORT

(a) Origins and Development

This chapter and the next two consider the role of the voluntary sector in transport provision, for which the usual generic term is 'community transport' (CT). Use of volunteer labour is clearly 'unconventional' where it is fulfilling the same role normally played by professional public transport drivers and managers. In this sense, the immediate advantage of volunteer labour is simply to avoid the expenditure on wages which normally constitutes 70% of operating costs. Beyond this, CT becomes necessary when a desired service is otherwise uneconomic, or when a conventional (commercial) operator does not have the resources to put on a particular service, or through official indifference to claims of local need, or where the community itself is able to identify specific local problems of a type soluble only by grassroots initiative. The role of CT relative to other public transport may be supplementary, or specialized, or more contentiously as a partial replacement.

Although defined on figure 2.2 as an 'unconventional' feature of public transport, community management and operation goes far beyond the context of normal everyday services for the general public which is assumed in other chapters. CT has to be identified separately from UCMs, as the great majority of the CT sector has a different emphasis and background, in its devotion to the travel needs of persons with impaired mobility such as the elderly and disabled. This vital distinction has been made in chapters 1 and 2 above, but should be reiterated. It was not intended that any form of transport using unpaid staff should automatically qualify as 'unconventional' through this fact alone. Within the UCM definition adopted in this book (chapter 2) most CT services are not particularly innovative or new. Voluntary assistance in the health and welfare fields, including related transport, has a long history. Voluntary associations and charities go back to the nineteenth century, well before the welfare state (e.g. the St. John Ambulance from 1877). Therefore, most CT efforts could be said to be 'normal' practice in supplementing the statutory health-related transport services, unless they show a degree of innovation on a par with other UCMs. The clearest case of this is the new generation of dial-a-rides, following on logically from those of the 1970s (chapter 3) but re-orientated as a CT mode towards the elderly and handicapped. This constitutes the overlap zone on figure 1.1. Further, UCMs were defined as being available to the general public, which would clearly exclude services provided specifically for clients with special transport needs. The latter overlap with 'welfare transport', which is that provided as a statutory obligation by public agencies for transport to hospitals and social services facilities. Few modes run by volunteers are without eligibility restrictions, but community buses (chapter 10) and many car schemes (chapter 11) are open to the public at large. These comprise the UCM/CT overlap zone in figure 2.1.

Still on terminology, any kind of transport which makes special provision for persons with impaired mobility is variously referred to as 'accessible', 'needs-based' or merely 'special' transport. These terms cover both the statutory and voluntary sectors (Bailey & Layzell 1983). 'Accessible' in this sense refers to the design of buses or other vehicles in a more sympathetic manner for people with physical handicaps, and in order to take wheelchairs. 'Needs-based' provision is counterposed to the normal commercial practice of providing facilities only if demand is high enough to justify them. 'Need' here is meant in two ways: where travel is essential to the passenger but demand is too small and specialized to be economic, and in a physical sense, where certain disadvantaged groups may be unable to use conventional public transport, however plentiful it may be.

The voluntary sector in general is a highly complex assemblage of individuals, local associations, charities, and national organizations which engage in work on behalf of the community not for payment but on grounds of generosity, principle and commitment. 'Voluntarism' and 'self-help' - helping one's own community by voluntary effort - have had an uncertain relationship with the public and private sectors and are often misunderstood by government. Motivations and roles can be summarized as follows.

(i) Genuine altruism, allied with a sense of pride and purpose on the part of the helper.
(ii) A mediating role, as a pressure group on behalf of the community towards institutional decision-makers.
(iii) Service provision, meeting the needs of local communities in a flexible, sensitive and informal way, as part of 'bottom-up' development.

The interest here is less in neighbourly mutual aid but more in semi-formal local organizations. Self-help activities have grown to encompass community cooperatives, job creation schemes, community housing and shops, conservation work, as well as transport (Wilmers et al 1981). But there is an ideological context which cannot be ignored. It could be argued that self-help is used as a means of legitimizing social and economic policies of government and its agencies, by perpetuating inequitable divisions, by compensating for public expenditure cuts or rates reductions. Hence, self-help will continue to expand as new needs are exposed. As if to endorse this, government has set up schemes to elicit further voluntary activity (e.g. in a rural context, McLaughlin 1987).

The origins and development of voluntary (community) transport services for people with special needs has been summarised by, among others, Bailey & Layzell (1983) and Sutton (1985, 1987), in each case relating the voluntary sector to statutory provision by public agencies in their mutual concern for the elderly and handicapped. Indeed, the changing relationship between the two sectors has been a recurring theme for many years, and the voluntary sector cannot be understood without reference to the public sector. It is important to affirm that in this context the 'voluntary sector' is not comprised wholly of unpaid volunteers. 'Voluntary work' for example is unpaid work for purely charitable purposes. A 'voluntary organization' is one that works on a non-profit basis, independently, for the benefit of the community; it may raise its own funds or be in receipt of grant. It frequently employs professional paid staff in addition to volunteer helpers; members of the management committee may be unpaid, as well as representatives of funding bodies. Many, such as small local associations, are entirely voluntary, including for example community bus and social car committees.

As hinted above, voluntary involvement in 'welfare' transport not only pre-dates the welfare state but also pre-dates the internal combustion engine. Prior to World War II ambulance services throughout the country were heavily dependent on the initiatives of the St. John Ambulance Brigade and the Red Cross Society. With the establishment of the welfare state in the 1940s and statutory provision of health, education and social services administered by local authorities, the voluntary sector came to assume a supplementary 'back-up' role. The new public authorities had transport responsibilities appropriate to their own clients. Education departments must provide or pay for home-to-school transport (chapter 6). Social services departments (SSDs) were obliged to provide transport for clients given places in certain institutions: for the elderly, travel to day centres, clubs, and day care in residential establishments; for the physically handicapped, travel to day centres; for the mentally handicapped, travel to adult training centres; for the mentally ill, to day centres and day hospitals; and for children, to day nurseries. Health authorities (taken out of local government control in 1974) provide ambulance services for 'accident and emergency' cases, and for out-patients where the doctor affirms that the patient is unfit to travel by other means. Outpatients may also be conveyed by the hospital car service, which is operated on behalf of health authorities by volunteers using their own cars (see section 11.5). Bailey (1979) provides a brief summary of the situation in the 1970s.

Voluntary transport assistance has been used traditionally to convey elderly and handicapped persons on a range of 'institutional' journeys similar to those provided by the state sector. Local organizations of all types frequently have private vehicles for use by their members on group outings, which would include social and recreational trips. However, facilities to enhance the general mobility (non-institutional trips) of disadvantaged groups were very rare. Statutory and voluntary sectors cooperate in various ways. Some voluntary organizations use social services transport when it is available, while others seek funds from SSDs to support their own transport for tasks that are congruent with local authority responsibilities. Health authorities and SSDs support voluntary car services, organized by local agencies such as the WRVS, when more convenient for institutional journeys (section 11.5). Specific hospitals may have their own coordinators of voluntary help. Local authorities may support a Council for Social Service or a 'Volunteer Bureau' which coordinates voluntary transport for various medical or welfare needs on an area-wide basis.

The current situation is dominated by a great expansion in demand for health and social services and an even greater relative increase in demand for transport. Increasing demand stems largely from the reorganization of both local government and the National Health Service in 1974, and the switch to 'community care' in the social services a few years previously. Decline in conventional public transport has compelled state agencies to acknowledge a wider role in providing for more general transport needs. Continuing growth in the elderly population has also put great pressure on public services and enhanced the importance of the voluntary sector in supplementing them. In the 1970s there was a substantial growth in the number of community groups concerned with elderly and disabled people.

Inflation in the 1970s and government budgetary cuts in the 1980s have added to the trend of shifting further responsibilities to the voluntary sector. The latter has now become a vitally important pillar of the welfare state, no longer just 'filling in gaps' - if it ever was.

Despite cooperation, roles of the various sectors are ill-defined and have remained fluid. Rapidly increasing demands and costs have directed greater attention to the matching of needs and resources. This could be achieved either by integrating services of different health and local authority departments, or by integrating the public and the voluntary sectors, deliberately developing the latter. Much blurring of institutional boundaries had already occurred. Comparative costs are revealing: in the late 1970s the average cost of one passenger trip was £7.79 by the outpatient ambulance service, £1.28 by social services transport, £0.17 by stage carriage bus, and £0.11 by a voluntary minibus scheme (Sutton 1985). This indicates the urgency of rationalizing the most expensive 'special' services, pooling vehicles among departments, and making appropriate use of the voluntary sector.

Coincident with these trends have been the emerging recognition of the concept of 'need', and attempts to cater for the *general mobility* of disadvantaged people rather than merely for institutional trips. Statutory agencies have the power to do this but not the resources, and hence the voluntary sector has taken a lead. CT schemes originated from a desire for community-based organizations devoted specifically to transport provision, responding to the needs of disadvantaged groups, intended to be widely available and to offer a range of service types. The first community minibus project is generally regarded to have been a scheme in Birmingham in 1966; the first dial-a-ride for the elderly and handicapped began in Edinburgh in 1975; the first rural community bus commenced in Norfolk the same year (chapter 10).

Success in the Birmingham scheme led to the establishment of Community Transport, a national charity aimed at promoting the concept. This has since given its name to the CT sector as a whole. Renewed concern for transport for the mobility deprived was part of a wider movement on behalf of the disabled by various pressure groups culminating in a powerful alliance known as the 'mobility lobby'. The International Year of Disabled People in 1981 was an influential factor. After financial problems, the Community Transport charity was reorganized in the late 1970s with government help, which encouraged a more professional management, paid coordinators and less reliance on volunteers, and new relationships with local authorities. In 1982 it was instrumental in setting up the National Advisory Unit for Community Transport with limited-term DTp funding. Based in Manchester, the role of the NAU was to provide information and advice to existing CT schemes and to local groups intending to establish transport projects (e.g. Taylor & Thompson 1986). Also in 1981-2 two promising developments were the first national CT conference and exhibition, and the launch of Community Transport Magazine. Both have continued successfully, and have done much for the coherence and credibility of a highly disparate field. The fragmentation of large numbers of independent locally based CT groups had been a handicap for years, and some regional associations appeared, but not until 1987 with the re-launch of the Community Transport Association out of two predecessors, was there a single national organization of CT operators likely to act as an effective pressure group.

The search for respectability has been partly successful. The 1985 Transport Act affirmed that CT came within the official definition of 'public transport' (although the great majority of schemes are not available to the public at large), which means they may claim from transport budgets at central and local levels. Also, local authorities were permitted to subsidize CT services, e.g. through supporting concessionary fares, although on the other hand some authorities were considering putting them out to tender (Masey & Hopkin 1986).

(b) Numbers and Types

CT is an enormous and complex field, in terms of the number and diversity of services available. Yet even this does not do justice to the potential demand comprised by that sector of the population with impaired mobility, especially if one also accepts that the range of permitted trip types should be equivalent to those enjoyed by the population at large. The usual rule-of-thumb for the number of disabled people in Britain is about 5 million or 10% of the population; this is about four times the number of registered disabled. Precision is impossible due to uncertainties over what constitutes disability. Much higher claims are often made: if all the elderly were assumed to have some problems of physical fitness then this immediately includes 15% of the population; disabled younger people, the blind and the mentally handicapped can be added; temporary disabilities such as pregnancy might be considered. A case could be made that the 'disabled' are nearer 25% of the population. In 1988 the Office of Population Censuses and Surveys produced more accurate figures indicating 6.2 million disabled adults in Britain, of whom 4.4 million were mobility impaired, although arguably *any* handicap might justify special transport.

Expected trip rates are another question. 'General mobility' provision, through its own success, generates higher expectations of travel in those people formerly almost housebound. Given the confidence to use special or CT services, the passenger may seek more independent 'optional' trips for social and leisure activities. Despite increasing burdens on services, they have become the catalyst for a great improvement in quality of life.

Uncertainty over the size of the 'needy' population is mirrored by uncertainty over the number of CT services available, save that it is very large and increasing strongly. There is of course no official list of CT services, and no comprehensive national survey has been done. What should be included? One could take only the present generation of community minibus services aimed at providing general transport for anyone with mobility difficulties. In addition, should one include the very large numbers of community groups who run minibuses or cars in conjunction with other general or specific functions - from national organizations such as Age Concern or PHAB to small local associations? An important distinction would be between those who hire out their transport to other groups and those who confine its use to members only. Should one include only transport for the elderly and handicapped, or also that operated by youth groups, churches, social or sports clubs? In practice it is impossible to distinguish.

However defined, recent expansion cannot be doubted. Compared with the few general purpose CT schemes starting in the early 1970s, a directory produced in 1983 by Community Transport Magazine listed over 300. A 1986 listing by Community Transport Services Ltd, supplemented by other known cases, produced a total of 1700. These are mapped in figure 9.1. The data exclude statutory welfare transport provided by local authorities, and include the 'general public' modes of community bus and social car, but otherwise there is no strict adherence to the categories above. The real national total is unknown, but much higher. Patterns on the map merely reflect differential knowledge of numbers in each county/ region, and therefore it is not a representative sample. The 180+ schemes recorded in Strathclyde might be a realistic figure. Essex and Surrey give complete lists in their PTPs. Excluding social services, education and health transport, there appears to be 145 schemes in the Essex list which could be described as CT, mainly using minibuses but also cars, vans and ambulances. At least 33% of buses and ambulances are available for hire by other groups. The Surrey list appears to have 125 CT schemes, of which 67 use cars, 30 minibuses and the remainder adapted vehicles including ambulances (Essex PTP 1985/6, Surrey PTP 1984/5). If the Strathclyde, Essex and Surrey lists are near complete, this implies one CT scheme for every 1600-2400 persons over age 60. Extrapolating over the whole country, this would suggest a national total of CT schemes anywhere between 6000 and 10,000 - probably nearer the

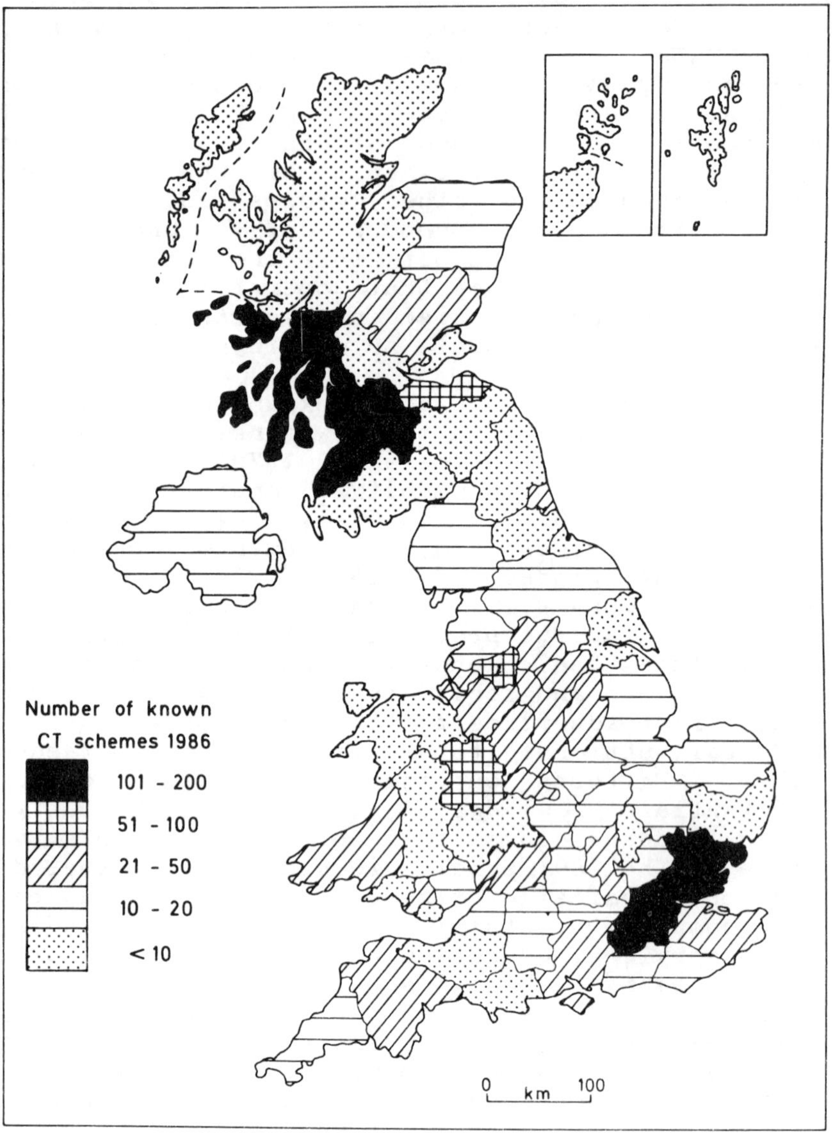

FIGURE 9.1 Community Transport schemes of all types, as known, 1986.

latter. There are thought to be at least 70,000 volunteer drivers.

As an example of the difficulties of gathering information in this field, but also of the importance of transport to the voluntary sector, a survey in Northern Ireland (Acheson 1988) provides useful data. The total number of voluntary organizations in the province is unknown, but a questionnaire was sent to 1600, of which 441 replied (27.5%). Of these, 259 used minibuses (58.7%). Even of those groups with their own transport, 43% still experienced 'considerable' or 'extreme' difficulty in meeting their transport needs. From the total number of respondents, 22% had no need for transport, another 22% had all the transport they needed, but 42% said they would start new projects or expand their work if they had more transport. If 'no reply' and 'no need' are excluded, then this figure rises to 65%. From a smaller sample, about one third were willing to hire their vehicles to other groups. It cannot be certain that Northern Ireland is typical of the UK, but in the absence of other regional scale surveys it may be reasonably indicative.

Transport provision for disadvantaged persons, primarily the elderly and handicapped, can be summarised as follows.

(i) *Statutory health and welfare services*, as described above; some vehicles may be hired to voluntary groups.
(ii) *Conventional public transport vehicles with suitable adaptations for the mobility impaired*; commercial operators are belatedly paying attention to the ergonomics of vehicle design and installing appropriate features.
(iii) *'User-side subsidies'*, i.e. subsidizing the person instead of the service, mainly implemented in the form of concessionary fares for the elderly, children, etc, but also the mobility allowance for the disabled and special subsidized taxi fares.

(iv) *Voluntary associations providing transport for their members in connection with organized activities*; some may allow their vehicles to be hired by other groups.
(v) *Voluntary associations formed specifically for general purpose transport for the disadvantaged*, including group hire of vehicles, minibus pooling, scheduled services (including rural community buses), demand-responsive buses (dial-a-ride), and social car schemes.
(vi) *Volunteer-run car schemes exclusively to back up health and social services transport.*
(vii) *Voluntary sector furniture van and storage services*, on behalf of disadvantaged groups. Not very well known, there are over 200 such operations in Britain.

The rest of this chapter concentrates on category (v). Community buses and car schemes, as 'general public' modes, are covered in chapters 10 and 11.

9.2 COMMUNITY MINIBUS SERVICES

Whatever the mode of operation, the use of minibuses in Britain is remarkably high, and relative to population is much higher than in any country in Europe. The use of a large proportion of the minibus sector in non-commercial operations for the benefit of the community is also highly distinctive. A comprehensive survey of minibus use in Britain was conducted by the TRRL in 1984 (Walmsley 1985). In that year there were 82,000 vehicles classed as minibuses. The normal understanding is of the 9-16 seat range, which comprised 63,300 vehicles or 77% of the total; those with 8 seats or less (sometimes called 'microbuses') comprised 14%, those with 17 seats or more (usually called 'midibuses') 2%, and others such as motor-caravans and type conversions 7%.

Further analysis was based upon a 3.3% sample. Perhaps the most significant result was that no less than 35% of the total were in use by the voluntary sector. This was broken down as:

schools and colleges 18%, youth and sports clubs 9%, community and charity 4%, others 4%. In addition, another 9% were employed by health, education and social services authorities, and most of these can probably be regarded as being 'in community service'. The remainder of the minibus sector is made up as follows: 9% other statutory bodies, 12% commercial operators (PSV stage carriage and contract hire vehicles), 23% industrial users and 13% private individuals.

Despite much concern over the conveyance of infirm and disabled persons, specially adapted vehicles are a small proportion of the total. The estimate was about 2000 adapted vehicles, or just over 2%. In response to a survey question, it appeared that 18% of vehicles regularly carried people with mobility difficulties, and 6% carried people in wheelchairs - the clear implication being that the number of adapted vehicles is sadly inadequate.

Other survey results showed that minibuses run by the voluntary sector tended to be older and to run a lower mileage than public authority and commercial vehicles. Only about 50% of voluntary sector buses were under five years old, and about 75% run under 10,000 miles per year. The latter figure suggests that there is great scope for more shared use of community group vehicles, and wide benefits would be achieved if regulatory and financial arrangements encouraged organizations with vehicles to hire them to those without.

It was precisely this point that influenced the first of the recent spate of statutes aimed at liberalizing UCMs and CT. The Minibus Act 1977 inaugurated a system of permits by which minibuses intended for use "for educational, religious or social welfare purposes" without any view to profit could be operated 'for hire or reward' without the vehicle needing to keep PSV standards or the driver needing a PSV licence. This meant effectively that a community group could charge fares on its vehicle and hire it out to others, on a non-profit basis. Minibus permits could be issued by the Traffic Commissioners, local authorities, and designated bodies in the voluntary sector. They were quite easy to obtain, and proved immediately popular. Subsequent legislation has amended points of detail; some larger buses (17+ seats) are now eligible. One major reservation was the exclusion of social and recreational travel purposes in the permit conditions: the 1985 Transport Act now allows permits for groups concerned with recreation. By 1985 about 12,000 minibus permits were in force, and this has become the main legal mechanism for CT groups to operate. According to the TRRL survey, 35% of minibuses in the voluntary sector are covered by permits (Walmsley 1985). A breakdown of applicants based on earlier data (Bailey & Layzell 1983) suggests that over 40% of permits are held by youth clubs, schools and student unions, over 10% by voluntary organizations; another 30% were issued by local authorities, some of which may also have gone to these groups.

Since the pioneering Birmingham project in 1966, the Community Transport charity was able to expand the number of local voluntary minibus schemes with others in Coventry (1972), Manchester (1973), Wolverhampton, Newcastle and Stockport (1974). Among others independent of the national charity were early schemes in Liverpool (1971) and Tower Hamlets, London (1974), although these were associated with inner city action projects rather than purely transport provision. As explained above, the present generation of CT schemes involves local non-profit organizations whose sole purpose is to provide a public transport alternative, for general travel needs, for a wide range of people. These are most often named after their locality, with the label 'Community Transport' (CT) or 'Voluntary Transport' (VT). Some schemes of this general type are provided by local authorities themselves. However, no data source allows these to be distinguished from the enormous range of other organizations with their own minibuses. Their rate of expansion may be deduced from overall figures estimated above, and growth in demand for and provision by individual schemes can be seen from figure 9.2. Newer schemes grow the fastest, and there is some

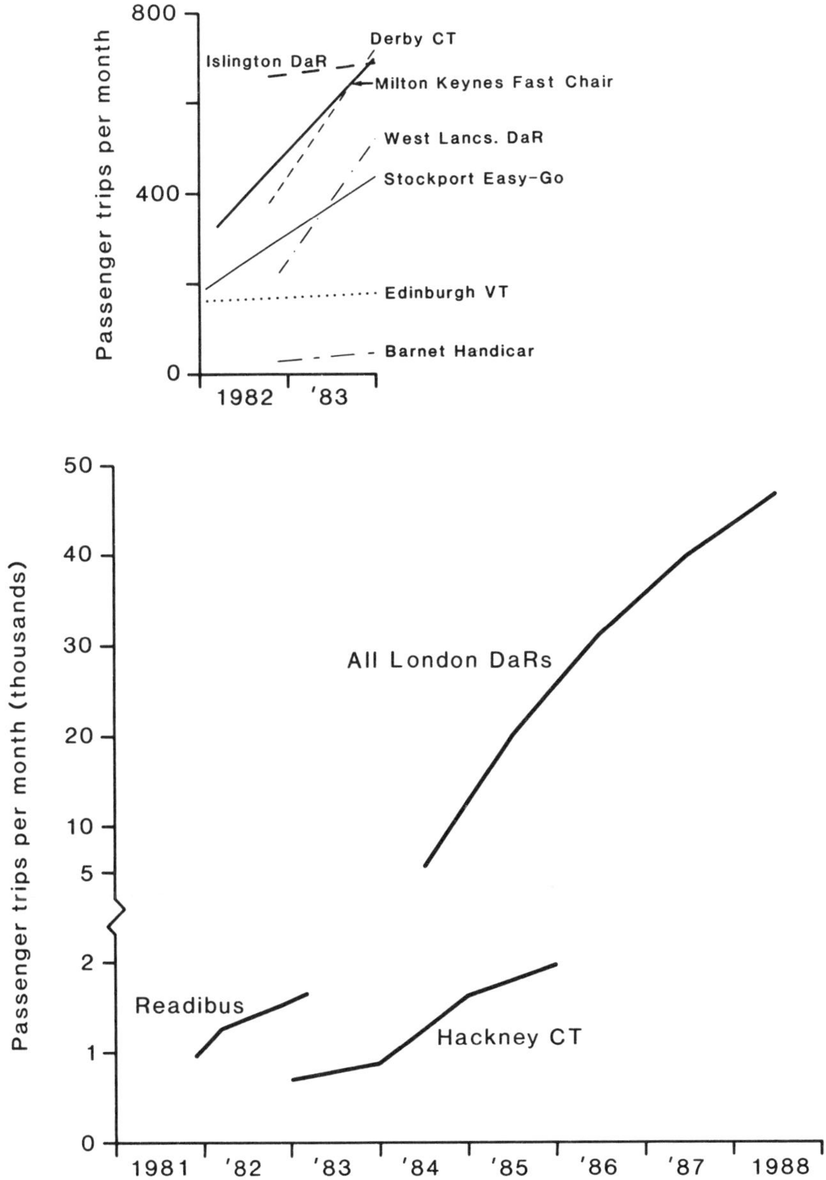

FIGURE 9.2 Growth rates for selected CT schemes (after: Bailey & Appleby 1986, Fulford 1987, Heraty 1984, Harkell 1988).

evidence that dial-a-rides (see next section) grow more rapidly than traditional CT.

The most common mode of operation, especially in the early years, was group hire of minibuses. Existing community groups without their own vehicles would be required to affiliate to the CT association, to be vetted according to appropriate eligibility criteria, and then to pre-book hire of vehicles when needed. Such groups would normally outnumber individual members. A hiring group may provide its own driver or use drivers and helpers employed by the CT organization, but in either case drivers would have to be registered and usually undergo an informal training programme (although this is not a legal requirement). Eligibility criteria may vary widely: narrower conditions would cause a greater proportion of disabled users and vehicles would have to be suitable. An example of more liberal interpretation is Hackney CT which accepts membership from, *inter alia*, arts-based and festival organizations, ethnic minorities, women's groups, religious groups, youth and sports groups, and housing cooperatives (Fulford 1987). The range of client groups handled can be seen from table 9.1. In both a large sample of voluntary organizations in Birmingham and a single CT scheme in Nottingham, the elderly, handicapped, youth and children's groups predominate. In the Birmingham case, passengers taken by CT organizations outnumber those transported by the SSD.

Modes of operation have diversified in recent years, and may include scheduled or semi-scheduled services suitable for individuals, many-to-one or many-to-few D/R services (e.g. collecting from individuals' homes and feeding into a shopping centre), or a completely D/R service (these are frequently hived off as a separate dial-a-ride organization). A successful CT association may operate different service types with different vehicles, which may include various buses, voluntary cars, a taxi arrangement, and one or two trucks. A convenient case study is Hackney CT (Fulford 1987).

TABLE 9.1 Client groups conveyed by CT operations (source: Sutton 1985).

	All voluntary organizations Birmingham 1977		Nottingham VT 1982-3	
	passengers (000s)	%	passengers (000s)	%
Elderly	299.3	27	8.5	39
Physically handicapped	343.6	31	1.0	5
Mentally handicapped	121.9	11	2.4	11
Mentally ill	11.1	1		
Children and families	254.9	23	7.7	36
Church groups			0.5	2
Community groups			0.3	1
Other	77.6	7	1.2	6
	1108.4	100	21.6	100
Number of vehicles	119		5	
passengers/vehicle/year	9300		4300	

A relatively new development is minibus pooling schemes. While the hiring of vehicles to local groups by a central agency, such as CT associations in the London boroughs, is more efficient than independent hiring arrangements, this could be further extended by pooling vehicles from several suppliers under one umbrella scheme. The obvious candidate would be the pooling of SSD and voluntary sector vehicles for clients of both. In Birmingham and Nottingham, for example, the voluntary sector uses SSD transport and SSDs use voluntary transport as appropriate (Sutton 1987). Previously, minibus pooling had only been considered for the journey to work, along lines familiar in the USA ('van pooling'). A TRRL investigation (Greening & Jackson 1982) found the possibilities limited by legal and economic obstacles, and a specific experiment at Ilford, London, found that volunteer drivers were strongly deterred by the need for a PSV licence (Jackson 1982). This journey purpose is still not eligible for minibus permits. Transferred to the welfare or community sector, however, the concept has generated a few schemes (Warrington 1987, Transnet 1987).

The best known is the Shared Transport Service in Birmingham. This was established in 1979 as a 'broker' to coordinate requests from groups without their own vehicle with the availability of minibuses held by other groups who are willing to share them. It was merged with Birmingham CT in 1983 and now operates its own vehicles as well as running the minibus pool. In London, Brent CT and Ealing CT operate similar schemes. Tower Hamlets CT coordinates minibus supply to community groups on behalf of the borough council. In some cases the pool or brokerage agency undertakes vehicle administration and servicing work on behalf of their end users. Among public agencies (not CT), shared vehicle use and the 'brokerage' concept is best known from the East Sussex ESCORT scheme, where social services clients and the disabled share with the general public and schoolchildren (see section 4.4).

Community transport was eventually seen as a potentially large market for vehicle and equipment suppliers. Once the CT sector had got itself organized and attracted attention, vehicle manufacturers and coachbuilders began to take notice of its requirements. The evolution of minibus design has been described above in the context of commercial PSVs (chapter 5), and details of vehicle types can be found elsewhere (Anon 1986, Hall 1988). Production minibuses adapted from van designs by major manufacturers, such as Ford, Vauxhall and Austin-Morris, were available in the 1960s and 1970s and were within the means of small organizations. While the range has since widened considerably, popular models such as the ubiquitous Ford Transit are still based on van conversions, and new models custom-built for 'special' transport purposes were slower to appear. Most multi-national car companies now produce minibuses with more generous space standards, e.g. Iveco-Ford, Peugeot-Talbot, Fiat, Dodge/Renault, as do new small firms such as Optare. Nevertheless, there is still a great reliance on body conversions to accommodate special fittings such as wheelchair lifts and ramps, special doors, steps and rails, and sometimes communications equipment. It is estimated that about fifty firms are involved in vehicle conversions for CT use.

For most, if not all, CT organizations the most persistent problem is *funding*. Possible sources are apparently numerous, but have proved unreliable, short-lived, prone to change, or demand too much influence on the transport business. Most projects survive on a combination of local fund-raising, donations, and grant-aid contributions from several different sources. Despite merciless cutbacks in central government support, local authorities are still among the most common sources of funds. The necessity here is to align CT objectives with appropriate budgets such as social services, education, public transport or general purposes. Central government sources are broken down into numerous limited-life programmes. Originally the most influential was the Urban Programme, stemming from the Inner Areas Act 1978. Many CT schemes have been set up as part of inner city development projects, under partnerships between the DoE and local authorities.

CT workers are often obtained from government job creation schemes, principally through the Manpower Services Commission (Department of Employment). From 1982 the Community Programme, aimed at the long term unemployed, has been available as a source of CT staff. Similarly, the Youth Training Scheme/ Programme can provide school leavers. The Voluntary Projects Programme provides support costs for volunteer workers. Although widely used by the CT sector, MSC schemes have caused much disquiet, mainly due to the imposition of their rules on the transport operation and the uncommitted attitude of temporary or part-time staff. The DHSS also has a fund for volunteers in the health and social services fields. In other cases, charitable trusts and transport-related businesses have made contributions (Taylor & Thompson 1986).

Fares obviously have to be set at levels that client groups can afford, and revenues typically cover operating costs only. Increasingly, the local authority pays for new or replacement vehicles. In 1988 the Urban Programme was reduced, local authority budgets cut further, the Community Programme abolished, and considerable doubts have been raised about the suitability of the new Employment Training programme. An atmosphere of renewed crisis has overtaken many CT schemes.

9.3 DIAL-A-RIDE FOR THE MOBILITY DEPRIVED

Traditional CT service modes, as they developed during the 1970s, were often believed to be falling short of the ideal of providing a fully personalized service for anyone with impaired mobility. Group hire of minibuses, for example, assumes that all the 'needy' population are members of affiliated community groups, and does not necessarily carry passengers for any trip purpose nor necessarily does it provide a door-to-door service that such people require. Other modes of operation are relatively flexible, but the ideal is for a totally D/R service with doorstep pick-up and set-down, available to any person in difficulties for any trip purpose, using a vehicle with suitable specialized equipment. Hence, around 1980 there was a re-evaluation of the 'dial-a-ride' concept, which at that time was about to become extinct in its earlier form as a commercial PSV stage bus mode. The unfortunate experience and apparent failure of the 1970s generation of DaRs has been documented in chapter 3. A comparison between 'DaRs of the 1970s' and 'DaRs of the 1980s' is offered by Bailey (1984). There was little or no continuity between the two. The mode was recognized as ideal for the client group concerned, while a closer targetting on one section of the population together with the use of volunteer labour might improve DaR performance with a well defined role and perhaps lower costs.

There has never been a clear-cut division between traditional CT and D/R services. A number of CT groups had always followed the D/R ethos, the first being a scheme in Edinburgh in 1975 - at least five years before the term 'dial-a-ride' was being used in this context. Not all D/R services actually call themselves a DaR - a number of these, including some earlier schemes, are included in the comparative study by Bailey & Appleby (1986). Causal factors in the renaissance of DaR in the early 1980s were similar to those for the CT field as a whole - the influence of the voluntary sector, permissive legislation (as a CT rather than a PSV mode), and the availability of funds (as described in the previous section). Banister et al (1984) also see the development as a valid example of needs-based planning, although lacking an integrated strategy. In the background, there was some influence from the continuing success of DaRs in North America and certain European countries, contrasting with earlier British experience. This is attributed by Sutton (1987) to the use of computerized scheduling techniques, although ironically the 'hi-tech' aspect was partly blamed for DaR failure in Britain.

As an alternative to adapted minibuses, needs-based D/R transport might be provided by the use of cars and taxis. The mobility allowance was frequently used by the disabled to pay for taxi journeys, but its value only permitted a small number of trips. In London, the GLC operated a 'taxicard' system which enabled taxi trips at a substantial discount. First tried in 1983 as a pilot scheme in Southwark, taxicard was soon extended to widespread coverage, and intended to complement the DaRs. These however should be classified as 'user-side subsidies' and do not qualify as CT. Unfortunately, taxis cannot normally accept wheelchairs, and adapted versions have only recently become available. Elsewhere, taxi-like services are provided by adapted cars run by voluntary organizations: examples are the Stockport Easy-Go and the Barnet Handicar (both in Bailey & Appleby 1986).

Two of the earliest schemes to call themselves a 'dial-a-ride' - Coventry (1979) and Islington (1980) - were both operated by local authorities and funded largely through the Urban Programme. Innovation proceeded rapidly after 1980. The Readibus service in Reading, starting in 1981, has been claimed to be "the prototype British transport service for disabled people" (Heraty 1984), which is unlikely to be true. It is probably fair to claim that the main centre of DaR innovation was London, due to the strong support for the mode displayed by the former Greater London Council.

Development of the DaR network in London is described in Potter & Heiser (1987), and more briefly in Banister et al (1984) and Harkell (1988). Services of this sort were first proposed by disability groups in the late 1970s, leading to the pioneering venture by the Borough of Islington in 1980. The support of the GLC was vital, and lobbying produced the offer of a start-up package for further DaR experiments. Each borough with a viable scheme was allocated two buses, seven drivers and 2.5 office staff. By 1982 there were five active services, the newcomers being Brent, Hammersmith & Fulham, Ealing and Camden. In the same year the GLC launched a major initiative with

£0.5 million for a London-wide programme of "experimental innovations" in transport for the disabled. This produced the taxicard scheme plus the extension of DaRs throughout the capital, at a time when, before the 1984 London Regional Transport Act, the GLC was responsible for all forms of public transport in the city. Grants rose to £3.6 million for 1985/6, and in June 1985 all 32 boroughs were covered by 29 separate DaR schemes. In addition, two specialist DaRs were established to serve the Asian community, the rationale being partly that language problems were experienced with telephone booking.

Management was by autonomous local committees in each borough, consisting of paid staff and volunteers. While generally successful, as DaRs ceased to be experimental and became a permanent resource across the capital, the need for some coordinated strategy was felt. A federation of London DaRs eventually became a central users' association. However, development was frustrated by the abolition of the GLC in March 1986, and responsibility for transport for the disabled was passed to London Regional Transport. £5 million p.a. was allocated to DaRs for 1986-9, two-thirds of which came from the boroughs. This marked a standstill in real terms and, as a body accountable only to central government, LRT has shown more concern for costs than for needs and put pressure on DaRs to improve cost-effectiveness. The system is back to a new period of insecurity.

Popularity of the D/R travel facility is attested by the rate of expansion. In 1984, the total in London was 29 buses and 4400 members with 5500 trips per month. By 1988 this had risen to 120 buses, 51,000 members and 46,400 trips per month (Harkell 1988). Individual London DaRs may accommodate 2000 or more trips per month. Birmingham Ring and Ride in 1986 was carrying 8000 passengers per month with seven vehicles - the largest in the country. Examples of growth rates are plotted on figure 9.2. Over the country as a whole it is difficult to estimate total numbers of D/R special transport facilities as the precise service type operated by any one scheme is not

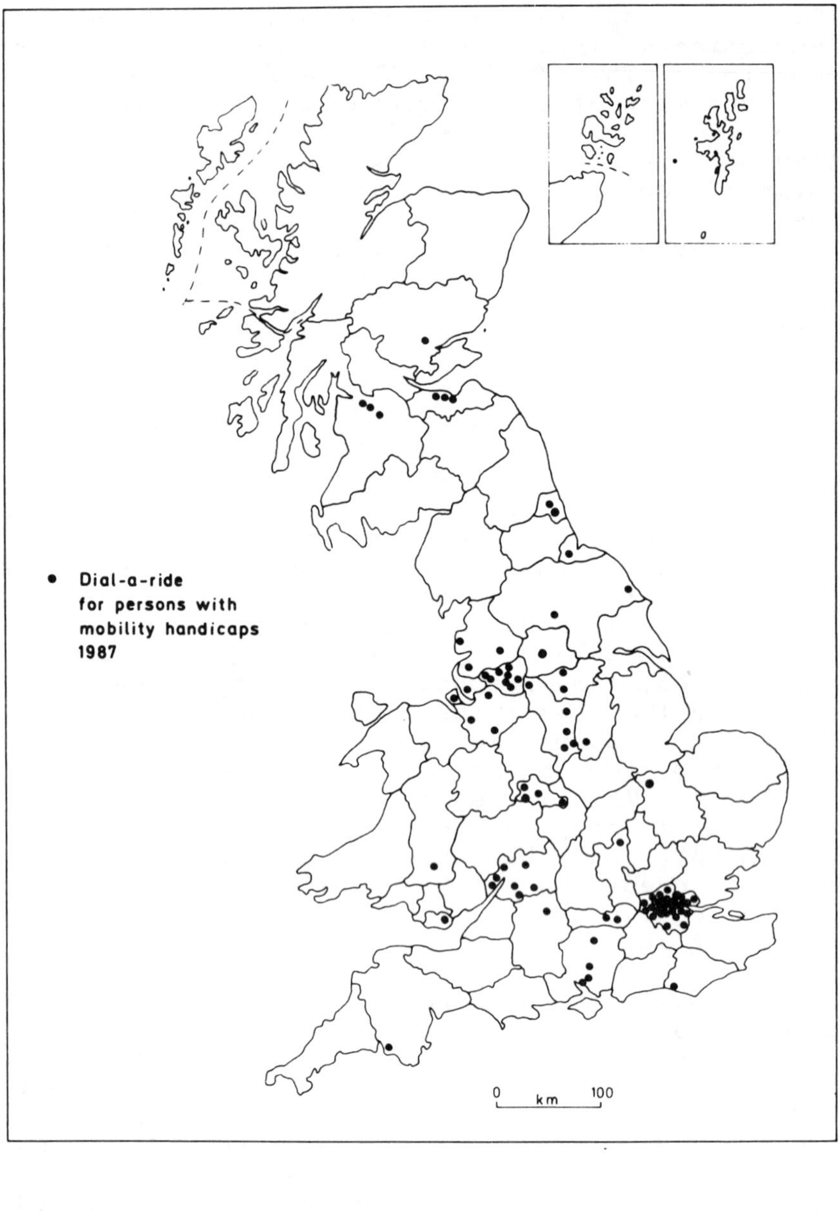

FIGURE 9.3 Dial-a-ride schemes for the mobility handicapped 1987.

always obvious. Figure 9.3 locates those which identify themselves by the names 'dial-a-ride' or 'ring-and-ride' plus a few others known to be D/R. The 90 DaRs shown here should therefore be regarded as a minimum. Most are urban based, simply reflecting the main centres of demand. Some non-metropolitan counties reflect the activities of sympathetic authorities, e.g. Derbyshire, Gloucestershire.

Aspiring passengers must usually become members of the DaR association so that eligibility can be checked against pre-determined criteria - normally restricted to people physically unable to use ordinary public transport. All services require pre-booking by telephone, at least a day and sometimes three days in advance. In some cases short notice bookings can be radioed to drivers. Arrangements for regular and block bookings are variable. Escorts and helpers are usually permitted to travel, and sometimes provided by the DaR agency. Service details vary from one scheme to the next, but generally, operations are completely personalized with no element of fixed routeing or timing, and therefore loadings per trip tend to be low - often just one or two people, more like a taxi than a bus. There is little attempt to reconcile a many-to-many service with passenger sharing of the vehicle, which proved so difficult for the 1970s DaRs. Also unlike the earlier generation, many DaRs operate seven days a week up to 2300 hrs. These aspects all raise costs (see below). Trip origins may be anywhere within the catchment area but schemes vary in terms of permitted journey distance, some setting a limit of *x* miles radius. Journey purposes are rarely circumscribed, the main restriction being that hospital or social services trips should be diverted to ambulance or SSD transport.

Much more data are available for DaRs than for the disparate field of traditional CT. This is primarily due to surveys conducted by the London Dial-a-Ride Users' Association. A wider but incomplete database was being collated by the NAU. Two London surveys have reported, one of six DaRs in early 1986 (Potter & Davis 1986) and another of eight DaRs in autumn 1986 (Potter 1987, Harkell 1988).

From these, it is at least clear that the right client group is being targetted. Eighty-four percent of the sample was over the age of 45 and 61% was over the age of 60. No less than 82% were women. While this is just an extreme form of the typical user profile seen in the rest of the UCM/CT sector, the under-representation of men is unusual. Also surprisingly under-represented were ethnic minorities, as only 2% of the sample were non-white; the reason for this is unknown. Particularly important for the DaR mode is to classify users by personal mobility standards. An 11-point classification is given in table 9.2. The two survey samples vary somewhat, but the biggest category (32-37%) was 'very poor mobility' which indicates frail or weak individuals able to walk only with a stick for support. The samples vary significantly in the proportion of passengers dependent on wheelchairs (19-31%). Unfortunately, comparative data from the general population are not available to test whether these distributions are representative. There is no doubt about customers' inability to use conventional public transport - 26% were barred through their wheelchairs alone, 18% by difficulties in getting on/off buses, 6% purely by distance to the bus stop or station; the remainder were deterred by *multiple* problems, e.g. another 23% by a combination of distance, difficulty in getting on/off, and other problems (Potter 1987).

Table 9.2 also classifies mobility by journey purpose, verifying the wide range of trips available on DaR. Overall, shopping is most important (38%) followed in equal measure by clubs/day centres and social/visiting (24-25%), a pattern to be expected from the nature of the client group. As the sample size for most mobility categories is small, it is doubtful whether any significant trends can be identified with confidence. Also, given the convenience of the minibus, there is no clear relationship between personal mobility and average journey length. These vary between about 4.5 and 7 kms -

TABLE 9.2 Physical mobility and DaR use. 1986 surveys, London (source: Potter & Davis 1986, Potter 1987).

Passenger mobility category	% in sample	Journey purpose %: Club	Visit/Social	Shops	Medical	Other	% in sample	Use DaR as much as would like? Yes %	No %
Wheelchair bound needs constant attention	7	31	6	57	4	3	5	64	36
Wheelchair bound needs pushing/can be left	7	23	25	51	-	-	11		
Wheelchair bound can self-propel	3	42	27	18	-	13	8		
Wheelchair user/can stand	2	65	-	-	-	35	7		
Minimum mobility	9	30	18	44	7	2	10	68	32
Very poor mobility	37	23	29	34	9	6	32	59	41
Poor mobility	12	19	23	33	19	6	6	53	47
Low mobility	8	30	16	36	16	3	9		
Low mobility due to other factors	6	19	21	50	5	10	5		
Medium mobility	5	33	37	21	-	9	4	68	32
Good mobility	4	11	55	34	-	-	3		
	100	25	24	38	8	5	100	61	39

N = 234
Survey Jan/Feb 86

N=381
Survey autumn 86

longer than conventional bus trips in London; 78% of trips were under 6.5 kms. There is more of a correlation with age: the younger age groups had a greater diversity of journey purposes and longer trip lengths. The proportion of shopping trips dropped significantly for the over-60s, in favour of social and club visiting.

Further evidence of the vital importance of DaR for the target groups is given in table 9.3, which presents results of an enquiry into alternative means of transport. Results from the two 1986 surveys differ somewhat, but there is no doubt of DaR's success in generating new trips with the proportion of the samples saying they would not otherwise have made the journey reaching the staggering levels of 52-65%. This is clearly a measure of suppressed demand due to the difficulties or inability in using conventional public transport. The only realistic alternative seems to be the taxi (23-32%). Joint membership of DaR and taxicard is possible, but taxis are still too expensive for many and of course can rarely accept wheelchairs. The survey reports were accompanied by many oral comments from respondents to the effect that they were totally dependent on DaR, praised the service, and would be virtually housebound without it (Potter & Davis 1986, Harkell 1988). Outside London results are less certain, as greater variation amongst schemes is to be expected. On Readibus the age/sex user profile was similar, but about 80% were able to use regular buses and only 5-10% of trips were newly generated (Bailey 1984, Heraty 1984). Most London users make one or two trips per week, the average being 1.2 (table 9.3). On a vehicle basis, the average number of bookings per trip is 1.3 and thus the occupancy (passengers/vehicle trip) would be slightly higher.

Currently, two severe problems are seen to afflict the future of DaR - first, its very high cost, and second, the impossibility of keeping up with the full potential demand and satisfying all needs. The mode is over-extended relative to resources and under-developed relative to demand.

TABLE 9.3 Trip rates and dependence on DaR. 1986 surveys, London (source: Potter & Davis 1986, Potter 1987).

Alternative in absence of DaR	Survey Jan/Feb 86 (%)	Survey autumn 86 (%)
Own car	0.8	1
Lift in other car	4.6	5
Bus	3.8	3
Taxi	32.1	23
Would not have made journey	52.2	65
Would have had home visit	1.2	1
Walk	2.1	} 2
Rely on neighbour	0.4	} 2
Other/no reply	2.5	} 2
	100	100
	N=240	N=380

Frequency	Actual use (%)	Desired use (%)
Daily	-	5
3-4 times a week	-	14
1-2 times a week	32	31
About weekly	37	25
Every 2 weeks	11	11
1-2 times a month	9	6
Irregularly	6	4
Other	5	4
	100	100
	N=371	N=359
Actual trips per week	1.2	
Desired trips per week		1.8

Survey autumn 86

TABLE 9.4 Comparative DaRs: costs and trip rates 1983 (source: Bailey 1984, Banister et al 1984, Heraty 1984, Bailey & Appleby 1986).

	Sale 1975 (1983 prices)	Reading DAR (1982/3)	West Lancs. DAR (1982/3)	Camden DAR (1983)	Ealing DAR (1983)	Hammersmith and Fulham Handicab (1983)	Southwark Taxi (1983)
Number of vehicles	9	4	1	6	2		
Passenger trips/vehicle hour	9.1	1.8	0.5	0.7			
Cost/vehicle hour £	5.98	7.28	3.80	8.33			
Revenue/vehicle hour £	2.98	0.66	0.45	0.27			
Revenue/cost %	49.8	9.1	11.6	3.3			
Cost/passenger trip £	0.73	4.08	8.41	16.42	10.86	15.03	5.35
Av. trip length (km)		4.5	14.4	5.1	12.6	9.9	7.7
Passengers/month	13,234	1,586	236	704	215	133	618
Trips/member/month			1.79	1.52	0.96	0.95	1.83
(trips/person/month	0.26)						

Cost data for several schemes are reproduced in table 9.4. Bailey (1984) updates information from the Sale (Manchester) DaR, which was probably the most successful of the 1970s PSV versions, for the sake of comparison. While the 1970s schemes were killed off because of their high costs (see chapter 3), their 1980s counterparts appear to be much worse. The West Lancs. DaR is cheaper per vehicle hour due to its high use of volunteers, but otherwise every index for each service tabulated is worse than Sale. Despite the partial use of volunteers, high costs are due to the inability to share overheads among a large vehicle fleet, the personalized service and low vehicle occupancy, higher average trip lengths compared with urban buses, and the high degree of dead mileage. The taxi-like service at bus equivalent fares is a major factor, such that the low throughput of passengers appears an inefficient use of the vehicle and produces extremely weak revenue/cost ratios. Operating ratios at 10% or less and costs per trip of £10-15, as in London, are appalling if judged on conventional criteria. Justification must depend on the 'social service' principle, and costs compared with the alternatives of ambulance or SSD transport. It is gratifying that this argument has up to now been generally accepted, especially by the former GLC, but the future looks insecure.

In another respect, DaRs have been the victims of their own success. It was quickly realized that, even though usage was increasing rapidly, it was only scratching the surface of the mobility levels required, bearing in mind the numbers of disadvantaged people and the 'normal' trip rate. From the examples in table 9.4, each registered member makes only 1-1.8 trips per month, yet membership continues to increase steadily. On-bus surveys found that users (the more active members) made an average of 1.2 trips per week, but about two-fifths of users said they would prefer to use DaR more often. The preferred trip rate was 50% higher at 1.8 per week (tables 9.2 & 9.3). Much latent demand is still unsatisfied.

At a more personal level, DaR has succeeded in unlocking the barrier of lack of confidence which afflicted many housebound disabled people. In opening up travel opportunities not previously existing, it has raised expectations and added further to demand. Many schemes have discovered that a publicity campaign attracts more users than they can cope with, which is counter-productive in terms of the disappointment caused. This is reflected in the problems of booking a ride. Asked why they did not use DaR more often, 35% of survey respondents cited difficulties in booking by telephone. Asked about booking in particular, various surveys reported 58-74% saying they had problems, with 41-55% specifically citing telephone congestion. Other problems were rules over minimum and maximum advance booking times and restrictions on where they could go (Potter & Davis 1986, Potter 1987, Harkell 1988). Many potential travellers have been deterred from using the service.

Resolving this is not simply a matter of more telephone lines. Microcomputers are now within the means of DaR agencies and suitable software is now available, through research projects into IT applications. About a third of London DaRs and several elsewhere were using computers by 1988. Thus it is hoped to make scheduling more efficient and relieve part of the bottleneck of demand.

The true scale of latent demand can best be assessed by starting from the total number of disabled people. The Greater London Association for the Disabled estimated there were 465,000 people in the capital with a transport handicap, and of these 235,000 were unable to use buses independently and ideally required specialized door-to-door transport. Compared with this, 51,000 were DaR members in 1988 of whom about three-fifths were active members. Many taxicard holders also belong to DaR but at least another 20,000 can be added. Therefore only 30% of the needy population is being reached, leaving aside the problems of telephone booking. Potter & Heiser (1987) calculate that if *half* the needy population is targeted for DaR (assuming the remainder travel on taxicard) then, assuming an average trip rate of one per month and a DaR minibus making 450 trips per month, at least 260 vehicles would be needed - over double the present number. But this assumption of one trip a month can hardly be regarded adequate, and the more active users travel weekly. On this basis a similar calculation shows the need for a fleet of 1000 minibuses! To put it another way, if existing resources were equally allocated among *all* the transport deprived in London, this would give each person roughly one return trip a year (Harkell 1988)! This kind of demand assessment led in 1988 to lobbying of government for sufficient resources to support "one trip a week".

Higher levels of DaR service would presumably operate in a different form to that running at present. However worthy the cause, the high costs per passenger cannot be ignored. The government's decision (through LRT) to freeze DaR funding in real terms was accompanied by predictable comments on the need for improved efficiency. There might indeed be a *prima facie* case for this, given the personalized, taxi-like service contrasting with the bookings problem. Cost variations - three London DaRs incur costs per trip of over £15 while another six keep such costs below £10 - might suggest scope for efficiency gains. However, partial explanations might be offered: for example, the most expensive services are in the central zone of the metropolis where traffic speeds are lower due to congestion. More attention needs to be paid to evaluation methods and to finding more appropriate indices (Hunter & Smythe 1988). A promising line of enquiry might be to encourage more vehicle sharing (occupancy rates) by greater coordination of trip requests, to make it more like a bus than a taxi, which should help both to reduce cost per passenger and relieve some of the excess demand. Computer despatching should help here. Alternatively, a semi-scheduled service might be considered, or D/R diversions from a fixed route, which would better accommodate higher flows on longer distances and improve resource utilization.

A more controversial alternative would be to turn over part or all of the operation to taxis. In the case of Readibus, a hypothetical comparison by Heraty (1984) suggested that taxis might be more cost-effective. At the same resource cost, it was estimated that taxis could provide about 200 trips per day compared with 42 on Readibus in the survey period. In London also, the unit cost of the taxicard scheme was much cheaper than the DaRs. The reason for this is the economies of scale achieved by taxi operators and the marginal extra costs of the 'disabled' traffic. Objections to such a plan centre on the inability of taxis to accept wheelchairs and the lack of 'care and attention' on a commercial service. When the new generation of wheelchair-accessible taxis becomes widely available, this option may have to be re-examined. In the general sense, planning of service types and funding commitments should not be excessively constrained by pre-determined objectives, but all the options with their implications should be studied on equal terms.

9.4 GENERAL ISSUES

Along with the growth in the 'general mobility' CT sector since the early 1970s and its improving level of organization, there has been a growing appreciation of the size of the market by outside suppliers. In turn there have been critical discussions on equipment and support services and legal and technical aspects of the CT environment. More appropriate vehicle designs and a wide range of specialized body conversions are now available; interior fittings, comfort and safety standards are critically reviewed. A code of practice for vehicle suppliers has been drafted. Insurance packages have been specially developed for CT. Emergency cover and safety matters are constantly under discussion. Training of both paid drivers and volunteers, although unofficial, is more comprehensive. Lobbying skills are better developed, at local and higher levels, essential in the light of constant funding problems and changing political/legal structures.

In this respect, problems are now coming from a more distant source. In 1988 the EEC reactivated a proposal to harmonize driving test standards in all member states, which would include an additional stringent test for all drivers of minibuses. This would obviously deter the vast numbers of volunteer drivers working with community groups, who have fairly recently been relieved of the need for a PSV licence in the UK, and many groups may have to abandon a transport function. It would be disastrous if the recent trend of permissive legislation in the UK were to be reversed by an alien bureaucracy.

Another symptom of the rising respectability of CT is its association with a developing corpus of academic research. This focuses on policy, management, design, consumer needs, costs and benefits (e.g. Meadows & Oxley 1988), plus information technology applications. A pressing research requirement at present is seen to be for better evaluation methods (Banister et al 1984), as simple indices such as 'cost per trip' do not show CT in a very good light. Other measures of productivity are conceivable (Hunter & Smythe 1988), but they do not do justice to the benefits received by passengers. The most comprehensive study of this issue is Bailey & Appleby (1986) who provide a full statement of costs and benefits, to users and to operators. Unfortunately, this does not succeed in solving the problem of developing a measure that adequately reflects user benefits in a non-subjective way and at the same time expresses value for money in terms that funding bodies can appreciate.

CT is a huge and complex field that deserves a book to itself. Indeed, the vast number of agencies involved raise questions about the need for greater coordination, about the efficiency of resource allocation, about the relative roles of the public and voluntary sectors and who should accept responsibility for transport for the mobility impaired. Answers do not seem any closer, while the successful growth of CT, originally to fill a vacuum, has allowed resource provision to be diffused more widely.

Voluntary sector transport has undoubtedly become better organized, more confident and more professional. Dependence of the public 'health and welfare' sector upon the supportive role of community groups was mentioned above, and it is arguable that over the course of time the distinctions between them have become purely institutional. CT operating practices now resemble the type and range of services provided by the statutory sector, but despite this blurring of distinctions CT has not gained recognition of equal status as a transport provider. It has never fitted comfortably under the aegis of either health authorities or local authority social services or planning and transportation departments, and has had difficulty being accepted as either a social service or a public transport facility (Sutton 1987). Hence, for example, the annual scramble for funding from diverse sources. A permanent recognized status is highly desirable. Definition of CT as 'public transport' under the 1985 Transport Act has not led to a secure financial basis, agreed objectives or incorporation in the transport planning process. Suspicion that the voluntary sector is being exploited by government as a partial substitute for state responsibility has not been dispelled. On the other hand, if special transport services at present run by the voluntary sector are a legitimate part of the public transport system, then perhaps they should all be run by professional operators. This could be done either by integrating with conventional public transport with a much greater use of vehicles accessible to the disabled, or by integrating general mobility services with health, social services and education transport.

Chapter 10

COMMUNITY TRANSPORT - COMMUNITY BUSES

10.1 COMMUNITY BUS ORIGINS AND DEVELOPMENT

(a) Concept and Terminology
In the context of UCM solutions to exclusively rural transport problems, the community bus has been one of the most significant and conspicuous developments. It is both 'unconventional' and a community transport (CT) mode, its essential raison d'etre being the avoidance of labour costs through reliance on volunteers for operation and management. Within the overall CT sector, however, its significance is relatively minor, being outweighed by the large numbers and diversity of urban-based services provided by the voluntary sector mainly for the benefit of elderly and handicapped people. In this latter context, the rural community bus stands out as being the *only* true 'public transport' mode, in the sense of being available to the public at large without any eligibility restrictions. Other CT modes provide alternative public transport for those unable through disability to use conventional forms; they circumscribe eligibility through various criteria of 'need', and usually require aspiring users to join some kind of local association. They are certainly not available to anyone. Rural community buses are *supplementary* public transport in the sense that no other conventional forms are adequate, or even exist, in the areas concerned. This mode was not founded primarily due to the physical incapacity of users (e.g. few have wheelchair facilities), but as a supplementary or replacement service for everyone. Social car schemes (another CT mode) are halfway between these two poles. The community bus overlaps the UCM and CT sectors, and has a special place within both.

It was strange that for much of their first decade, community buses achieved a higher public profile than the rest of the CT sector, which had been relatively overlooked, such that confusion arose over the terms 'community bus' and 'community transport'. The latter term was often applied to UCMs in rural areas which made use of voluntary labour, and therefore was taken to refer to community buses, social cars and other lift-giving schemes. For example, Banister (1982) states that community transport schemes are of these two basic types. More seriously, the Department of Transport in 1978 issued *'A guide to community transport'* as an advice booklet aimed at encouraging local groups in the countryside to set up community bus and social car schemes, apparently oblivious of the rest of the CT sector. This is symptomatic of much official ignorance of the traditional CT field, the mobility lobby, and the role of volunteers, in contrast to the higher profile achieved by rural transport problems in the late 1970s at the time of RUTEX. Minibuses used by urban CT groups are often loosely called 'community buses', and there was some resentment when the first rural projects appeared that both the name and the concept had been stolen from where they truly belonged in the urban 'elderly and handicapped' context (Sutton 1983).

To illustrate the confusion, the following service types have all from time to time been described as 'community buses'.

(i) A minibus run by a local group for its own members.
(ii) A minibus available for group hire but owned by a local authority, bus company, or other local voluntary organization.
(iii) Minibuses available under pooling arrangements.
(iv) A bus run by a professional bus company, under contract to a local group or committee composed of community representatives, e.g. the Bassetlaw Community Bus and the Horncastle Bus Club (described in chapter 4).
(v) A dial-a-ride for the elderly and handicapped.

It would appear that any degree of local consumer involvement qualifies for the 'community bus' label. To ensure consistency, the term is not used here in any of these contexts, but only in the 'rural' case where a small bus is operated and managed by volunteers, used on a range of services designed to meet local needs, and available to the general public.

(b) Development - The First Phase

The first rural community bus in Britain was inaugurated in November 1975 in an area of north Norfolk around the village of Sharrington. It achieved immediate fame, and the 'North Norfolk (or Sharrington) Community Bus' has become probably the best known single UCM service of any kind, and also the one most written about. The origins, development and first year's progress of the scheme have been described in Norfolk (1976), Orriss (1976), National Norfolk (1977) and Wilkes & Madgett (1977), and in a broader context also Harris (1978).

As in so many other new schemes, the first community bus was born in circumstances of conventional bus withdrawals and the threat of complete service deprivation in a number of villages. Unlike many subsequent schemes, the main impetus in north Norfolk came not from the community itself but from an alliance between the regional public sector bus company and the local authority (i.e. 'top-down'). The original idea emerged from growing concern at national level within the NBC about the short term prospects for public transport in rural areas, especially in the early 1970s climate of severe reductions in bus services, inflation, loss of passengers and rising car ownership, which induced a belief that something radically different had to be tried. Norfolk was a prime candidate for new experiments, as an overwhelmingly rural region with very dispersed population for whom transport and accessibility problems were endemic. There was also a history of fruitful cooperation between the local NBC subsidiary, the Eastern Counties Omnibus Company, and Norfolk County Council in planning imaginative service revisions to meet changing circumstances. A likely test area was also available. Recent service reductions in villages south-west of the market town of Holt, and the imminence of complete withdrawal, was producing much vocal opposition.

The solution - to allow the local community to run its own bus services with its own small vehicle, using voluntary labour to reduce costs, and able to respond to specific needs through grassroots contact - was developed jointly by NBC's Marketing & Research Department, ECOC, and Norfolk CC. Community organizations could not be expected to finance their own bus purchase, and would have to be given a suitable vehicle by the NBC. This would have the advantage of keeping Eastern Counties' flag flying and reassure the population that they had not been abandoned. Trade unions had also to be assured that this was not the beginning of a trend to take away bus drivers' jobs in favour of unpaid volunteers. The County Council's commitment was maintained by the need for initial subsidies.

The project was therefore based on a three-way partnership between the bus company, county council and community representatives. Eastern Counties provided the bus (paid for by the council) and appropriate professional support, such as maintenance, fuelling, cleaning, ticketing system, permits and licences for the routes and vehicle, training to PSV standard for the volunteer drivers, publicity and operational advice. The County Council was responsible for policy guidance, e.g. on legislation and relationships with other services, and for ensuring financial viability by guaranteeing deficits up to an initial £2000; this permitted the council much influence on service levels and fares. The Council acted as banker, paying costs to ECOC, receiving fares revenue, and reimbursing bus committee expenses. Operations and day-to-day management were entrusted to the Village Bus Committee, comprising representatives of all the villages served. Their responsibilities were to organise a pool of volunteer drivers, and bus routeing and scheduling in accordance with local demands. In addition, a steering committee was enacted as a channel of liaison among the three parties.

Six villages were original members of the scheme, with a total population of 630. This was regarded as an absolute minimum catchment size for

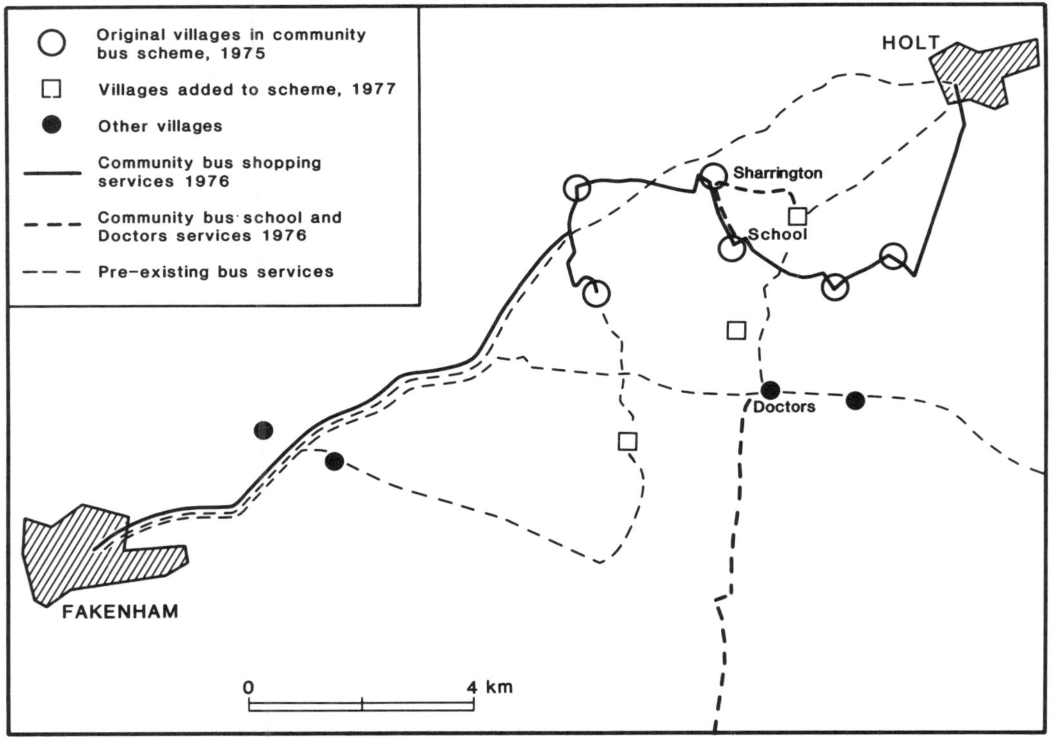

FIGURE 10.1 The North Norfolk community bus, original operations 1976 (after: Norfolk CC 1976).

a community bus, and the number of participant villages was later increased. The vehicle was a 12-seater Ford Transit certified to PSV standard. Service patterns were based upon off-peak stage carriage routes, primarily for shopping, to the towns of Holt and Fakenham at frequencies of 2-4 per week; there was also a weekly doctors' service and a schools service in term (figure 10.1). These were licensed by a permit, appropriate to small buses, under Section 30 of the 1968 Transport Act, and this confers eligibility for new bus grant and fuel tax rebate. It soon became apparent that stage carriage was insufficient to meet financial targets, and thus a programme of excursions (to Norwich and elsewhere) and private hires was encouraged to enhance revenue. In the first years a pool of 15-20 volunteer drivers was established.

Also known as the 'Norfolk Village Minibus' or locally as the 'Sharrington Bus', this first community bus had a great impact, not necessarily in terms of trip generation or economic indicators, but more due to the nationwide fame it attracted and the example set for future forms of public transport. Its form of organization became the model for further community bus schemes in the late 1970s.

Early experience of the North Norfolk scheme was sufficiently promising to encourage further examples of the genre. By the end of the 1970s there were twelve. There was also an unknown number of unlicensed minibuses avoiding the PSV requirement by not charging fares, or possibly running illegally. Considerable obstacles remained in the way of significant expansion, principally the need for generous cooperation from NBC and the county council (assuming the Norfolk model), the need for PSV licensing of drivers and vehicles, and the heavy responsibilities placed on local people giving up their time without payment. There were doubts about the durability of community public transport, whether the novelty would wear off or volunteers lose interest. A brief review of the first phase can be found in Banister (1982). Three of the pioneer community bus committees gave evidence to the Select Committee on Nationalized Industries (1977-78). In addition, a large amount of information was collected in a private survey by Kaye & Kite (1983) which contains case histories of 17 bus projects.

The second community minibus was set up in the Uwchaled area of Clwyd in September 1976. This was part of a thorough review of public transport undertaken by Clwyd County Council in which numerous experimental schemes were tried out, including voluntary or essential car schemes (in which the county was also a pioneer), reduced fares experiments and one postbus (Clwyd 1978). Planning and organization of the service was almost identical to that of Norfolk, with the same three-way partnership of local committee, county council and bus company, with the bus - a 16-seat Dormobile body on Ford Transit chassis - being provided by Crosville Motor Services. The location was a cluster of half a dozen villages around Cerrigydrudion, which had been deprived of bus services for some years, with a catchment population of 1400 - significantly higher than in north Norfolk. The main purpose of the minibus was to provide off-peak shopping services to four small towns, plus monthly excursions to Llandudno and Wrexham. A spin-off for the bus company was timed connections with its own services for onward travel.

A month after the Uwchaled CB was the establishment of a third project in East Sussex. In this case the initiative was 'bottom-up', originating from a travel survey undertaken by Sussex Rural Community Council. Fifteen villages (population 2800) in the Cuckmere valley between Seaford and Hailsham were identified as having no or few bus services, many of these being scheduled for withdrawal by early 1977. Publicity from the Norfolk scheme had suggested the solution, and the necessary collaboration was arranged between East Sussex County Council and Southdown Motor Services - the regional NBC operator. A bus management committee was constituted in the usual way. It is interesting that the minibus used was the same vehicle that had been deployed on the former Eastbourne dial-a-ride (see chapter 3), and thus lower depreciation costs were charged to the community bus. The schedule of the 'Cuckmere CB' was based on trips to Hailsham and Seaford four days per week, with rail connections, plus a programme of excursions and hires (Southdown 1977).

Northamptonshire was the next county to institute a community bus, with the 'Lilbourne CB' starting in June 1977. Again, this was established by a tripartite of local committee, county council and NBC subsidiary, in this case the United Counties Omnibus Company. Remarkably, this service reversed the trend by serving just three villages on the Northants/Warwickshire border with a total population of only 450. Nevertheless, this seems to have been successful from the outset, and encouraged Northamptonshire, like Norfolk, to set up several other community bus schemes.

It was at this stage that the community bus mode became involved in the RUTEX programme. Although not by this time a completely new idea, one example of the mode was an obvious candidate for RUTEX, and seemed to be an appropriate solution for one chosen experimental site in Devon. An area to the north-west of Tiverton was suffering from very poor transport - two 'market day only' services both facing withdrawal. Demand was large enough for a minibus rather than some kind of car service, but costs required the use of volunteer drivers; routeing had to be flexible to cover a dispersed area. As an official experiment, this was clearly a 'top-down' approach, but the advantage was that the vehicle - a 16-seat minibus with customized body - was provided on loan by the Department of Transport. The Western National Omnibus Company remained the operator in the legal sense, but the bus was volunteer driven and managed by a local committee. Management was more complex, with a RUTEX working group and TRRL project officers also involved. More data was therefore collected for this bus scheme but, strangely, the 'Exe Valley Market Bus' is the only CB to have had a TRRL report (Devon Rutex 1978).

Various difficulties were presented by the nature of the test area. It lay in the upper Exe valley, with the main demand being for shopping access to Tiverton; hence the name given to the community bus. The population of 1500 was very dispersed and there was no coherent pattern of nucleated villages and main routes which would allow most demand to be tapped by a fixed route scheduled service. It was therefore decided to operate the bus in a demand-responsive (D/R) mode, with pick-up and set-down times and places being subject to pre-booking by passengers, and with a loose timetable of arrivals at and departures from Tiverton. The Exe Valley Market Bus is therefore quite unique as a hybrid between a community bus and a public dial-a-ride. Experience of urban DaR services in the 1970s had firmly shown that they were more likely to succeed as a less ambitious, informal and non-technical, 'many-to-one' version (see chapter 3). Oxley (1980) counted the Exe Valley bus scheme as a 'third generation' DaR and, although the general concept might be doubted, this particular CB has most justification to be regarded as such.

The catchment area was regarded as a quadrant with its apex at Tiverton. This was then divided into northern and southern zones, each of which would have one separate return service to town on two weekdays. Fares were based on a concentric zoning pattern (figure 10.2). Reservations were taken by telephone the previous evening by volunteer controllers. A notional timing system for each stretch of road allowed controllers to tell passengers when to expect the bus. A passenger list, arranged in order, could then be given to the driver. A Saturday service was also planned, following a fixed circular route, but this was soon discontinued for lack of use. Private hires and excursions were possible. From its commencement under the aegis of RUTEX in October 1977 the Exe Valley Market Bus turned in good results, and at the conclusion of the experiments in 1979 it was allowed to continue with financial support from Devon County Council.

(c) Further Development and Variations

It was soon evident that significant expansion of

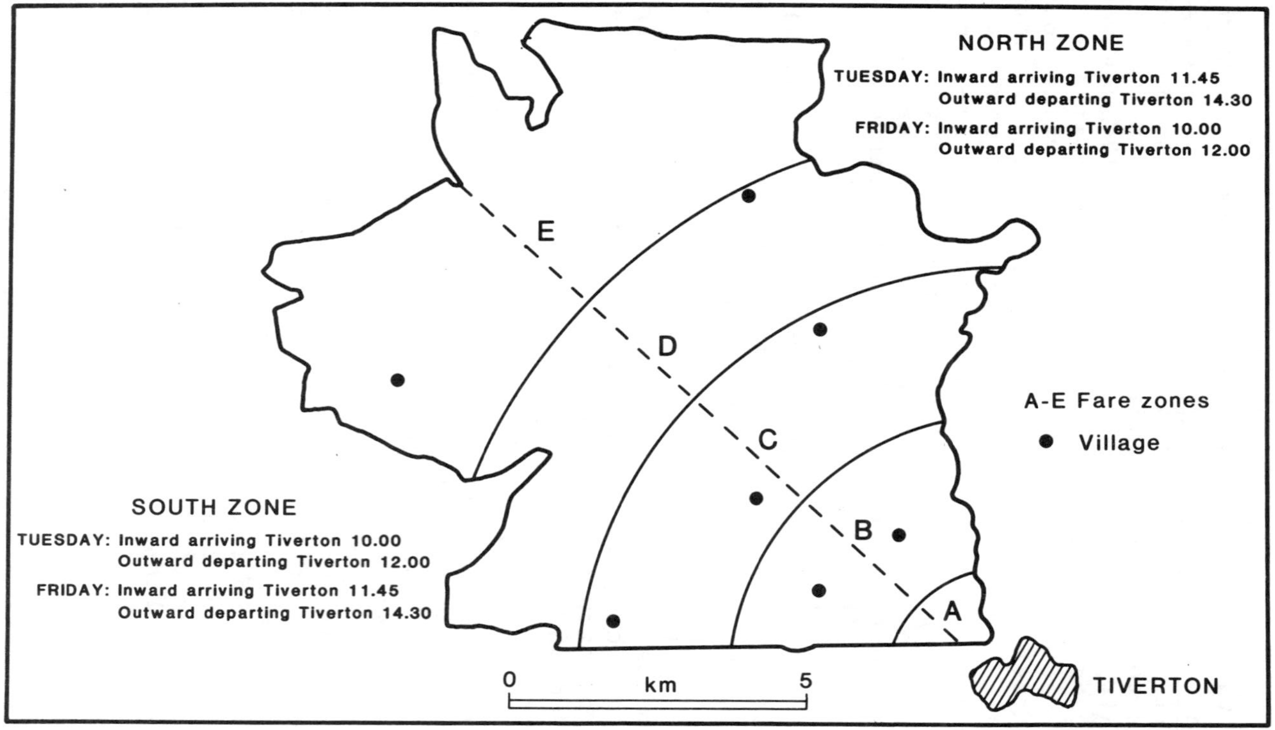

FIGURE 10.2 The Exe Valley Market Bus 1977-9 (after: Devon Rutex 1978).

the mode required legal changes. The 1977 Minibus Act had instituted a permit system so that local groups and welfare organizations could run minibuses on a non-profit basis without PSV licensing. This was not specific enough for community buses which were running relatively formal stage carriage routes, and so the Transport Act of the following year gave full legal recognition to the community bus (one of the few UCMs with a legal definition) with appropriate concessions. Hence, volunteer drivers over 21 with a full driving licence no longer needed a PSV licence; this removed a very severe impediment and extra costs in setting up new services. Similarly, vehicles with 8-16 passenger seats no longer had to be maintained to strict PSV standards, but merely to conform with new Community Bus Regulations. For their stage carriage routes, a streamlined procedure was established to obtain road service licences from the traffic commissioners. Community buses may operate contract carriage or express carriage services. Special CB licence discs had to be displayed on the vehicle. Rules were revised slightly by the Public Passenger Vehicles Act 1981. The definition of a 'community bus service' is "a stage carriage service provided (i) by a body of persons....concerned for the social and welfare needs of one or more communities; and (ii) without a view to profit...."

With new legislative support, community bus development proceeded at a greater pace to 1985. The scheme at Meopham (Kent) got under way in 1978/9 immediately upon passing of the latest Transport Act, and is claimed to be the first to use volunteer drivers without PSV licences (Kaye & Kite 1983). It is also remarkable for being located so close to London, verifying that pockets of need can exist on the fringes of the metropolis. Other launches in 1979 included the Soar Valley CB in south Nottinghamshire, and the Langwathby CB (later renamed 'Fellrunner') in Cumbria, which had evolved from a social car scheme.

Counties which had pioneered community buses in the early years pushed ahead with follow-up systems. Norfolk promulgated its second, based at Aldborough in the north-east of the county, in 1978 in advance of the new legislation. This was followed by three more - Denton/Alburgh, Docking and Wells & District - in the next five years. Northamptonshire followed the Lilbourne CB with new minibuses at Rushton, Grendon and Gayton/Tiffield. After RUTEX gave a lead in the Exe valley, Devon followed with another four schemes, one at Tavistock, and three in the South Hams area - Coleridge, Harbertonford and Blackawton. After launching the Cuckmere CB, East Sussex had a second at Westmeston, originally very small, but more recently (under the name 'Downland') integrated with the ESCORT Village Rider service (see section 4.4). Neighbouring West Sussex encouraged the upgrading of the 'Harting Minibus', which had been running informally since 1973, to full CB status. A second followed at Sutton Bignor & Barlavington. With the proven example of Norfolk, the innovation spread to other parts of East Anglia, with two in Suffolk - the Waveney Valley and High Suffolk CBs - and one in Essex (Coggeshall). Nottinghamshire added to its diversity of community transport provision with the 'Countryman' in the rural areas between Mansfield and Newark. Remaining English community buses are all in the south, with two each in Hampshire (Broughton & Mottisfont and Meon Valley), Gloucestershire (the Cotswold 'Villager' and Mickleton),and Somerset (Wedmore and Dulverton), and one each in Oxfordshire (Stanford-in-the-Vale), Wiltshire (Minety) and Cornwall (St Breward).

In Scotland community bus development has been slower, perhaps due to the existence of numerous other UCM types, but interesting variations have appeared. The first was a little-known scheme at Edinvillie (Grampian) where, in 1975, the local community association felt impelled to take action into its own hands in order to relieve the village's isolation and secure access to the nearest town. A 12-seater minibus was contracted from a local operator at a very low rate, and driven on behalf of the community association by a volunteer with a PSV licence. Fares could not be charged on contracted services, and thus the venture was paid for out of the association's general fund-raising activities. The service comprised a weekly shopping trip to the larger settlement of Aberlour (McLaren 1979). As a grassroots initiative, this owes nothing to the 'Norfolk model' or the 'NBC model' which initiated community buses south of the border, and yet got under way only one month later than the Sharrington scheme.

Other community buses in Scotland have all been started since 1980. The three services most closely resembling the CB norm are all in the south of the country and were all started on the initiative of community councils. Two are in Borders Region - Ettrick & Yarrow, and Hobkirk & Southdean (near Hawick) - both of which connect with Border Courier services (see section 8.2). The other is the Colvend & Southwick CB, near Dalbeattie in Dumfries & Galloway Region.

In Wales, the Uwchaled CB remains the only example of the mode, and elsewhere other UCM types have been found more appropriate.

All the community buses described above have been mapped on figure 10.3. According to a fairly strict definition, there were 39 such services by the end of 1985. As there can never be any absolute certainty as to what is meant by a 'community bus', and taking account of the 'possible' cases and type variations also shown on the map (and described below), then the best estimate of numbers on the eve of deregulation is 40-55. It is impossible to say whether the apparent southern bias to the map has any significance.

Later community bus projects often exhibited significant differences in origins and organization from the earlier examples, which might be described as the 'Norfolk model' or the 'NBC model'. The tendency was to become less dependent upon the local NBC subsidiary and the county council, and new schemes were more likely to be initiated by local organizations, with a less active input from the local authority and none from the NBC. It had been hoped that the more

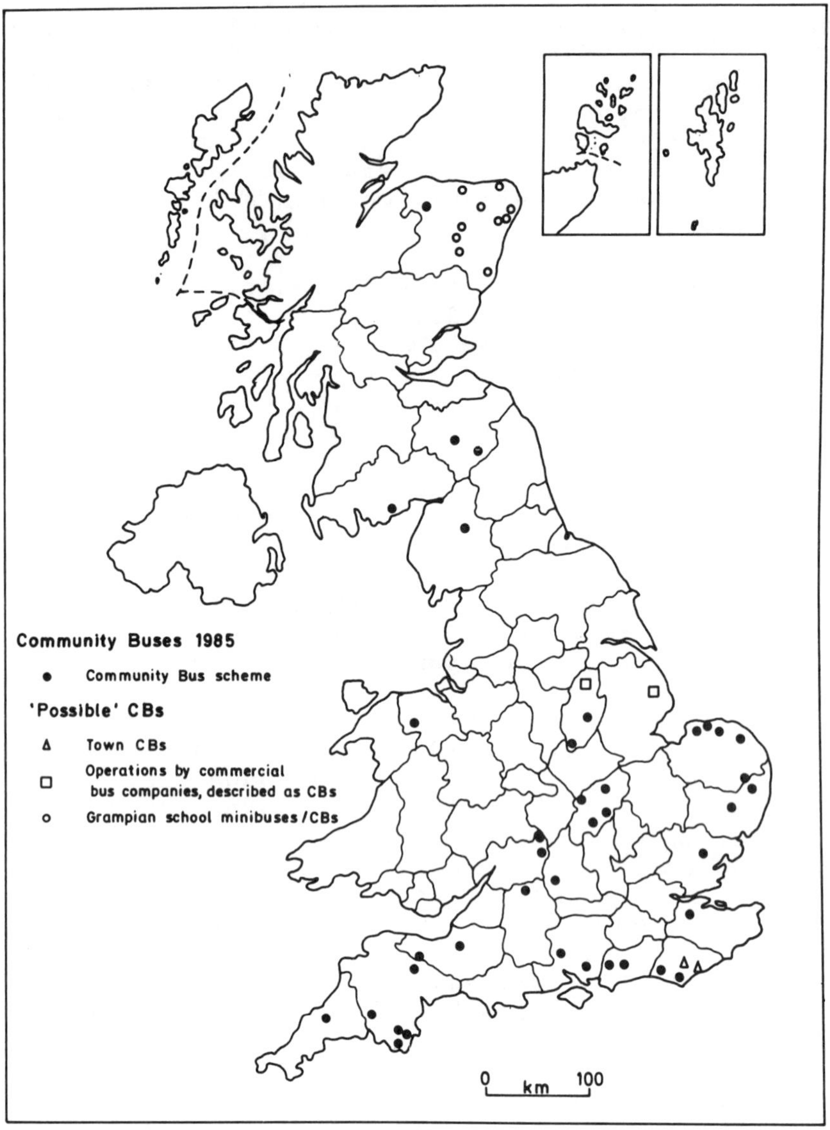

FIGURE 10.3 Community buses 1985.

formal early schemes would serve as demonstrations whose influence would spread, and after appropriate permissive legislation, public agencies could withdraw into a promotional role. Even where the tripartite of bus company, county council and local committee still prevails, the onus is now on the latter to make a case for a new CB project, as in the later Norfolk services. The earliest example of 'community action' was the Edinvillie experiment cited above, which made no demands on public agencies. Complete independence is rarely possible, but increasingly, the prime mover in getting community bus schemes off the ground is one or more parish councils or the community council; in Meopham (Kent) it was the welfare committee of Age Concern; for Stanford-in-the-Vale (Oxfordshire) it was the Faringdon Local Transport Group (Kaye & Kite 1983). In nearly all cases, the cooperation of the county council is still essential, although the vehicles themselves are now acquired from a variety of sources.

Alternatives to the formal procedure of obtaining a new minibus via the NBC would have the great advantage of reducing costs significantly for the local group and the council, and would present much less of a barrier to setting up new schemes. A further trend is therefore to acquire vehicles by contracting or leasing from commercial operators, by buying second-hand, or by borrowing them from local authorities for part-time use. An extreme case is that of Meopham CB which started with an old converted ambulance bought for £30! Some are offered the use of vehicles owned by other social organizations. Colvend & Southwick (Dumfries & Galloway) hired their minibus from a local businessman who used it to transport his staff. Private commercial operators may assist with vehicle maintenance, e.g. Coleridge (Devon). Vehicles for Westmeston (East Sussex) and Wedmore (Somerset) CB activities were minibuses owned by the councils' education departments and made available for community use when not needed for school transport (Kaye & Kite 1983).

This mechanism has been exploited to the full by Grampian Regional Council. After a 1983 review of rural accessibility and possible 'community' solutions (rather later than many other local authorities), the council began a programme of establishing community buses with education department vehicles. At times when not required for school transport, the minibuses are taken over by volunteer drivers for use on essential local services. There were ten of these by 1985 (Grampian 1985; figure 10.3).

As community bus numbers increased, there arose greater variety in the type and scale of services offered. It was once thought that the 600-700 population catchment of the Sharrington CB was an absolute minimum for viable operation, and subsequent schemes generally increased this. However, the tendency to get vehicles from alternative sources, much more cheaply, allowed local groups to exist with smaller budgets, and hence it was possible for community buses to provide for the needs of one village only with a low level of service. Edinvillie, Westmeston and Meopham, for example, began with one stage service of less than 5km, and in two of these cases this operated only once per week. On the other hand, some of the Norfolk CBs felt obliged to expand further as other villages in adjacent areas were cut off by subsequent conventional bus withdrawals. Most schemes are based on essential off-peak village-town shopping trips, while the other most common service types are bus or rail feeders and doctors' surgery connections; many assist with schools services. Variations are: connections with London commuter trains (Coggeshall), superstore visits (Soar Valley), a playschool service (Aldborough), and Sunday church trips (two in Grampian Region). Another trend is for community buses to become more dependent on hires and excursions, with some earning a reputation for imaginative tours programmes.

Yet another variant is the urban community bus. In Hailsham and Bexhill (East Sussex), the community bus legislation has been used to authorize volunteer-run buses on routes within the towns and their housing estates. The Hailsham Town Bus is run by a committee from the town council. It comprised a converted Mercedes van with 16 seats and space for shopping trolleys or wheelchairs. There is little comparison with rural counterparts, being much more intensively used, running five trips per day on two urban routes, and attracting a much higher passenger load. One problem is the great commitment required from drivers. Also unique was the tolerance of some competition (before deregulation) between the CB and commercial public buses (Kaye & Kite 1983).

10.2 TRIP GENERATION AND SOCIAL EFFECTS

(a) Passengers

Reports produced by community bus sponsoring agencies - county councils and bus companies - on the setting up and early progress of each scheme were generally enthusiastic and favourable. Outstanding needs existed, and passenger trips generated by the new services showed that most were being satisfied (Wilkes & Madgett 1977, Clwyd 1978, Southdown 1977, Devon Rutex 1978). While not wishing to disagree with this, data do not exist in the appropriate form either to establish prior 'need' in a systematic manner or to allow an analysis of passenger loadings in sufficient detail. In (probably) all cases a preliminary survey had been done on local travel patterns and problems, and bus committees normally monitor passenger loadings and keep in touch with local opinion, and thus, despite the lack of published data, there might be sufficient grounds for claiming success.

Table 10.1 reveals the fragmented nature of the data that are available. Unfortunately, reports tend to cover the first year of operation, and once the scheme is on a secure footing the sponsoring agencies withdraw into the background and no further reports are produced. Hence, more recent data are sparse. Nevertheless, great variation among CB schemes is evident, in terms of passenger numbers and mileage.

The North Norfolk (Sharrington) CB was rather slow to build up momentum, being monitored over the first five months at 120 trips per week.

TABLE 10.1 Community bus patronage - examples (source: Norfolk 1976, Clwyd 1978, Banister 1982, Kaye & Kite 1983, Banister & Norton 1988).

Scheme	Trips per yr.		(date of info.)	Trips per week	(date)	Mileage p.a. or week		(date)
Sharrington				120	(21 wks 1975/6)			
				250-300	(1977)			
	17000		(1979/80)	330	(1979/80	14800	stage	
						+10600	non-stage	(1981/2)
						26000		(1986/7)
Uwchaled	1494	stage	(1977/8)					
	+416	non-stage		37	(1977/8)			
	1778	stage	(1981/2)					
	+1636	non-stage		66	(1981/2)			
Cuckmere	7000		(1980/81)	140	(1980/81)	15000		(1980/81)
Lilbourne	3500		(1977/8)	70	(1977/8)			
Exe Valley				85 stage only	(1977/8)	9600	stage	(1977/8)
						+1800	non-stage	
Meopham				5	(1982)	20/wk stage only		(1982)
Coleridge				50	(1982)	140/wk stage only		(1982)
Hailsham				600 one-way	(1982)	250/wk		(1982)
Aldborough	5747		(1981/2)	110	(1981/2)	20000		(1981/2)
						26000		(1986/7)
High Suffolk						16000		(1986/7)
Soar Valley	12-15000		(1982)	230-300	(1982)	25000		(1982)

However, nearly two-thirds of this was school traffic, and average loadings on off-peak shopping journeys were only 3-5 per trip or 16-18 per day (Norfolk 1976). Even this was regarded as fulfilling a need. Consolidation over the next few years to over 300 per week can be verified from the table, but the mileage data suggest this may be largely due to exploitation of spare vehicle time for use on excursions. That 42% of vehicle mileage is attributed to non-stage purposes is not untypical. The latest mileage figure of 26,000 (42,000 km) also takes account of expansion of the service's catchment area.

A problem with available data is that they rarely distinguish stage carriage passengers or mileage from excursions and hires. The Uwchaled data illustrate the growing relative importance of the latter (table 10.1). Including this, the combined passenger totals are still low. Available data make it difficult to determine whether low loadings refer to the early stages of a CB development process, or remain fairly constant over time where the bus is deliberately targetted at a small market. The latter is probably true in the cases of Meopham, Uwchaled, Lilbourne and Coleridge, which have survived over the years; a relatively large excursions and hires business might be inferred. Larger operations might be summarised as those doing 24-40,000 km (15-25,000 miles) p.a. and carrying 7-15,000 passengers, on all types of business. The 600 passengers per week carried by the Hailsham Town Bus is exceptional, reflecting the short distance and intensive nature of its urban duties. From the few cases with adequate data, trip rates can be calculated at 0.04 - 0.44 trips/person/week.

Disappointingly, there are very little data on passenger characteristics and journey purposes on community buses. Even the TRRL report on the Exe Valley Market Bus (Devon Rutex 1978) has very little information, save that "virtually all" journeys to Tiverton were wholly or partly for shopping. There were some newly generated trips in that about one third of passengers travelled to town more frequently than before the service started. A "significant proportion" had diverted to the CB from previously getting lifts in the household or other car. Although most bus committees conduct user surveys, these are not published, and there is nothing, unfortunately, to corroborate or expand on the morsels of information revealed above.

(b) Social Implications

Social benefits are wider than the ability to make trips not previously possible. Several CB organizations report that their bus scheme has activated a higher level of social intercourse among the community, and where a number of villages are involved, has enhanced relationships between them. Excursion programmes have also been a vital benefit in raising the quality of life for rural dwellers without access to cars. Community bus schemes have enhanced locality self-confidence and elicited loyalty to popular effort. Sutton (1983) argues that rural CBs have, through development within the community itself, produced a much greater responsiveness to need than has been achieved by urban CT projects (chapter 9).

Such benefits are hard earned, for various external costs and problems are imposed upon organisers of community bus services. The main problems are the level of work and commitment required, and the difficulty of maintaining an adequate pool of volunteer drivers. As with any kind of self-help scheme, success depends upon a few key individuals with the motivation and administrative skills to turn an idea into action. The chairman, secretary and treasurer of the bus committee might well find they have taken on a full-time job. Retired professional people are more likely to have both the time and the aptitude. Individuals volunteering for such tasks no doubt have a genuine willingness and enthusiasm to help their communities, but the result is often excessive responsibility put on the shoulders of a few, with possible problems of continuity if a key member decides to give up the job.

The scale of responsibility involved can be gauged from the list of tasks explained in *'A guide to community transport'*. This was a handbook produced by the DTp in 1978 (after the

Transport Act of the same year) aimed at encouraging local groups to set up CB and social car schemes. Any group concerned about transport is enjoined first to define the problem as precisely as possible by doing a local survey of existing travel patterns, and ascertaining outstanding needs in terms of desirable journeys which are not currently possible. As academics and planners have discovered, 'need' is an elusive concept extremely difficult to capture in a simple questionnaire survey of the type likely to be within the means of voluntary workers. If a realistic transport idea emerges from the survey, the group is advised to consult the local authority and the main bus company in the area. Planning service type, routes and frequencies is a meticulous, semi-technical exercise likely to be done by non-experts on a trial-and-error basis. Acquiring a suitable vehicle also needs considerable expertise, some knowledge of law and finance, personal contacts with local suppliers and familiarity with 'the system' for obtaining grants. Arrangements for licensing, insurance and maintenance have to be made. Finding enough volunteer drivers to cover the necessary operational hours might need an informal survey, although local knowledge should help here. Driver training should be organized. Fares and a ticketing system must be decided upon. Excursions and private hires might be essential for viability, but these present a major additional workload. Publicity in the form of printed leaflets and timetables is vital to raise awareness of a new service. Finally, the need for an accountant within the organization is obvious, especially as such schemes are unlikely to cover all costs and responsibilities are due to various funding bodies.

Apart from the sheer workload, the other persistent problem is that of maintaining sufficient numbers of drivers. It was originally feared that after the novelty had worn off, volunteer drivers would lose interest and replacements would be hard to find. Community buses would be 'transient'. Generally, this has not happened, but there is a fairly high turnover of drivers and this remains a constant source of anxiety to CB committees. Volunteers available in evenings and weekends only may not be so useful if the bus operates mainly during the daytime. A major deterrent in the early years was the need to obtain a PSV licence, although the number of people failing the PSV driving test was not high. After this requirement was removed by the 1978 Transport Act, it was normal for CB committees to provide some kind of driver training, and a few insisted on PSV standards continuing. It seems unlikely that lack of drivers has caused any CB scheme to collapse, but there are some cases where driver shortage has led to temporary cutbacks in services.

The only community bus definitely known to have failed was that at Brendon (Devon), which was withdrawn after a couple of years for lack of use (Devon TPP Part 2 1982/3).

The widely held impression of community buses concurred with that of the Select Committee for Nationalized Industries (1977-78), that the overall contribution of the mode was inevitably limited. On the other hand, the popularity of the schemes which got off the ground led to excessive optimism in some quarters. A past chairman of the National Bus Company was quoted as saying that community buses could be run by "every village that has a cricket team". Banister (1982) summarised the opposing viewpoints as "panacea or palliative?"

There was never any serious suggestion that CBs (or CT in general) were a panacea. Considerable concern has been expressed that community bus development is not occurring in places of greatest deprivation or need, but in places where the necessary social and administrative skills are most likely to be found. In the light of the considerable workload and commitment required from the CB committee, and the type of experience necessary, it is commonly believed that those with most aptitude for this are professional people, probably retired, with middle class values, and most likely to be ex-urban migrants into the rural area. Locations with the greatest propensity to set up CB schemes are those with this type of population structure, and perhaps also they are likely to be the larger villages, closer to towns, and better endowed with other facilities. On the other hand, locations with the greatest need for supplementary transport are likely to be the smaller, remoter settlements with an older population structure, and fewer people with the initiative or know-how to inaugurate a self-help scheme.

Consequently, the possibility remains that community buses together with other voluntary sector activities might actually exacerbate the inequalities in rural service provision. Until recently there was little data to verify this, but a survey of Norfolk and Suffolk by Banister & Norton (1988) goes a long way to confirming it. Of 134 parishes surveyed, 41% had some kind of voluntary transport, including 27% with access to a community bus and 25% with a social car scheme. Average number of services per parish correlated positively with parish population. A wide range of other voluntary sector activities showed the same trends so that overall, the distribution of self-help ventures is extremely patchy with most parishes devoid of most services, and these are least likely to exist in the smaller parishes.

The survey also provides a profile of typical volunteer workers. Over 50% of volunteers in all transport schemes (and over 60% of organizers) are from social class I (professional and managerial) while this group represented only 24% of the study area population. 48% of volunteers were 'urbanites', either retired or commuting. 49% were 'newcomers', having lived in the area for ten years or less, but another distinct group of long term residents also contributed with 34% having lived in the area for 20 years or more. Looking at community bus schemes alone, no less than 88% of organizers were retired! There is a distinct polarity among drivers, with 63% of stage service drivers being retired, but 88% of excursions/hires drivers being working people (i.e. available in evenings/ weekends). Overall, these findings confirm the preconceptions and give weight to questions about equality of service provision.

There are many areas of need where voluntary transport does not exist, and yet a reading of county PTPs uncovers many cases where encouragement from the authority meets with little interest from local communities. Even a community bus is seen as a 'last resort' not to be considered while there is still a slim chance of keeping the conventional service. Increasingly, responsibility is passed to the community to set up its own transport project, and if local people are disinclined to do this then the assumption is that there is no need. Quotations from PTPs are given in section 14.3(c) below to exemplify local authority attitudes in this field.

The final word can be left to Cheshire Community Council which, in consultation on the county's PTP, retorted that they

> "...would not feel it reasonable to ask rural people to take on the considerable investment of time and effort required to run a voluntary transport scheme simply to enable the County Council to reduce its commitments.....many people in Cheshire's own isolated rural areas already feel that they receive few enough services in return for the rates which they pay". (Cheshire PTP 1982/3).

10.3 ECONOMICS

Insofar as it is possible to generalize about community bus economics, one can conclude first that fares receipts are insufficient to cover operating costs unless a significant number of excursions and hires are also provided. Second, if excursions and hires are exploited to their full potential then it is possible for CBs to more or less break even on their operating account, or to make a modest surplus which could be set against vehicle depreciation. Third, overall revenue is unable to cover depreciation in full, and it is this element which usually demands subsidy. Thus, the vital point is the arrangements made for acquiring the minibus, whether it was purchased new, second hand, or borrowed from another organization, and whether it was paid for by the county council (perhaps via NBC). Similarly, the need for a replacement vehicle after a few years' operation produces periodic searches for supplementary funding.

Financial data on community bus operation are reasonably abundant but disappointingly fragmented, such that it is rarely possible to compare one bus scheme with another, or to be certain whether one's results are representative of others. Sufficiently detailed cost breakdowns could only be found for the Exe Valley Market Bus and the Cuckmere CB, which give the 'setting up costs' for the original projects in the late 1970s. Often, historic data are available from reports on the projects' early stages but suitable recent data are rarely published. Table 10.2 lists the start-up costs and first year operating accounts for these two services. The above points can be verified from the Exe Valley data where income is close to operating costs so long as depreciation is excluded; otherwise, revenue is only about half total costs. The Cuckmere results are better, but it is unclear whether the cost of the vehicle is covered. New bus grant was normally available (until phased out in 1984) for schemes with stage carriage routes, but in neither of the above cases was the minibus acquired new.

Early results encouraged some deceptive optimism. At Sharrington, the first three months produced fares revenue of £325 against direct costs of £316; the first six months at Exe Valley brought in £765 against costs of £602 (Orriss 1978, Devon Rutex 1978). These of course ignore the repayment of setting-up costs, and had the cushion of local authority and RUTEX guarantees. More realistic were the £134 revenue against £436 costs for the first three months at Uwchaled, and the £464 revenue against £1189 costs for the first five months at Cuckmere (Clwyd 1978, Southdown 1977). Including excursions and hires, the 30% operating ratio at Uwchaled was raised to 56% at the end of the first year, but this remained around the 50% mark for the next few years. Also in the late 1970s, the Select Committee (1977-78) attempted to produce comparative data for the CBs then in operation. Average weekly results were

TABLE 10.2 **Community bus economics - two early examples (source: Balcombe & Dredge 1980, Southdown 1977, Select Committee 1977-78).**

Exe Valley Market Bus (1977-8)		Cuckmere Community Bus (1976-7)	
Setting up costs	£	**Setting up costs**	£
Purchase of vehicle	3600		
Preparation of vehicle	105	______	156
Driver training (PSV)	1455	______	1252
Insurance, licensing, etc.	105	______	71
Publicity	190	______	160
Administration	105	Depreciation	173
	5560		1812
Operating costs	£	**Operating costs**	£
Tax, licensing & insurance	295	______	282
Fuel	560	______	882
Maintenance	830	______	998
Drivers' expenses	35		
Spare bus hire	55		
Additional driver training	40		
Publicity	15		
Miscellaneous	80	______	307
	1910		2468*
Income	£	**Income**	£
Stage carriage fares	970	______	1057
Private hires and excursions	300	______	382
Rebate of fuel duty & VAT	350	______	448
	1610*	Donations, etc.	165
			2052
Total costs and income	£	**Total**	£
Equivalent annual cost vehicle	1040		
Equivalent annual set-up costs	300		
Annual operating costs	1910	Annual costs	2468
	3250		
Annual income	1610	Annual income	2052
Deficit	1640	Deficit	416

*Rounding discrepancy

given as: Sharrington £48 revenue against £44 costs; Uwchaled £15 against £29; Cuckmere £29 against £74; Lilbourne £34 against £82. Weekly figures should however be regarded with great caution.

Dispensing with the PSV requirement after the 1978 Act removed significant driver training and some licensing costs. Attempts to improve the financial position subsequently focused on further exploitation of excursions and hires and cutting the cost of vehicle acquisition. It became normal for excursions and hires to bring in 50% of total revenue, and to aim at 100% profit on a full bus. Some have attracted a tourist market (e.g. the Fellrunner CB). Absolutely crucial was the ability to finance a replacement vehicle, as many minibuses on CB duty only lasted for three years. The need for a reliable means of periodic vehicle replacement was argued by the Sharrington group in evidence to the Select Committee on Nationalized Industries (1977-78). Those with established arrangements with their county council and NBC subsidiary (like Norfolk) were generally able to continue this, although not always with new vehicles. In return, such schemes had to cover costs on operations and sometimes to repay a proportion towards the vehicle.

Various alternative means of acquiring vehicles have been summarised above - second hand purchase, contracting or leasing, shared use with other organizations. Balcombe & Dredge (1980) estimated that the lower cost of a used vehicle would be offset completely by higher maintenance bills, and new bus grant would tip the balance in favour of purchasing a new one. Schemes without council or NBC support have chosen not to take this risk, and preferred to buy second hand. Meopham's £30 ambulance was mentioned above. In other cases where local authority funding is obtained, the bus committee may have to contribute 25% or so, perhaps through a repayable loan. Sharing a vehicle with other agencies is perhaps the best way to spread the costs, especially if the partner is a local authority department willing to take overall responsibility. Shared school/community minibuses are apparently successful examples. In the case of Wedmore CB, *all* financial arrangements are taken over by Somerset County Council. The loss of new bus grant in 1984 could have been a serious blow, but since 1986 there has been the Rural Transport Innovation Grant as a possible alternative recourse. This does allow the part-funding of replacement vehicles, as well as the setting up of new operations.

Unfortunately, there is very little evidence on the comparative performance of community buses and conventional buses when run under the same conditions. The only relevant indication of this is a calculation made for the Exe Valley Market Bus that the deficit incurred was almost identical to that of the pre-RUTEX bus services in the area, but the CB operated at twice the frequency and carried twice the number of passengers (Devon Rutex 1978). More evidence is needed on this aspect.

10.4 COMMUNITY BUSES AFTER 1985

While specific mention of community buses in the 1985 Transport Act was very minor, its wider implications could have been much more serious. Since road service licences were abolished, the Act replaced this with a system of permits for CBs. More importantly, local bus committees and sponsoring authorities were extremely concerned that deregulation would cause serious disruption to existing community bus services, or possibly cause their extinction, after having put much effort into fostering their development over several years. Competition of a predatory nature by commercial bus companies, even if it only affected part of a CB route, could have wrecked the viability of a volunteer-run operation. It also seems inappropriate that CB stage carriage routes should be put out to tender, in that their original *raison d'etre* was that no other bus company was willing or able to run the desired routes at acceptable subsidy. In any event, a lower tender from an independent bus company would be extremely unrealistic and there would be doubts about its durability. Consequently, community buses, together with postbuses, were exempted from the tendering procedure until 1991 (under SI 1985/1921).

Since the 1985 Act came into force, the main instrument for sustaining community buses and encouraging new schemes has been the Rural Transport Innovation Grant (in England, the Rural Transport Development Fund). New CB initiatives were a specific target for the pump-priming assistance of the RTIG/RTDF, and it would therefore have made no sense to expose new schemes immediately to tendering. Funding was primarily intended to help the inauguration of new transport projects by contributing to start-up costs such as vehicle purchase, but would also be considered for a limited degree of running costs. This has been interpreted to include the funding of replacement vehicles for existing CB groups. Often the county council is expected to match the RTDF share, and district councils might also contribute.

Rural transport projects of all types which had been assisted by the RTIG/RTDF up to November 1987 are mapped in figure 14.2 below. Community buses can be extracted from this pattern and compared with the 1985 distribution in figure 10.3. In this short interval, fifteen new community bus schemes had been established (although it is uncertain if there are any more without funding from this source), and assistance had been granted to a further eight services already in existence. Of the latter, the six in England were all given capital grant towards the purchase of 16-seater minibuses. These are: Mickleton, Stanford-in-the-Vale, High Suffolk, Meopham, Harting and Tavistock. The new buses thus acquired allowed the communities to offer new or extended services; two had wheelchair lifts. Two schemes in Scotland (Southdean & Hobkirk, and Colvend & Southwick) were granted funding from RTIG for replacement vehicles, apparently not for new services but to ensure the continuation of existing ones.

As far as new CB schemes are concerned, it is necessary to accept the classification of modes given by the Development Commission (administrators of RTDF). In at least one case, a project called a community bus appeared on other evidence to be a different UCM type. Despite this, there appeared to be fifteen new CBs by November 1987, of which eight were in Cornwall (figure 14.2). Appearing simultaneously, the latter are the first example of anything resembling a coherent policy of instituting CBs in areas of local need after a systematic search of the whole county. The groundwork was done by Cornwall Rural Community Council in formulating the original proposals and coordinating volunteers towards setting up local community bus associations. Their vehicles are identical 15-seater Freight Rover minibuses. Operation is at a parish scale, and devoted jointly to the needs of local primary schools, the adult non-car population and voluntary groups. The whole project was joint-funded by the council's education and transportation committees (25% each) and the Development Commission (50%) through the RTDF.

The remaining new community buses established in 1987 (figure 14.2) are: Burton Fleming and Woldsward (both in Humberside), Malpas (Cheshire), Shipston (Warwickshire), Woodford & Addington (Northants), Sharnbrook (Bedfordshire) and Amberley & Slindon (West Sussex). The most novel feature arising from these is that the Sharnbrook scheme is run by the local school/ community college on behalf of the community.

Others will no doubt arise until the RTIG/RTDF budget is exhausted. This only serves to reiterate that raising finance for replacement vehicles is the achilles heel of community buses. Expiry of RTIG/RTDF can only pass more of the burden back to county councils, with a consequent slowdown in the innovation rate and periodic difficulties for existing schemes. Further, there remains the problem, as with any voluntary transport, that resources will come to those communities best able to help themselves, and not necessarily to those in greatest need (Banister & Norton 1988). In order to help the weakest, perhaps a top-down approach is best. A systematic search for transport-deprived areas and active marshalling of local volunteers by a county-wide agency, with vehicles acquired from external funding, as in Cornwall, may be the most likely means of securing equitable transport provision.

Meanwhile, a further threat comes from Europe. The proposed harmonization of vehicle and driver licensing has led to EEC draft regulations to compel minibus drivers, including volunteers, to obtain a special higher standard licence (i.e. to pass a second driving test). This would strongly deter volunteers, have financial implications, and pose a severe threat to the viability of all community transport services.

Chapter 11

COMMUNITY TRANSPORT - CAR SCHEMES

11.1 TYPOLOGY OF CAR SCHEMES

Within the context of providing alternatives to conventional public transport in rural areas, the use of private cars has been most often invoked in terms of what are usually called 'social car schemes', and this chapter is primarily devoted to these. The basic rationale is that the use of private cars for the public at large, with or without eligibility restrictions, constitutes an increase in overall public transport resources in areas where they might otherwise be lacking. Private motorists, as drivers, obviously incur no wage costs, and thus car schemes of all types share the essential advantage of all CT modes in such a significant cost reduction. It is 'supplementary transport' in a real sense, as it is used to convey individuals or small passenger loads on specific journeys which could not possibly justify the provision of specialized transport, or which could not be made by any existing conventional bus. Most types of car scheme are also locally based and are thus able to exploit the network of informal social relations and fund of goodwill that exist at community level.

Like community buses, most car schemes are located in the overlap zone between UCMs and CT. They are 'unconventional' in that cars are being deployed to fill an unfamiliar role and supplement the supply of conventional public transport (figure 1.1). To change the emphasis slightly, they are unconventional means of serving the *general public* where suitable alternatives are absent, as distinct from the great majority of the CT sector which is devoted to passengers with special needs such as the elderly and disabled (figure 2.1). Having said this, car schemes are not such a clear-cut case as community buses (chapter 10), as there is in reality a transition between those with few limitations on who can participate (e.g. regular lift-giving and car pools) and those restricted to specific urgent cases (e.g. social services or health authority back-up arrangements). Overall, a large majority of car schemes *do* have eligibility restrictions.

Within the context of the wider CT sector and its concern for mobility-deprived groups, the true range of volunteer-run car services can be appreciated. Sutton (1987) gives a four-fold classification.

(i) Centrally organized services with volunteer drivers working in collaboration with a public agency, such as the Hospital Car Service run on behalf of health authorities and similar arrangements for transporting clients of local authority social services departments, in each case as a back-up to statutory 'special needs' transport provision.
(ii) Services run by volunteer drivers recruited by local agencies, such as district-wide Councils for Voluntary Service or 'Volunteer Bureaux' in order to provide transport for a range of welfare or medical purposes over a fairly wide area.
(iii) Car sharing or pooling schemes organized locally, e.g. at a parish scale, by community groups such as the Women's Royal Voluntary Service, to provide for 'essential' travel needs.
(iv) Non-organized informal lift-giving, such as between neighbours, for any purpose.

Others have suggested classifications which clarify the means of widening car availability to the general public. Types (iii) and (iv) can be expanded further. Green (1978) offered a series of definitions, some of which are based on American usage.

(a) *'Lift-giving'*, as in (iv) above, is offering a trip in a private vehicle to a passenger on a non-organized, occasional basis. The passenger may or may not make a monetary contribution. A *'lift-giving scheme'* implies a minimum organization with at least a modicum of advertizing and perhaps some agreement on cost sharing.
(b) *'Car sharing'* means giving lifts on a regular basis, usually involving the same passengers and destinations, and the passenger may make a

contribution towards the car running costs. Sometimes, however, 'car sharing' is used as a generic term for all types of car schemes.
(c) *'Car pooling'* is an arrangement among car owners who make regular journeys between the same origin and destination. By mutual agreement, each owner takes it in turn to use his own car for the journey with the others as passengers, the remaining cars being left behind. With a regular rota all costs should be equal, and no money need change hands.

The last is mainly used for the journey to work. The classification is extended by Green (1978) to include (d) *'van pooling'* and (e) *'subscription bus'*, which are other means of collective commuting more common in the USA.

There is much confusion amongst such terminology, and it is rarely used consistently. For example, excluding the completely informal types, the best known 'public transport' substitute is the rural social car scheme. These derive from categories (iii) and (a) above, but with a rather more formal organization than mere lift-giving. They fall between the informality of 'friends and neighbours' lift-giving and the 'institutional' trip requirements of hospital and social services cars. Unfortunately, different titles have been used in different areas. 'Voluntary car', 'community car', 'essential car' are the most common names. A RUTEX scheme of this type was given the cumbersome title of 'Voluntary Essential Transport Service'. Dyfed has 'country cars'. Oxfordshire has a 'car ferry service'. The most commonly used generic label is 'social car', and henceforth this term will be used for all such schemes.

As stated above, there is a transition among car-based services with respect to passenger eligibility. This was declared in chapter 2 as one criterion distinguishing UCMs, on the basis that certain voluntary modes are run on such low budgets that they could operate only if passenger demand was kept down to low levels. Thus the mode is to be regarded as a 'last resort', where there is no suitable alternative transport and where the journey is 'urgent' or 'essential'. Restrictions are therefore imposed, based upon the 'type of person' requesting the trip and the desired destination. It is possible to classify car schemes by these two variables, each with a scale of relative specialization or generality. This was done above in figure 2.3. The 'grey area' is delimited in the centre of the figure, between the ambulance and hospital car for clear-cut medical cases, and at the other extreme various car sharing/pooling arrangements which have no limitation. Social car schemes at least insist on passengers being non-car owners, but most interpret 'essential' trips to mean those for health and welfare purposes, mainly targeted on the elderly. Other car schemes, whether run by social services departments, volunteer bureaux or charities, run largely 'institutional' trips for clients of various welfare centres or otherwise at the discretion of the agency concerned. Also, passengers are usually confined to members of the community organization offering the facility. Eligibility restrictions can cause difficult problems which will be discussed below.

11.2 SOCIAL CAR DEVELOPMENT IN RURAL AREAS

It is impossible to date the origin of rural social car schemes, as they evolved naturally from pre-existing habits of lift-giving at community level. Informal lift-giving has been a sociable tradition in the countryside for generations, and is as old as the motor car itself. Declining bus services since the 1950s has, despite the rise in car ownership, put more people in a position of dependence on obtaining lifts from friends and neighbours. The proportion of trips made in another person's car can be quite significant, for example, typical rural travel surveys have shown 7-8% of all trips to be lifts, compared with 10-16% by public transport and 57-70% by household car. Over 50% of rural car owners routinely give lifts. Care is needed in drawing conclusions from this. It should certainly not be assumed that all the travel problems of isolated car-less rural residents can be solved by neighbourly lift-giving. Hardship, deprivation, immobility, trips foregone, are regularly reported. There are problems of people in need not knowing anyone willing to give a lift, to the right place at the right time, and potentially helpful motorists might not know the persons who need assistance. Some people are reluctant to ask or to accept 'charity'. It was gradually recognized that the village 'grapevine' was not enough, and a slightly more organized system was needed to put lift seekers in contact with drivers.

It was also recognized that, to be a realistic public transport alternative, a new scheme had to be more widely available than anything currently running. Charities of various types often provided lifts, as did local community associations, but these circumscribed tightly the passengers and journey purposes that were allowed, limited their services to members only, and did not exist in all places. Similarly, hospital cars and agents of the social services such as volunteer bureaux were restricted to passengers who had specific health or welfare appointments or who had 'places' in day centres, etc. The predecessor of social car schemes may be the general purpose lift services organized by the Women's Royal Voluntary Service on the home front during World War II. Since then the WRVS has regularly provided lift facilities on behalf of the health and social services. With this experience, and as a locally based organization, the WRVS has been most frequently asked to take on board the responsibility of running rural social car schemes. The main initial problem was that the former arrangements were funded by health authorities and social services departments, and this might be withdrawn if the WRVS proposed to cater for a wider range of journeys. Local authority support was therefore essential.

Implicit in this is an appreciation of the role of the motor car in rural areas. Regardless of the status of public transport, by far the best mode of transport for the countryside in terms of convenience, flexibility, speed, load carrying, etc, is the private car. It is also ideally suited to the small scale and highly dispersed pattern of trip demands. Congestion and parking pose few difficulties in rural environments. The only problem is that not all people have access to a car, and consequently it makes sense that public transport alternatives are based on the premise that existing vehicle resources should be exploited to the full and 'shared' amongst the community.

Early promoters of social car schemes were Lincolnshire, Shropshire, Cumbria and Clwyd, which had all launched experiments by 1975. It is not certain which was the first. The earliest records are of one being established in Bridgnorth (Shropshire) in November 1971 and of others in Caistor and Horncastle (Lincolnshire), beginning the same month. The Bridgnorth example was set up by the former Rural District Council in conjunction with the WRVS, and was financed 50% by central government and 25% each by county and district councils (Fearnside 1980). The Lincolnshire schemes were organized by the WRVS and funded by the former Lindsey County Council and the Department of the Environment (National Consumer Council 1978). These are the only cases of direct central government support for social cars. Cumbria began similar experiments in 1975.

Most available information at this stage refers to Clwyd (Rhys & Buxton 1974, Clwyd 1978). The first scheme commenced in Wrexham Maelor District in 1974, and the second in Glyndwr District the next year; others were to follow. In outline, the organization is similar for all such schemes. A committee of local representatives is formed, most often but not always from the WRVS. A survey produces a list of potential volunteers willing to drive their own cars for the benefit of less advantaged neighbours. Initial publicity is required. Requests for transport are processed by the coordinator who then contacts an available driver. After the journey the driver completes a log sheet, and at regular intervals suitable expenses based on a mileage rate are reimbursed by the county council which is underwriting the scheme. Under contemporary legislation it was not possible to levy a fare on the passenger without the car becoming legally a PSV. Committee expenses such as telephone calls were also reclaimable.

The Clwyd case is interesting in that it developed into a two-tier form. Originally, the 'Voluntary Social Car Scheme' was aimed at the elderly, handicapped, housewives and mothers with children, for visits to nearby towns, hospitals, doctors and chemists. Later, there was felt to be a need for a more restricted service capable of being implemented over wider areas. Consequently, a 'Voluntary Essential Car Scheme' was inaugurated in the Glynceiriog area in 1977. Eligibility for this was defined as the elderly and infirm for visits to hospitals and clinics, the collection of prescriptions, and the physically or visually handicapped for visits to social clubs. The latter task replaced duties previously undertaken by social workers. This distinction is symptomatic of the dilemma facing organizers of whether to permit a broader range of trips and risk heavy demands on resources, or deliberately to assist the most needy by allowing urgent cases only - a policy which would also keep costs to the minimum. Social car schemes around the country vary widely in this respect, but there is a strong tendency for most to adopt the latter.

A survey of rural UCMs was undertaken in 1977 by the National Consumer Council (1978). This found 98 car schemes in Britain, of which 81 were by the WRVS. It is not clear however whether these are all comparable in type and scale.

Under RUTEX social car schemes were implemented in all four experimental areas. In North Yorkshire, the Hackforth & District car sharing service was designed to enhance the existing informal pattern of lift-giving in an area where public transport was almost totally lacking (North Yorkshire Rutex 1980c). This was the smallest of the RUTEX projects, serving only three parishes with a population of 240. The scheme was run entirely by a single volunteer coordinator in each parish. Any trip was allowed within the experimental area, but outside this cars were required to connect with local buses for onward travel. A further unusual feature was that the legislation which facilitated RUTEX - the Passenger Vehicles (Experimental Areas) Act 1977 - had authorized the payment of fares by passengers on car schemes, on condition that volunteer drivers were vetted by the local authority. In the Hackforth experiment there were two fare scales, a lower one for journeys which drivers would have made anyway and a higher one for special journeys made at the passenger's convenience.

In Devon, the South Exmoor Car Service was similar in type although it was based in a larger area - five parishes with 1500 people (Devon Rutex 1980b). Limited bus services in the area were supplemented by habitual car sharing, but a more widespread facility was needed. An *ad hoc* committee of local representatives was formed to recruit volunteers and coordinate trip-making. Lifts within the project area were generally allowed, as were trips to the local towns of South Molton and Dulverton. Trips further afield were only allowed in the cases of passengers with hospital or clinic appointments. Again there was a two-tier fare system, depending on whether or not the passenger was carried on one of the driver's regular journeys. One odd feature of this scheme was that it became localized, with participation almost exclusively from two of the five parishes, partly reflecting variations in publicity and enthusiasm.

The biggest RUTEX car scheme was that in the Dinefwr experimental area (Dyfed) with a population of 7000. Given the incongruous title of 'Voluntary Essential Transport Scheme' (VETS), it was organized on the typical WRVS pattern, with coordinators at village and central levels (Dinefwr Rutex 1979). The main need here was for feeder services from the dispersed rural settlements into the towns of Llandovery and Llandeilo where there were consumer outlets and better transport connections. In this case the car service was specifically for 'essential' journeys, at the discretion of the organizers, which could not be made by other means. In practice this meant that most trips were for medical purposes, as people were put off from requesting shopping and other trips. Here there were no fares, but a reclaimable mileage rate for drivers.

Different again was the Scottish RUTEX car scheme, known as the South Ayrshire Hospital Transport Service. In this case lifts were offered only for access to hospitals and other medical facilities in the vicinity of Ayr. The service was available to three villages for which transport to hospitals had proved a particular problem (Scottish Rutex 1981b). Coordination was done by three local volunteers in collaboration with the community council. Fares were paid in advance direct to the organizers, the drivers being reimbursed a mileage rate. At the conclusion of RUTEX the South Ayrshire scheme was abandoned, in the belief that the few people who used it did not justify the organizational effort, and informal lift-giving would suffice. The South Exmoor and Hackforth schemes continued with certain reorganizations and/or extensions, while the Dinefwr services were adopted willingly by Dyfed County Council and later expanded throughout the county.

As with other UCM types, innovative ventures forced the pace of legal change. The question of payment to voluntary drivers was important, not just for the economics of individual schemes, but also to encourage passengers that they were paying for a legitimate transport service instead of receiving 'charity' and feeling under an obligation to others. Surveys had indicated that many passengers would actually prefer to pay fares. Accordingly, this was legalized in the 1978 Transport Act so that casual passengers might offer a contribution if there was no prior advertising. In the case of social car schemes, passengers could be carried at separate fares with the agreement of the local authority and traffic commissioners if the arrangement was without a view to profit and aided the social and welfare needs of the community. Advertising was however only permitted in a restricted number of places. These concessions were broadened further in the 1980 Transport Act such that a vehicle with less than eight seats was not a PSV and not operating for hire or reward if fares were received on a non-profit basis and agreed with the passenger in advance of the journey. Advertising restrictions were removed. This liberated not only social car schemes but any form of lift-giving where payment was sought. These provisions were endorsed in the Public Passenger Vehicles Act 1981.

Insurance was another problem requiring solution if social cars were to flourish. Volunteer drivers would appear to be running their cars for hire or reward, and participation in a social car scheme had to be cleared with insurance companies. In some cases a supplementary premium was demanded. Coincident with the passing of the 1978 Act the insurance companies jointly declared that giving lifts for payment on a non-profit basis would not be a breach of motor policies.

From the late 1970s social car schemes expanded prodigiously. As far as can be discovered from available evidence, the situation in 1985 was as shown on figure 11.1. This map was particularly difficult to compile and has to be interpreted with great caution. First there is the usual problem of incomplete information, with some county PTPs listing each individual scheme and others referring to them only vaguely or without giving their exact location. A major problem is the inability to distinguish clearly between social car schemes of the relatively broad type described above, and other car services operated on behalf of hospitals or social welfare agencies for 'institutional' journeys. The latter may be assumed to operate in all areas, including urban. The WRVS, for example, organizes many rural social car schemes but also acts as agent in many other areas for the social services, providing back-up transport for their clients. Some confusion is inevitable. Different scales of organization is another factor which prevents consistency in mapping or enumeration. In many cases social cars are implemented in specific localities only, as a deliberate policy aimed at isolated pockets of transport deprivation. Other authorities take an area-wide 'broad brush' approach and try to arrange contingency car schemes to cover a whole district or county. Comparability is most uncertain.

It is realistic to take a few examples briefly. Six years after the pioneering Bridgnorth

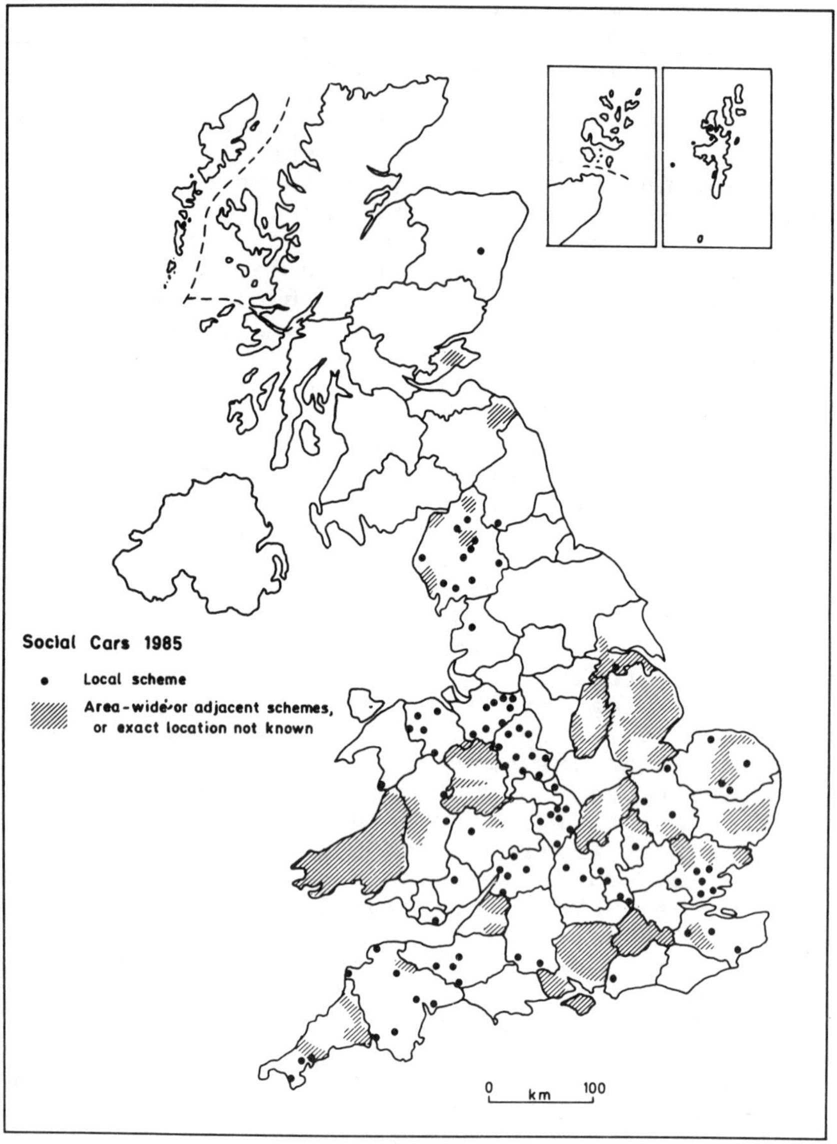

FIGURE 11.1 Social cars 1985.

experiment, Shropshire initiated a programme of supporting social cars in any parish which declared a need for one and which could find sufficient volunteer drivers. A round of local surveys and meetings had concluded that gaps in the county's public transport were very localized and specific, such that any larger mode would be inappropriate. Schemes are administered by community councils or district councils although the coordinator must be a local resident; the county paid 75% of costs (Fearnside 1980). As in some other counties, fares were not charged despite being allowed by the 1978 Act, but passengers were encouraged to make a voluntary contribution. In 1977-78 an additional 10 schemes were started; by the end of 1979 there were 30 in total; the next year there were 45 and in 1981 there were 61; the 1984 total was 68 which covered two-thirds of the county (Shropshire PTPs 1979/80 - 1985/6).

Few counties can match this rate of expansion, although these are all parish based and hence the numbers seem impressive. Other authorities have policies of making social car facilities available throughout the county area. After RUTEX, Dyfed CC resolved on expanding the 'Voluntary Essential Transport Scheme' in Dinefwr District to the whole county. Renamed 'country cars', the service was replicated in 30 mutually exclusive zones, organized as before by the WRVS recruiting local coordinators. The whole county was covered by 1984, by which time the system had recruited about 800 volunteer drivers and over 200 organizational staff. A blanket approach means there is some variation in service type in different environments: country cars also operate in towns, but here only the disabled are likely to qualify. Dyfed country cars have achieved considerable fame, but it is realized that even the 20,000 journeys a year merely expose the full scale of mobility needs among certain vulnerable groups (Dyfed PTPs 1979/80 - 1985/6, Pettit 1987).

Other examples of substantial area-wide coverage are: the Lindsey and South Holland areas of Lincolnshire, large parts of Nottinghamshire and Northamptonshire, Breckland District of

Norfolk and central Suffolk, rural parts of Avon. Hampshire and Cornwall announced their intention to promote car schemes in all rural areas of the counties, but information on these is sparse. Surrey on the other hand provides a complete list of car schemes of all types; the 63 listed cover the county and are run by almost as many organizations (Surrey PTP 1984/5). In all probability most of these are for members only and restrictive in permitted usage, and not comparable to 'public transport' social car schemes, but it is quite likely that a similar range of organizations offering transport exists in all counties without being widely publicised. In addition, most other counties have specific localized car schemes, notably Cumbria, Cheshire, Staffordshire, Warwickshire and Essex (figure 11.1). Finally, the apparent absence of social cars in Scotland is most odd, for which there is no obvious reason.

Deregulation does not appear to affect social car schemes. It was thought they might lose traffic to the newly liberated shared taxis and hire-cars, but this does not seem to have happened. By 1987, the Rural Transport Development Fund had assisted only one social car (in Suffolk).

11.3 TRIP GENERATION AND SOCIAL EFFECTS

(a) Passengers

Data on social cars are fairly abundant, mainly reflecting the large number of schemes around the country, but extremely fragmented and inconsistent. Table 11.1 attempts to summarize data from various sources, and illustrates the wide differences that exist among services. The first problem is the different scales of organization, in that some are localized parish schemes while others are area-wide at a district or town hinterland scale. Total passenger loadings cannot therefore be interpreted without reference to the catchment area population, and expressed as a trip rate. Another problem is that it is not always clear whether a 'trip' means a car journey or a

TABLE 11.1 Passenger numbers on social car schemes (source:Shropshire PTPs 1979/80 & 1985/6, Sutton 1985, N. Yorks Rutex 1980c, Devon Rutex 1980b Dinefwr Rutex 1979, Dyfed CC Powys, other PTPs).

	Number of schemes	date of information	passenger trips range	total	mean(one scheme)
Shropshire	11	1978/9	24-1055	3505	319
	44	Apr-Dec 1984	10-6115	16214	368(9 months)
(excluding	42	Apr-Dec 1984	10-938	9592	228(9 months)
Volunteer Bureaux)	42	(estimate for year 1984)		12444	296
Lindsey	area-wide	1977		3808	
South Holland (Lincolnshire)	area-wide	1977		3687	
Hackforth RUTEX (N. Yorks.)	1	1978/9			156
S. Exmoor RUTEX (Devon)	1	1978/9			555
Clwyd	2	1977	182-422	604	302
Dinefwr RUTEX (Dyfed)	1	1977/8			c.1000
Dyfed-Powys	15	1981/2	12-721	3132	209
Dyfed	30 county-wide	1986/7		c.20000	c.700
Powys	6	1982/3		764	127
Bawburgh (Norfolk)	1	1980/81			189
Nottinghamshire	5	1983		2412	482
Leominster (Heref & Worcs)	1	1983/4			>600

passenger journey, or is counted as a one-way or return journey.

Annual trip totals appear to vary from little over 100 to about 1000. The Dinefwr RUTEX case was at a larger scale than most, while other data from Dyfed and also Lincolnshire refer to area-wide administration, not comparable with the other examples, in reality operating at a more local level. The remaining range of 100-600, or 2-12 trips per week is more realistic. The Shropshire data in particular emphasize the enormous range in passenger loadings that is possible among different schemes in the same county. Here, parish organizers were not working to universal eligibility guidelines (unlike the WRVS), but were free to respond to local circumstances and interpret the needs as they thought fit. Even among WRVS schemes however, as in the Welsh examples, patronage varies widely despite national guidelines. It is notable that the volunteer bureaux among the Shropshire data have significantly higher loadings, as their cases are referred to them by the social services. A survey by the National Consumer Council (1978) discovered a much higher patronage on WRVS car schemes than on non-WRVS services - an average of 1082 passengers (from a sample of 22 schemes) against 266 (sample of 10). This may reflect different scales of organization, or the greater experience and resources of the WRVS compared with local 'amateur' efforts, or the sample may have included cases where the WRVS is acting as agent for the social services.

Overall, it is very difficult to draw general conclusions from such data other than that passenger carryings are very low. This is of course expected given the 'essential' or 'last resort' nature of the mode, and also the small capacity of motor cars compared with buses. The average number of passengers per car journey is 1.1-1.3. This is much closer to a taxi service, and the cost per passenger is therefore high (see below). Catchment areas are much wider than that of any fixed-route bus service, and hence social car patronage seems even lower when compared to the populations they serve. Trip rates have been calculated for a sample of 15 car schemes in Dyfed and Powys, and come out as low as 0.027 trips/person/year. This is equivalent to one person in 37 making just a single trip in a year, or one person out of 1,926 making one trip a week! Where not limited to essential journeys, the trip rate can rise to the 1.47 trips/person/year found on the South Exmoor scheme.

Data are similarly inconsistent with respect to journey purposes. Available information is summarized in table 11.2, which attempts to deal with highly variable trip classifications. A medical emphasis would be expected, and is universally revealed except for the two anomalies of the South Exmoor and Hackforth RUTEX. 'Work' and 'education' trips would not normally be permitted on social cars, but these have caused a quite different distribution in these two cases. Here, the car service was available to non-car owners for any local journey which could not conveniently be made by other means, without any requirement that the trip should be 'urgent'. The high percentage of work trips in South Exmoor were made however by just two people travelling regularly, and shows how the trip pattern can be biased on a low-intensity service. Similarly, the high proportion of education trips in the Hackforth scheme was accountable to one person regularly attending evening classes. Elsewhere, the trip distribution is biased towards health purposes, which reflects the normal interpretation of 'urgent' or 'essential' journeys. The proportion of health-related trips varies between 51% and 87%. There are some social car schemes where *only* trips to hospital or to a GP's surgery are permitted (in a few cases the doctor actually administers the car service). Day centres and other social clubs, mainly for the elderly, are often legitimate destinations, but their sparse showing in table 11.2 probably reflects local factors. The validity of shopping trips is a cause of some debate, as their proportion of social car journeys is variable but generally low. Some kind of selection or rationing process must be going on, to limit those shopping trips regarded as 'essential'.

TABLE 11.2 Journey purpose on social car schemes (source: Nat. Consumer Co. 1977, Dinefwr Rutex 1979, Devon Rutex 1980b, N. Yorks Rutex 1980c, Sutton 1985, fearnside 1980, Shropshire PTP 1985/6, Moyes 1987).

%	Lindsey (Lincs.) area-wide 1977	Clwyd 2 schemes 1977	Dinefwr RUTEX 1978	S.Exmoor RUTEX 1979	Hackforth RUTEX 1979	Nottinghamshire 5 schemes 1983	Shropshire 37 schemes 1979/80	Shropshire 68 1984	Aberaeron (Dyfed) area-wide 1984/5
Hospital visiting	33	29				19	11	12	35
Hospital treatment							41	41	
Doctor/dentist/optician	33	22	60	7	7		14	11	9
Other medical	15					41	14	7	19
Chemist/prescription	6							2	
Shopping		13		10	6		4	5	18
Personal business			30	9	7	25			5
Social visits	12	10		3	20				7
Social club/day centre		23				15			
Education				3	29				
Work				69	13				
Escort					14				
Transport connections			10						4
Other/unspecified							16 (includes day centres)	23	4

A detailed analysis of social car trip patterns has been made in a private study by Moyes (1987), which examined results of the Aberaeron scheme, part of the Dyfed 'country cars' service. In 1984/5, 985 trips were made to 37 destinations, although eight of these accounted for 83% of the total. The biggest single destination (28%) was a hospital, the others being two more hospitals and five town centres. 78% of all trips were of distances below 50 miles (80 km), but there was no bias towards short trips, with journeys of 30-50 miles outnumbering those of 1-20 miles; this reflects average distances to town in rural Wales. The 985 trips were made by 178 people, although just over half of these made only one or two trips during the year; about 13% of users made more than 10 trips. A similar pattern obtains with respect to drivers. The 985 trips were conducted by 34 drivers, but over 60% undertook only one or two trips per month while a few individuals gave lifts every three or four days on average.

Examined at the level of the twelve areas into which the Aberaeron scheme is subdivided, the data are characterised by extraordinary variations among areas, by journey purposes, and from month to month. Some areas showed little or no usage, and others quite a lot. Comparing 'observed' trip numbers with those 'expected' from the scheme totals, trips for each purpose varied by area from zero to 200% or 300% of the norm. Organizers' workloads may vary by a factor of three in successive months. While the dominance of a few users is sustained at area level (indeed, the single heaviest user may account for one quarter of all trips), the individuality of overall trip characteristics does reflect the range of needs that the system was designed to meet.

Data on the social characteristics of users are surprisingly rare. It is commonly believed that most patrons are women and/or elderly, with some handicapped and children with escorts. RUTEX reports provide brief verification of this, but few reliable statistics. During the Dinefwr experiment, 56% were elderly, 15% handicapped and 5% children (Dinefwr Rutex 1979, Walden-Jones 1979).

Finally, it is of interest to know how people would have made necessary trips in the absence of a social car service. The only evidence on this comes from RUTEX surveys. Where trips were supposed to be 'essential' they would presumably be made by some other, less than adequate, means. In Dinefwr, two-thirds of trips would have been covered by private lift arrangements. Where trips do not necessarily have to be 'urgent', as in South Exmoor and Hackforth, 19-29% would not have made the journey in the absence of the social car, and 18-33% would have got a lift from someone else. Otherwise, inadequate buses, expensive taxis, or even walking were the only alternatives. In these respects, the new facilities are clearly meeting at least a proportion of local needs.

(b) Social Implications

That there are social benefits to individuals is beyond doubt, despite the low level of most social car activity. There are also benefits to the wider community. It has extended, and not competed with, existing traditions of neighbourliness, and has provided help to those isolated in a social as well as a geographical sense. It has given many public spirited individuals a means of helping the community, which otherwise they might not have the resources to do. Many volunteer drivers are drawn into other community activities. Doctors have been relieved of unnecessary travelling since patients can now be brought to them (Walden-Jones 1979). There is no competition with existing public transport, and operators gain a few extra passengers through the car feeder service. For the local authority these are cheap means of filling gaps in the public transport network which otherwise would remain empty.

Compared with community buses (chapter 10), social cars have caused fewer problems but have had less impact. As the vehicles used are merely private cars without any particular distinguishing marks, they are 'invisible' when in service. As a discreet, diffuse and low intensity operation, it is less easy to generate the local identification and 'team spirit' seen in community bus projects.

On the other hand, the administrative workload and financial headaches imposed on organizers of CBs are much less onerous for social car schemes. There are similar problems in recruiting and keeping volunteer drivers, but in car schemes a temporary shortage is unlikely to prevent the service running. Due to the inconspicuous nature of the facility, considerable importance attaches to publicity. This usually takes the form of 'shop window advertising', posters and leaflets, and informing health visitors and social workers, yet lack of awareness remains a problem. Organizers do not want to solicit custom and thereby encourage non-essential use. The difficulty is how to keep a low profile and at the same time reach all those persons in genuine need.

After the 1978 Transport Act had relaxed the rules on social car operation, the Department of Transport (1978) tried to encourage more local groups to start services by issuing a handbook of advice called *'A Guide to Community Transport'*. The tasks required of local organizers are laid down in a simple manner. Assuming a 'bottom-up' approach, local groups concerned about transport are invited to consider social cars vis-a-vis community buses (the only two modes included). No matter how well they might think they know the problems, group members are urged first to conduct a survey of travel needs to define the issues precisely. Approaches to the local authority and main bus company in the area might result in suitable service adaptations, or at least assistance in setting up a voluntary transport scheme if a viable proposal emerged. It would be expected that a social car service would seem an easier option than running a community minibus, primarily to avoid the daunting problem of how to acquire the vehicle. Theoretically, this decision should be guided by the type of demand in the area (dispersed and essential, or predictable group demands for commonly used facilities). Although a larger pool of drivers may be needed for a car service, this may still put less pressure on local people's time and resources than that needed to run a minibus. Planning does not have to be so meticulous as for a bus service; it is only necessary to avoid duplication with existing transport and to circumscribe permitted journeys to selected destinations or to a reasonable radius, and bearing in mind the likely availability of drivers. Local contacts should be sufficient to recruit a pool of volunteer drivers, whose only preparation is to check their insurance policies. Passenger eligibility is a tricky problem, and the only recourse is to set simple guidelines and trust the discretion of coordinators on duty. Publicity must be organized. A fares system (recently legalized) must be decided upon, and related to any financial support arrangements which may be offered by the local authority. A coherent system for maintaining drivers' logs is needed, and the group Treasurer has responsibility for meeting claims and general accounting.

As a putative public transport alternative, social cars suffer by their inability to provide for the full range of trip purposes necessary for the pursuit of a normal way of life. They are ideally suited to a 'topping up' role for urgent trips where a conventional bus still exists to provide for everyday needs. Most often, however, they are the last and only resort in areas without any alternative, but despite this, they can only operate if trip demands are kept down to very low levels. Hence, crucial importance attaches to the vexed issue of *eligibility* - of passengers and destinations. As shown above, the limitation to 'essential' trips is usually interpreted in a medical sense - journeys to hospitals for outpatient treatment or visiting, doctors, dentists, opticians, chiropodists, and chemists for prescription collection being normally allowed. Possibly this emphasis may be partly due to some misunderstanding on the part of potential passengers, being reluctant to ask for any other kind of journey. The obvious problem is the implication that other types of journey are 'inessential'. Under what circumstances would a village coordinator refuse a request for a lift which s/he thought was not essential? This must place such people in an invidious position, with a responsibility to the drivers and to the sponsoring authority to keep trip demands within bounds, and a responsibility to be as helpful as possible to one's own community. The restrictive nature of eligibility guidelines can be judged from two examples from Lincolnshire.

(i) "Visit doctor, dentist, optician; collection of prescriptions and hearing aids; health centre visits; pensioners' occasional visits to friends where neither party has access to alternative transport; visit friends or relatives in hospital; short journeys to trunk route bus stop where bus journey is essential and no alternative transport is available; visit MIND or PHAB club".

(ii) "Visit to hospital for treatment or consultation; visits to relatives or in special circumstances friends dangerously ill in hospital; collect patients discharged from hospital; to link with public transport for any of previous types of journey".

(Lincolnshire PTP 1983/4)

These are hardly designed to encourage normal levels of mobility! While the first allows some social visits, neither seems to allow even shopping trips or personal business, and presumably other social, recreation and entertainment trips would be out of the question. Depending on local circumstances these may or may not be possible by other means, but the implication that these journey types are not essential suugests that rural residents should tolerate a lower standard of living than anyone else. Arguably, there is no such thing as an inessential trip, and any journey that is desired should be possible in order to ensure an acceptable quality of life in rural areas.

As with any other type of voluntary transport, there is much uncertainty whether schemes are being set up in the right places, where the need is greatest. It is consistently believed that the places most likely to establish social car schemes are those whose inhabitants possess the necessary social and administrative skills and the motivation to use them. These are most likely to be found amongst professional or middle class people, probably retired, and perhaps recent in-migrants to the countryside. Further, such a population is more likely to exist in the larger, less remote villages, which may have some facilities already. Thus, an inverse care law might operate, where those in greatest need are least able to undertake self-help measures, and existing inequalities might widen.

Results of a survey of voluntary services in Norfolk and Suffolk provide data in support of this (Banister & Norton 1988). This was reviewed in the context of community buses (chapter 10), but it is relevant here also. Out of 134 parishes surveyed, 41% had some kind of voluntary transport, including 25% with a social car scheme and 27% with a community bus service. Together with a range of other voluntary sector activities, there was a positive correlation between the average number of services per parish and population size. Overall, most parishes are devoid of most services, and the smaller parishes are more likely to be lacking.

The social characteristics of volunteer workers were also examined. In all transport schemes, over 50% of volunteers (and over 60% of organizers) were from social class I (professional and managerial), in comparison with 24% of the study area population. Forty-eight percent of volunteers were 'urbanites', either retired or commuting, and 49% were 'newcomers' with less than ten years' residence in the area. However, it is interesting that another 34% of volunteers were long term residents who had lived in the area for 20 years or more. Looking at social car schemes alone, 55% of organizers were retired and 27% housewives. Unlike on community buses, drivers were evenly spread throughout the community, with 28-37% from each of the retired, housewives and working populations. Generally, the data confirm the unequal distribution of self-help activity, although the polarization with respect to social car services is not so sharp as with community buses.

11.4 ECONOMICS

Social cars were seen as a cheap means of plugging gaps in the 'conventional' network, adding very small and politically acceptable increments to the subsidy bill, and certainly much cheaper than supporting additional buses. Low aggregate costs are due entirely to the low passenger loadings reported in the last section, and do not result from any intrinsic efficiency of the mode. On the contrary, it could be argued it is very inefficient with no more than 1.3 passengers per vehicle journey so that the cost per trip is high. The bespoke, individual nature of journeys resemble those of taxis, and costs are inevitably high although not at commercial levels. Hence the restrictions on eligibility are essential, apart from other reasons, to keep total costs under control.

All these points are verified by the data in table 11.3. The variability in trip numbers from one scheme to another has been noted above, and there is consequent wide variation in costs, revenues and deficits. There are some cases in the table of zero revenue, i.e. 100% subsidy, referring to the period before charging fares was allowed by the 1978 Transport Act. Fare scales are usually simplified, either a set fare for common destinations or a crude zoning system; mileage-based fare scales pose problems of calculation, shortest route, etc. The most common arrangement was for drivers to hand in all monies plus their journey logs to the scheme coordinator or treasurer, and then receive reimbursement of expenses (which *are* based on a mileage rate) every month. Attempting to cover operating costs by making fares equivalent to motor car running costs is rarely done, as this makes fares too high for the kind of people requiring lifts. Fares are often defined as 'bus equivalent', in agreement with the local authority. The deficit or subsidy on a trip basis is therefore high.

Expenses paid to drivers are usually based on local authority norms as used in, for example, the meals-on-wheels service. Clwyd in 1974 paid

TABLE 11.3 Costs of social car schemes
(source: Shropshire PTPs 1979/80 & 1985/6,Balcombe & Dredge 1980, Clwyd 1978, Dyfed CC, Powys CC, other PTPs).

Shropshire 1978/9 11 schemes

	Trips	Cost(£)	Revenue(£)	Deficit(£)	Deficit/Trip(£)
Range	24-1055	56-1081	0-31	56-1050	0.42-3.76
Average	319	428	18	410	1.29

Shropshire Apr.-Dec. 1984 42 schemes

	Trips	Cost(£)	Revenue(£)	Deficit(£)	Deficit/Trip(£)
Range	10-938	39-3233	10-990	29-2243	0.80-6.70
Average	228	799	261	538	2.36

Dyfed-Powys 1981/2 15 schemes

	Trips	Cost(£)	Revenue(£)	Deficit(£)	Deficit/Trip(£)
Range	12-721	69-2695	0.683	53-2012	1.47-7.12
Average	209	774	163	611	2.93

RUTEX schemes 1977-9

	Trips/week (one-way)	Revenue/week (£)	Subsidy/week (£)	Subsidy/Trip (£)
South Exmoor (Devon)	11	3.00	2.50	0.23
Hackforth (N. Yorks)	3	2.00	0.50	0.17
Dinefwr (Dyfed)	40	-	30.00	0.75
South Ayrshire Hospital Service (Strathclyde)	2	0.60	1.40	0.70

Others

		Trips	Subsidy(£)	Subsidy/Trip(£)
Lincolnshire (S. Holland)	1977	3687	6091	1.65
Clwyd (2 schemes)	1977/8	625	594	0.95
Hampshire (area-wide)	1981/2	1700	2320	1.36
Norfolk (Bawburgh)	1980/81	189	30	0.16

5 pence per mile, and rates in the late 1980s are generally about 25 pence per mile. The fares differential can be exemplified by the Warwickshire rates in 1984, when passenger fares were 8 pence per mile and drivers were reimbursed 20 pence per mile. Where a lift is given on a journey already being made for the driver's own purpose, then the fare could be on a 'shared cost' basis and there need be no subsidy from any central agency. This was done on the Hackforth and South Exmoor RUTEX services: in Hackforth, a higher fare was charged on 'special' journeys and there was no subsidy; in South Exmoor, drivers received a supplementary allowance for 'special' journeys only. The former arrangement is similar to that for regular car sharing schemes (section 11.5(a) below). Some services were reluctant to charge fares even after the 1978 Act. Shropshire suggested all passengers make a voluntary contribution but, perhaps inevitably, was disappointed with the initial response and later tried to raise this to cover one-third of costs.

Over the course of time, most car schemes have tried to recover a greater proportion of costs from revenue. Charging fares at bus rates is more common, and there are a few examples of fares being set between bus and taxi levels, to reflect the higher quality of service on a car journey. Only the Shropshire data allow a comparison over time, and table 11.3 shows revenue rising much faster than costs; the operating ratio has increased from only 4.4% in 1978/9 to 33.7% in 1984.

From the more recent data in table 11.3, it would seem that a deficit per trip of £2-£3 can usually be expected. This is much higher than for other modes. For example, the average of £2.93 per trip calculated from 15 social car schemes in Dyfed and Powys (1981/2) compare with £0.38 per trip from a sample of 21 independent stage bus routes, and £0.31 per trip from nine postbuses in the same area (see section 14.1 below). Lower deficit per trip values in the table derive from older data, but nevertheless there are some cases where low subsidy rates are sustained. These are generally very small scale schemes with firm eligibility restrictions, localized journeys and realistic fares. The example of Bawburgh (Norfolk), which cost only £30 in the year 1980/81 at a mere 16 pence per trip, is one such case. West Sussex cites the Marden/Chilgrove car scheme, which carried only three or four passengers per month at an annual cost of just £50 (1980/81); in the same area a conventional bus carried only five people per week and lost up to £400 p.a., but if this were removed it was estimated that transferring this traffic to the car scheme would raise the latter's subsidy ten-fold (West Sussex PTP 1980/81). This is a neat illustration of the 'supplementary' role of social cars. Generally, patronage at such low levels suggests that this mode of provision is influenced more by cost reduction than by local travel needs.

11.5 OTHER MODES BASED ON PRIVATE CARS

(a) Car Sharing and Car Pooling

It is often difficult to distinguish other car-based UCMs classified in section 11.1 above. 'Car sharing', although a very imprecise term, is defined as the giving of lifts on a regular basis, such as helping a regular passenger to reach a common destination at frequent intervals. 'Car pooling' is also based on sharing cars for frequent trips, but in this case it is a means of cutting down on aggregate car use by arranging for groups of car owners to share the use of one member's vehicle for the journey, each member taking it in turns to drive the others in his own car. What these have in common is that they both apply to regular trips, particularly the journey to work. Where they differ is that car sharing assists non-car owners to get lifts, while car pooling is restricted to car owners. Also, in neither case is any kind of organization necessary, although it is possible to have a formal 'scheme' if required, in order to inaugurate and coordinate lift-giving. This could take the form of anything from advertising on the company notice board to a computerized car sharing 'information system'.

Most of this activity (perhaps nearly all of it) is completely informal, among relatives, friends and work colleagues. Because of this informality, there is very little information about its prevalence. National aggregate data suggests that passengers who travel in cars belonging to other households account for about half as many trips as on public transport. If measured in terms of passenger-kilometres, lifts account for a share roughly equal to that of stage carriage buses (10-12% of passenger-kms). Some data from rural surveys were given in section 11.2. Normal data on journeys made in 'other car' cannot distinguish car sharers and poolers from casual lifts. The best estimate that can be made is that about 3% of those who drive their car to work are engaged in car pooling (Vincent & Wood 1979). Perhaps such arrangements are entirely satisfactory for drivers and passengers, but as mentioned above in the case of rural social cars, it can never be certain that a person in need of travel assistance can always rely on finding a driver willing to give a lift. Some sort of organization seems to be necessary to cover these circumstances. Informal car sharing, where there is no organization, is not appropriate to be classed as a UCM.

First, one could examine the 'minimalist' approach of encouraging greater car sharing merely by disseminating publicity about the need and opportunities, and clarifying the legal situation. One such policy was implemented under RUTEX in Devon. An area of ten parishes was identified in which, although existing bus services were poor, remaining trip demands were too dispersed and small scale to support any new form of public transport. Car ownership was already 85% of households and lift-giving widely prevalent. Accordingly, the '<u>mid-Devon lift-giving scheme</u>' attempted to mop up the remaining need for lifts by allowing motorists to accept payment from passengers. This was not legal at the time the system began in early 1978, but was permitted under the special arrangements of the Passenger Vehicles (Experimental Areas) Act 1977 by which Devon CC took out a 'General Authorisation'. This procedure was made redundant when the 1978

Transport Act allowed payment on a non-profit basis. There was no formal organization (thus differing from a social car scheme), but a leaflet distribution in order to raise awareness.

The assumption here was that inability to pay deterred lift-giving and the new arrangements would remove this inhibition. In fact, the scheme had almost no effect on local mobility patterns. A few extra payments were made but there were hardly any additional journeys (Devon Rutex 1979). Nevertheless, a subsequent survey disclosed there were still problems of matching lift seekers with potential drivers, and therefore an organized scheme of car sharing or social car would have a greater chance of solving residual problems.

As regular arrangements with the same passengers and destinations, car sharing and car pooling are best suited to the journey to work in urban environments. In spite of the lack of data, this is probably where most activity occurs. Subjectively, there might indeed be considerable advantages if a significant proportion of urban commuters could be induced to engage in car sharing or pooling - a reduction in traffic levels, lower aggregate energy consumption, reduced private motoring costs, and reduced public costs of traffic, accidents and environmental hazards. Such advantages have long been realised in the United States, where car sharing and pooling has attracted much attention and appears to be widespread. The concepts are collectively known as *'ridesharing'*, and this may be extended to embrace also minibus pooling, subscription buses, dial-a-rides and taxi sharing on urban commuter journeys (Transportation Research Board 1986). Hence the emergence of the 'paratransit' phenomenon was overwhelmingly urban based, unlike UCMs in Britain. Most information on car sharing and pooling is therefore of American origin (Green 1978, Bonsall 1980a).

In the United States, ridesharing activity is strongly influenced by high urban traffic levels, longer commuting distances and the special problems of low density suburbs. Individuals participate primarily to reduce private car running costs and to relieve the stress and aggravation of urban motoring. It is a large scale activity: commercial agencies exist to organize systems, computers are frequently used to match drivers and riders, in some areas it is a legal obligation for companies to have employee car sharing schemes.

A hypothetical study by Vincent & Wood (1979) estimated the effect of a 10% increase in car occupancy (normally 1.7 persons/car) at peak periods. In the case of car pooling, this would lead to a decline in car numbers by 9%, reduced traffic flows and shorter journey times, but no effect on public transport. If the 10% increase in car occupancy were achieved entirely by car sharing, then the effects would be mainly felt by other modes, with bus, train, taxi, etc, each losing about 20% of their custom, but with little effect on traffic congestion. In the first case, fuel conservation would be significant in absolute terms, but only about 1.5% of aggregate consumption. Remarkably, both scenarios might produce a net social benefit of about £100 million p.a. (1976 prices).

In the absence of any tradition in Britain of formal schemes of this sort, there was no reason to believe that there was any great propensity among the public to adopt them, despite the theoretical advantages. A survey in West Yorkshire by Bonsall (1980b) attempted to assess attitudes and potential. About 7% of commuters said they might make use of some kind of information or matching system for car sharers. There was more interest in simple lift-giving than car pooling. Most people were not impressed by potential cost savings, and preferred to keep their privacy and flexibility.

Few real attempts have been made in Britain to stimulate car sharing/pooling activity. After payment for lifts was generally authorized in the 1980 Transport Act, the government conducted a half-hearted leaflet campaign for more car sharing in the naive belief that it might have an impact on traffic congestion or aggregate energy consumption. The only specific experimental scheme to get off the ground was one mounted by the University of Leeds in West Yorkshire between 1979 and 1981. Using the name YORKSHARE, the project tried to stimulate car sharing and pooling at four sites by an information campaign aimed at employees and households, questionnaire and telephone surveys to trace potential sharers and monitor travel behaviour, a manual driver-passenger matching exercise, and certain incentives such as free reserved car parking. The target sites were an engineering firm in Wakefield, an out-of-town office centre at Boston Spa, an office complex in Leeds city centre, and a residential zone in Garforth (Bonsall et al 1983).

Application rates to join YORKSHARE varied from only 0.8% to 6.8%. The largest group were applications to ride, while pooling was least popular. A scheme based on a single workplace is likely to have the highest response. A simple geographical pattern emerged, with potential passengers having the shortest journey length and potential drivers and poolers the longer. Matching of applicant passengers with drivers was possible in most cases, but for various reasons actual participation was lower, with the proportion of the target population making firm arrangements ranging between only 0.3% and 2%. These are about 30% of applicants. The higher rate was for Leeds city centre, where reserved car parking proved a significant incentive. Most arrangements involved financial compensation from passengers - the suggested rate was 6 pence per mile (in 1980). Follow-up surveys indicated a lapse rate of 10% of arrangements terminating every six months.

Impacts of the scheme were low in proportion to the low level of activity, but perversely, the biggest impact was on public transport. Monitoring showed that public transport use declined abruptly when the scheme started, and did not recover. Solo car driving also declined sharply, but then recovered to some extent as dissatisfied sharers reverted to their previous mode. Overall, the main effect of YORKSHARE was to reduce public transport use - by 1.8% of previous usage by the target populations.

Reductions in vehicle-kms driven and in parking space requirements were insignificant. A tentative cost-benefit analysis concluded that after one year the schemes had produced a net benefit of £1500 over a total set-up cost of £2600. However, the system with the best rate of return is not necessarily that with the highest participation rate, due to the cost of the parking concession in the Leeds scheme. General community benefits were few as most savings accrue to the participants, in the form of reduced cost of public transport fares (50% of benefits), time savings (25%) and reduced motoring costs (25%). Loss of fares revenue to public transport operators is potentially serious, unless the impact can be concentrated on to one sector where the operator could rationalize services and make savings. Low participation rates and low positive impacts suggest there may be little purpose in pursuing car sharing/pooling any further. Results of YORKSHARE were broadly equivalent to those from American systems, and so it was unlikely to be under-performing (Bonsall et al 1983). Social factors primarily determine the low level of interest, as they did in the rural cases described above. Future encouragement of car sharing must be considered in relation to overall transport system objectives, unless it is to provide purely private benefits to a few people and undermine other modes.

(b) **Hospital and Social Services Cars**
The largest service (although by no means homogeneous) that uses volunteers and private cars is deployed as a back-up facility to statutory welfare transport. This is the sector provided by or on behalf of health authorities and local authority social services departments (SSDs) for the conveyance of patients or clients on 'institutional' trips, that is, for medical appointments or attendance at day centres, training centres, etc. This is often referred to as 'special transport' as it is not available to the public but confined to approved client groups, although many vehicles also have specialized facilities for certain categories of passenger such as the disabled (Bailey & Layzell 1983). Cars are used in support of this for less urgent

cases for whom public transport is not available or suitable, for small passenger loads and in sparsely populated country, and where special facilities are not needed. They are also much cheaper than ambulances!

The hospital car service (HCS) is operated on behalf of health authorities by volunteer drivers using their own cars, who are then reimbursed a mileage rate. Drivers are recruited by a local agency such as the WRVS. Alternatively, the term 'hospital car' is sometimes applied to an independently organized social car scheme which is only available for journeys to hospital. In this case, most passengers are likely to be hospital visitors rather than patients. Similar arrangements are made on behalf of SSDs, where volunteer drivers are recruited by the WRVS or some other agency. This is sometimes called the 'social car service', but to avoid confusion with rural social car schemes this term is best not used. The social services may also call upon the HCS. Taxis are also hired for these purposes, but this is of course expensive. Car transport for medical or welfare purposes might also be organized on an area-wide basis by a Volunteer Bureau or a Council of Social/Voluntary Service. The main problem confronting these and other voluntary sector organizations is the extent to which they attempt to fulfil a wider social role by offering 'general mobility' in addition to 'institutional' journeys, in which case they might risk losing their funding from the public sector (Sutton 1987). Further, many voluntary associations exist which provide car transport of a 'welfare' nature for their members, not referred to them by SSDs - taking elderly people to a social club is a common example, which might be done by local Red Cross branch, Senior Citizens' or Good Neighbours' Club.

As the HCS and SSD voluntary cars are not available to the general public, it is arguable whether they should be counted as UCMs and therefore no detailed analysis will be attempted. In England alone, the HCS carries about 3.5 million passengers annually, which is about 16% of patients using authorised transport. The high cost per passenger found with respect to social car schemes (section 11.4) is of less significance when compared to the even higher cost of ambulances. Some health authorities of a largely rural character show high vehicle mileage per patient, but costs per patient remain about average due to their greater dependence on HCS (Bailey & Layzell 1983). To take one example, the East Sussex HCS in 1980 accounted for 41% of patients carried and 56% of health transport mileage, but only 11% of costs. Nevertheless, there is still much under-capacity and dead mileage, and this has been brought into the debate on the efficiency of health and welfare transport services. The fact that HCS has been administered separately from the ambulance service has led to accusations of waste, and many health authorities are making savings by coordinating HCS with ambulance control.

(c) **Shared 'Green' Cars**
Other ways of sharing cars are theoretically possible, which if engaged in by large numbers of people could potentially show dividends in reduction of traffic, accidents, pollution, energy consumption, etc, but which in reality are likely to remain too rare to make any impact. Shared cars, sometimes called 'Green Cars' in the UK due to their attraction for the conservation movement, depend on the sharing of ownership as well as usage. Sharing car ownership means sharing the cost, which may mean ownership could be extended to those at present unable to afford it. The responsibilities of ownership, such as maintenance, could also be shared, and less private parking space is needed. It is also likely to produce behavioural changes, such as more sparing use of the vehicle by any one individual or household, and use of other modes when appropriate.

Shared car schemes could operate in various ways, but there are two basic levels - the technological and the informal. At the technological level, one might envisage a fleet or pool of cars owned in common by the members of a club in, for example, a residential neighbourhood. Users would be able to take any vehicle available when needed, and usage would be recorded by a meter in the car which logged the mileage, computed the cost, and identified the driver for subsequent charging. No advance booking would be needed. This 'pay as you drive' system would eliminate the distinction between perceived and actual costs and discourage inessential use. The technology exists for such a system, but a plan to mount a demonstration project in a part of London had unfortunately to be aborted on financial grounds (Pharoah 1987).

Alternatively, cars could be shared on a small scale, informal basis, with a single vehicle jointly owned and run by two or more households. Inclusion of more vehicles and households is conceivable, but the maximum feasible size of such a scheme would be quite low. An essential pre-requisite is for participating members to cooperate fully with each other and to show great tolerance and a public spirited attitude. Households with complementary lifestyles would make the most compatible members, e.g. by needing the car at different times of day. Cousins (1979) explained how this might work. Capital and recurrent costs would be paid according to use on a mileage basis, recorded by a log book or perhaps a meter. Some kind of informal booking system would be necessary, at least for non-regular requirements. Joint ownership would have to be cleared with insurance companies. A formal agreement is desirable, to clarify mutual obligations and financial arrangements.

Pharoah (1987) claims there are "at least 17 Green Cars operating in the UK more or less successfully". There is very little information about these. The largest seems to be the 'Inter-Action Shared Car Fleet' reported in Cousins (1979). This was run by a charitable trust operating a cooperative in Kentish Town, London. Its 40-45 members were able to share the use of eight or nine vans and four cars. Vehicles were purchased second-hand when required. They were available for private or Inter-Action business use, and controlled by a simple booking chart. Remarkably, all journeys within London were free to the users, but the scheme was believed to work well, depending on mutual trust among members.

Chapter 12

SHARED TAXIS AND HIRE-CARS

12.1 SHARED TAXIS AND HIRE-CARS UP TO 1985

(a) Operations

Passenger sharing of taxis and hire-cars is one of the most recent UCMs to develop, as the mode was not fully legalized until the 1985 Transport Act. Its 'unconventional' status depends on its being contrary to traditional British practice, and normally illegal, without special dispensation, up to 1986. Taxis in general, as they date back to the horse-cab era, must be regarded as 'conventional', based on the exclusive hire of the vehicle by a single customer. It should be noted that in the USA and elsewhere in the world taxis are regarded as a 'paratransit' mode, in that they are halfway between public and private transport. Taxi sharing is also normal in other countries, especially in the Third World and in many parts of the USA under varying conditions ('jitneys' originated in the USA in 1914).

A taxi is distinguished by its ability to "ply for hire" by which it is allowed to "stand" at likely places and solicit custom, and it may be hailed on the street. For this privilege a hackney carriage licence is required, issued by local authorities and sometimes by the police. Traditionally, taxi operation was controlled by District Councils under powers dating back to the Town Police Clauses Act 1847; bye-laws could be made subject to Home Office approval. This was extended by the Local Government (Miscellaneous Provisions) Act 1976 and the Civic Government (Scotland) Act 1982. A third (rarely used) form of control is a Local Act. A hire-car, by contrast, is unable to ply for hire or solicit custom in any way, but has to be booked in advance through an office or agency. This obviously differs from the 'self-drive' hired (rented) car. Terminology is often confused by inaccurate usage and local custom. In some areas, hire-cars are known by the unhelpful term 'minicabs'. Commonly, hire-cars call themselves 'taxis', so that it is difficult to distinguish the two modes operating in the same area. Hire-cars *may* be regulated by local authorities by the means cited above, but may still operate in places without regulation; taxis on the other hand may not ply for hire without licensing. There has always been great variation in the extent to which local authorities license vehicles, or drivers, over whole or part of their jurisdiction. In 1980 about 80% of district councils had some kind of licensing for taxis and about 60% for hire-cars. The cumulation of all relevant legislation as adopted in any area is known as the 'Taxi Code' (Open Tech Transport Project 1986).

Traditionally, taxi and hire-car operation was based on the principle of exclusive hire, under which groups may travel collectively on the same journey for one fare. It was gradually realized that although these modes had great potential in areas that were 'difficult' for ordinary public transport, their flexibility and convenience were outweighed by their high cost to the user. They were perceived as a rich man's mode and as a town mode. Many social and accessibility problems might be solved if they were made available to a wider market, which means reducing cost to the passenger. On a commercial basis, this could only be done by allowing passengers making individual journeys to share the vehicle at separate fares. This had not been authorized in the past presumably to protect the bus industry from 'unfair' competition; deliberate market segmentation is implied. Taxi sharing could be accomplished in several ways.

(i) *By operating like a bus, on regular routes and times*, in which case passengers could board and alight at any place without pre-booking (in this case taxis surrender their time-space flexibility).

(ii) *By operating regular routes on a shuttle or 'fill up and go' basis* (flexible in time but not in space, as on Third World 'jitneys').

(iii) By an office coordinator *matching telephone requests for journeys* and arranging routes and times on a 'best fit' basis to suit the passengers

(suitable in areas of lower demand, flexible, but passengers have to compromise).
(iv) Having received an advance booking, *by advertising spare seats on the journey to other people* for whom the route and timing might be convenient (suitable in areas of lower demand, no flexibility for later passengers).

Variations might be conceived: (i) and (ii) might have restricted stopping places or D/R diversions from the regular route.

Except in Northern Ireland (see next section), it is doubtful if there were any shared taxis prior to the RUTEX programme in 1977-9. The only candidate was Dennis Radio Taxis of Maidstone, which operated a dial-a-ride facility from 1972 and which was later developed into a full DaR minibus service (chapter 3). It is conceivable that shared taxis could have operated at this early date by licensing as a bus, by taking advantage of the small bus permit provisions of the 1968 Transport Act. Nevertheless, the key to full experimentation was the Passenger Vehicles (Experimental Areas) Act 1977 which allowed the minister to relax licensing regulations in designated areas, preparatory to the RUTEX programme. Four shared hire-car services were inaugurated in rural areas under RUTEX.

In Devon, the '<u>Taw Valley Car Service</u>' was designed to cater for residual demands in a large rural area around Chulmleigh (Devon Rutex 1980a). This zone straddled the Exeter-Barnstaple railway line, while contemporary bus services were complex, fragmented and irregular. The experiment was intended to complement these with essentially a feeder service to existing routes plus a supplementary local service into Chulmleigh. The scheme was to be based on commercially operated shared hire-cars, one in each of the sub-zones into which the area was divided, but insufficient local operators were found and one part of the area remained unserved. For each zone, a schedule of timed connections was advertised, which intending passengers would pre-book, and routeing would then be flexible to allow doorstep pick-up and return. Additional workings were provided into Chulmleigh, without onward connections. Although the system worked satisfactorily, patronage was low with little new trip generation or passenger sharing of the vehicle, and consequently the financial performance was also poor.

A similar but smaller scale experiment was mounted in the Scottish RUTEX area. The experimental zone was the single parish of Stair, east of Ayr, which was surrounded by small towns and main roads but had no internal public transport or even any shops (Scottish Rutex 1980). Anticipating very low demand, a hire-car firm was contracted to run connecting services to three peripheral villages on two days per week. Routeing was demand-responsive, depending on advance booking, and timing determined only by the schedules of connecting buses. Patronage however was only about one third even of the low level anticipated; vehicle utilization and degree of sharing were low, and revenues fell far short of costs.

Two shared hire-car services were attempted under RUTEX in North Yorkshire. The '<u>Colsterdale Car Service</u>' aimed to link a remote valley on the edge of the Pennines with the small town of Masham where there were bus connections to Ripon and Bedale (North Yorkshire Rutex 1980b). Again, operation was D/R, depending on pre-booked passengers, subject only to arrival times in Masham to meet selected buses. In the first phase a daily service was offered, but this was later extended to allow a choice of return journeys from Masham and connect with another weekly bus. Patronage was generally low but doubled in phase two; car sharing remained very low. Financial performance was very poor.

Rather different was the '<u>Northallerton Hospital Transport Service</u>'. In this case, shared hire-cars were used to transport visitors and out-patients to three hospitals in Northallerton, originating from two large rural sectors of the catchment area to the south and west of the town (North Yorkshire Rutex 1980a). Some conventional bus services were available, but journey times were long and did not suitably match visiting hours, and remoter parts were unserved. From each sector, two daily services were offered for outpatients and a Monday-Thursday evening service for visiting. Four local operators were used, and because of this a central coordinator was required. The schedule was based upon projected arrival times at the hospitals, with routeing and timing dependent upon advance bookings from passengers. It was aimed to avoid conflict with bus services. Benefits were widely spread but demand was lower than expected, leading to high subsidies.

Three of the four RUTEX shared hire-cars were terminated immediately after the experimental period, and only the Colsterdale scheme survives. The operations all worked well, but all were in areas of high car ownership where lift-giving was common, and people were reluctant to change established habits. Low and dispersed demand did not encourage high rates of passenger sharing. Results cast severe doubt on the usefulness of shared hire-cars in rural areas, which would have been impossible without RUTEX subsidies, and suggested that such environments might be better served by volunteer-run social car schemes.

Despite this disappointing experience, the impression remained that taxis and hire-cars were an underdeveloped resource with potential for expansion. Removal of certain legal barriers would both help the industry and extend a flexible mode of transport into territory or circumstances that had hitherto posed problems. Thus, the TRRL was encouraged to mount a research programme on the taxi and hire-car industry which produced a spate of reports. A basic survey was done throughout England and Wales in 1980 (Coe & Jackson 1981). Detailed case studies were done of taxis in Northampton and Basingstoke. A review of research up to 1983 was then presented (Jackson 1984), and the possibilities for taxi sharing were re-evaluated (Greening & Jackson 1983). Finally, the 1980 survey was repeated five years on, as a benchmark before deregulation (Coe 1986).

The main result of this research was to refute preconceptions that taxis were a minority mode used mainly by the wealthier classes. In England and Wales outside London, the number of licensed taxis had increased by 37% between 1972 and 1980, and by a further 11% to 1985, to give a total of nearly 19,000 vehicles. For private hire-cars the rates of increase were 46% between 1972 and 1980, and 33% to 1985, when there were almost 26,000 vehicles. Hire-cars show a steeper increase but are more likely to fluctuate with cycles in the economy. In addition, there were estimated to be about 4500 unlicensed hire-cars. Such an expansion, clearly contrary to trends in other public transport, indicates new markets have been found. Taxis have become relatively more affordable, with fares falling in real terms between 1961 and 1979, in sharp contrast with bus fares, although this pattern has been reversed since 1980. User surveys show there is little restriction on the range of passenger characteristics and journey purposes, and hence taxis and hire-cars can be seen as an integral part of the public transport system.

Sharing of taxis, however, had not proceeded very far by 1985, as shown in figure 12.1, suggesting that the general expansion of the mode might be even greater if legal barriers to shared use were removed. Until the 1985 Act, passenger sharing of taxis and hire-cars at separate fares was possible by two mechanisms.

(i) By continued use of ministerial designations under the Passenger Vehicles (Experimental Areas) Act 1977, later incorporated into the Public Passenger Vehicles Act 1981, and with the agreement of the local authority.
(ii) By operating the vehicle as a PSV, and obtaining a road service licence for the route (i.e. as if it were a bus).

This latter course was made easier by provisions of the Public Passenger Vehicles Act 1981 for a new 'restricted' PSV licence, under which taxis and hire-cars could be run as PSVs and minibuses could be run by taxi firms under certain circumstances, e.g. for contract hire work.

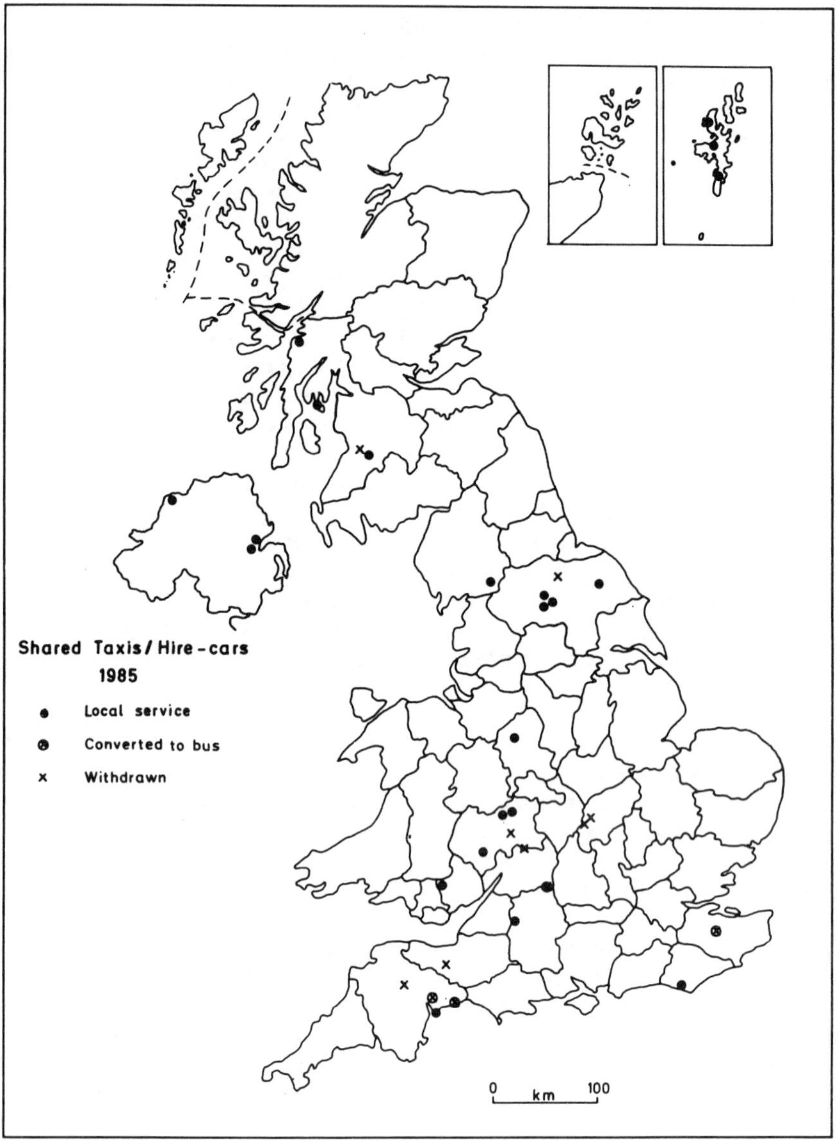

FIGURE 12.1 Shared taxis/hire-cars 1985.

Unhelpful restrictions remained on sharing, such that the arrangement had to be initiated by the passenger and not by the driver, and that it should not be advertised except with local authority approval as meeting the social or welfare needs of local communities.

Schemes mapped in figure 12.1 display very mixed fortunes. If there is any local authority showing a particular predilection for shared taxis, it is North Yorkshire. After RUTEX, the Colsterdale scheme was made more viable by expanding to replace a conventional bus service to two villages. By reducing subsidy for the latter, the shared hire-car was made to appear more acceptable as its deficit was not large in absolute terms. Folllowing this, three more were introduced in 1983/4: from Kirkby Malzeard (to Ripon), in Nidderdale (to Pateley Bridge), and from villages north of Pickering. Hereford and Worcester county has also seen some experimentation, although with mixed results. Two shared taxis had been withdrawn by 1979, but three later ones survived: at Stourport and at villages nearby, and south of Hereford. Figure 12.1 might suggest a high failure rate, but the variation in fortunes was extremely wide with some falling victim to their own success and being replaced by buses. Despite the failure of the Taw Valley RUTEX scheme, two other services in Devon were successful and converted to PSV minibuses. This also happened to a service in Gloucestershire, but two in Northamptonshire were discontinued. In Scotland, Strathclyde introduced a few shared taxis to solve specific local problems, while the mode has also found favour in the Shetland Islands.

Amongst the rural ventures, a distinctive scheme is that at Dent (Cumbria). Using the experimental areas legislation, this worked by advertising the number of spare seats on a journey which had been booked in advance, and then dividing the fare by the number of passengers. Hence people would make private arrangements to share, or use the operator as broker. Most journeys were to Kendal or to Oxenholme for rail connections, but more distant destinations were also authorized. In 1981-4 it carried almost 200 passengers per year with just over one trip per week, achieving much higher levels of sharing than other rural schemes; the council paid no subsidy but only publicity costs.

In common with most UCMs, shared taxis were first seen as expedients to tackle problems of rural environments, and legislation proceeded accordingly. Early results were variable but suggested that shared use would be more likely to succeed in higher density environments, as well as offering a better commercial prospect for the operator. Excluding Northern Ireland, urban shared taxis were rare before 1986. Taxis had been run as buses in the Stoke-on-Trent area by Potteries Motor Traction (chapter 5). Other town services were at Exmouth, Tredegar and Polegate (East Sussex); the latter, with another at Peacehaven, were later incorporated into the ESCORT project (chapter 4).

The best documented urban shared taxi scheme is that at Stourport (Hereford & Worcester). Started in 1982 on the initiative of a local taxi proprietor, this takes the form of two fixed routes, each covered by a single vehicle at times governed by passenger demand. Described as a 'taxibus', this is sometimes claimed to have been the first urban shared taxi service in Britain (Coe 1985). It was operated under PSV and road service licences with local authority support, as the routes agreed were not well served by buses. One route was within the town itself and the other was to the nearby town of Bewdley (5 km). In each case, bookings were made in advance on the understanding that the greater the notice given, the greater the chance of sharing at a reduced fare. Passengers could be picked up and set down anywhere on the route. Hence the service is as demand-responsive as it could be, consistent with the need to attract shared use. One route was used for school transport with pupils sharing in groups of four. After a TRRL survey and a fares revision, patronage continued at a satisfactory level to produce a 80% operating ratio with a modest level of subsidy. Passenger sharing caused no problems but the average car occupancy rates were little higher than for ordinary (exclusive hire) taxis.

(b) **Patronage and Economics**
Most statistical information on shared taxi trip generation and economic performance comes from TRRL reports, and this constrains the sample of services examined. Available data are summarised on table 12.1 in which the RUTEX schemes predominate. Certainly, there is little evidence from these four schemes that shared taxis and hire-cars are an economical and realistic means of solving rural transport problems, unless this performance is better than any conceivable alternative. It is interesting that the large size of the Northallerton scheme produces a high trip rate despite being restricted to a single journey purpose (to hospitals). The large Taw Valley catchment area produced a low trip rate. Generally, the number of passenger trips per week are of the same level as those on social car schemes (chapter 11) and in some cases appear to be lower. Unless there is no pool of voluntary labour in the areas concerned, it is difficult to see a suitable role for shared hire-cars in rural zones.

Low trip generation rates inevitably circumscribe the possibility of passengers' sharing the vehicle. Car occupancy rates, meaning the average number of passengers per car outing, rarely exceed the 'normal' figure of 1.7 for exclusive hire taxis, although the latter would be an urban-biased result. Theoretically, a more valid measure of car occupancy would be to calculate the ratio of total passenger-kilometres travelled to the total passenger seat-kilometres operated (i.e. taking into account the number of spare seats and distance covered). For the RUTEX schemes, this worked out at only 0.10-0.25 (Watts 1980). This severe under-capacity is due not only to lack of sharing but also to the very high rates of dead mileage incurred. Especially in rural areas, hire-car operators may be located at some distance from the origin and destination points of their customers, so that more than half of the round trip distance may be spent running empty to

TABLE 12.1 Shared taxi/hire-car services: patronage and costs (source: Watts 1980, Jackson 1984, Coe 1985).

	date of information	passenger trips per week	car occupancy (degree of sharing)	dead mileage %	total cost per week £	subsidy per week £	operating ratio £
Taw Valley RUTEX (3 zones, phase 2)	1978-9	2.4-11.0	1.4-1.7	50-68	5.30-17.59	4.39-14.92	4.4-17.2
Stair. RUTEX	1978-9	3.1	1.1	82	14.92	14.22	4.7
Colsterdale. RUTEX (phase 1 & 2)	1978-9	2.5-5.0	1.3-1.7	50	1.46-5.43	0.93-4.29	19.8-33.3
Northallerton hospital service. RUTEX	1978-9	30.1	1.6	33	86.71	67.69	21.9
Dent	1981-4	1.2	3.2				
Stourport (phase 2)	1983-4	c.90	1.8		c.50	c.10	c.80

and from base. Finding suitable operators has been a particular problem for rural schemes.

Despite an extremely low demand in the Dent area (Cumbria), an impressively high car occupancy rate was achieved, implying a concentration of the demand on to a minimum number of car trips. Hence the *modus operandi* of this service (explained in the above section) appears to be cost-effective, although it might need greater organisational and advertising effort.

A customer profile is difficult to determine as two of the four RUTEX services were used by a small number of individuals, and one other was exclusively for hospital trips. Nevertheless, the users seemed typical of other rural UCMs, being dominated by women and the elderly, mainly for shopping or health-related purposes. In the absence of these services, most people would have sought a lift or postponed the journey.

Financial results of the RUTEX schemes are uniformly poor, as expressed by the low proportion of costs recovered from revenue, reflecting the problems described above. Cost per passenger trip varied between £0.57 and £4.81 (1978/9 prices). The lower end of this range compares reasonably with social car schemes (cf. table 11.3), which is perhaps surprising given the commercial nature of hire-car firms. If the operator is required to keep a vehicle on stand-by at certain times for a possible call-out, then a guaranteed minimum revenue might be required (included in subsidy under RUTEX contracts). VAT is payable on taxi hires, unlike public transport fares. Costs are a problem for this mode in rural environments, but it was noted that in absolute terms the deficits were not too great, especially if a conventional bus had been replaced (Watts 1980).

A comparison between the costs of shared taxis and conventional buses was made by Greening & Jackson (1983), to find out under what circumstances the former might be justified. The main conclusion was that where demand is low it would be more cost-effective to provide a service with up to three shared taxis than with a conventional bus. This is not sustained when higher demand levels require a substitution ratio of more than three to one. Above this level taxis would become increasingly unfavourable in terms of fuel consumption and the effects of road congestion would come into play. Taxi fares at break-even levels would be higher than on buses in all environments.

All these results suggest that shared taxis and hire-cars are more likely to find a competitve niche in areas of higher demand than in most rural districts but avoiding city centres. Small town, suburban, or off-peak urban environments might be more promising. The Stourport example is the only urban shared taxi service to have disclosed operational data. That given on table 12.1 reveals some success in economic terms although the degree of sharing is unimpressive. It is highly significant that since deregulation most of the new 'taxibus' services are urban, but no operational data have been forthcoming.

12.2 NORTHERN IRELAND

(a) Operations

It is in the field of shared taxi operations that Northern Ireland makes its greatest contribution to the UCM sector, essentially through the 'Black Taxi' system of Belfast. For various reasons UCM and CT development has been very weak in Northern Ireland, and the shared taxis comprise the vast majority of the 'unconventional' activity there has been. In the absence of any legislative provision or official encouragement, they are an example of local people taking matters into their own hands to compensate for perceived inadequacies in public transport. They began much earlier and totally independently of any interest in shared taxis in Great Britain; the operation is larger by far (375 vehicles) than anything contemplated elsewhere, and they have achieved widespread fame outside the British Isles. The negative side is that they have been unable to attain full respectability due to their association with criminal activity in Belfast and their dubious legal status.

Such an enterprise could only originate in the unique circumstances of Northern Ireland in the early 1970s. The development and workings of this phenomenon are described by Lavery (1986a,b). Conventional urban bus services in Belfast were forced to endure intolerable conditions amidst civil unrest and frequent street disturbances. Buses were often attacked, or taken over to be set on fire and used as barricades; at other times, drivers were intimidated. Regular services were frequently suspended or re-routed such that large areas of the city were deprived of satisfactory public transport for long periods. In these circumstances, residents started to carry neighbours into town in private vehicles, and this developed into a semi-organized shuttle service which could perhaps be described as 'jitneys'. This changed to its present recognizable form when taxis were acquired from the second-hand market in London for use in Belfast, and organizations were set up to supervise the operations. Austin FX4 taxis of the type familiar in London but highly distinctive in Belfast give the system its popular name of 'black taxis'. Inevitably, organization mirrored the highly polarized state of community relations, with services running along particular roads to specific housing estates identifiable with each of the segregated social/religious groups. The first association of owner-drivers was formed in 1971 to connect catholic/nationalist areas of the inner city and outer suburbs with the city centre; originally the 'Falls Taxi Association', this is now known as the 'West Belfast Taxi Association'. In the following year a corresponding group was formed to serve protestant/loyalist areas in the north-western sector of the city; first called 'Shankill Taxis', this is now the 'North Belfast Mutual Association'. This original structure has survived to the present, and the associations feel confident enough to consider developing services further.

The service pattern has remained fairly stable over a 15-year period, and routes existing in 1985 are plotted on figure 12.2. The mutually-exclusive nature of route patterns and catchment areas between the two organizations is

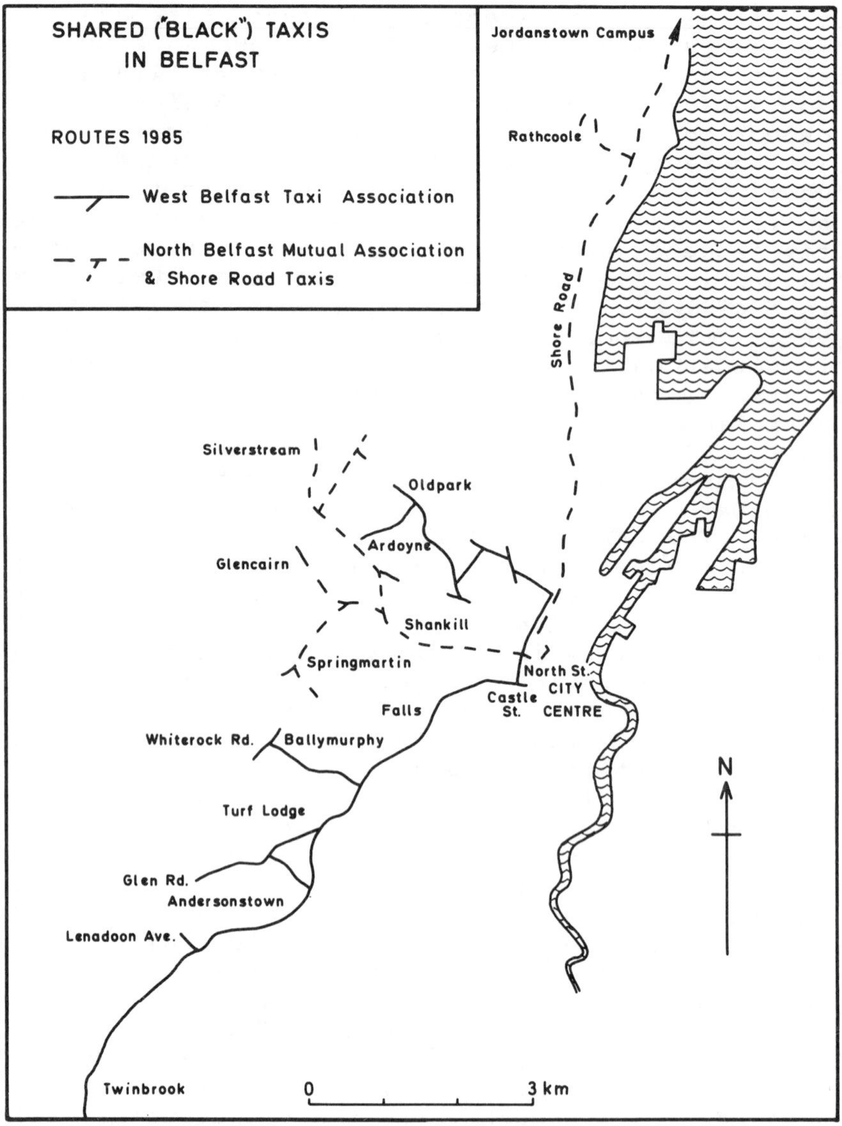

FIGURE 12.2 The 'Black Taxis' of Belfast 1985 (after Lavery: 1986a, b)

apparent. Routes are basically radial, serving high density, public sector housing estates characterized by high levels of unemployment and social deprivation and low rates of car ownership. They focus on informal termini at the edge of the central shopping zone. Although the problems experienced by buses in the early 1970s have receded, they have lost much ground to the taxis and the main operator - Citybus - now competes at a severe disadvantage along the arterial routes. The long Shore Road route to the large estate of Rathcoole and the University of Ulster campus at Jordanstown is operated by a subsidiary of the North Belfast association (figure 12.2). A new development was the Ardoyne route which started in 1985, and differs from the others in taking a devious route via minor roads to give a door-to-door service.

Black Taxi services are flexible in time but not in space. Routes are relatively fixed, taking the form of arterial routes (along the Falls and Shankill Roads) with various branches into the housing estates. Passengers wait at traditional places acting as termini at the outward ends of routes. Taxis stand and depart with unlimited frequency on the 'fill up and go' principle. The maximum capacity is six passengers. Services are available between 0600 and 0200 hrs, and during the daytime the average wait is no longer than five minutes. Minor deviations may be made on the inbound trip to pick up groups of people at pre-arranged times and places. Each association has its own traditional street as a city centre terminus, which is usually crowded with waiting taxis and passengers. Taxis carry destination boards on their windscreens and passengers queue up at a 'stop', just like a bus. The congestion caused is disapproved of by shopkeepers, police and the city council, but over the years a reluctant toleration has been maintained. Recently a few people have been employed as taxi dispatchers. On the return, outbound trip, taxis might deviate from the route to drop certain people, such as the elderly, at their doors.

Originally, the governing taxi associations were conceived as workers' cooperatives and are run by elected management committees. They assist drivers to purchase second-hand vehicles which are then run solely by their owners. Each association operates a monopoly within its own areas, strictly controlling the number of drivers and not allowing any freelance intruders. Each driver is free to operate as he sees fit, but in cooperating with other drivers, the system is self-regulating. The association does not impose central control on operations, although many cabs have citizens' band radios for inter-communication about passenger loads, traffic, etc. The West Belfast association has about 275 taxis and its North Belfast counterpart roughly 100. As well as attesting to the popularity of the services, the taxis have considerable employment potential in areas where jobs are desperately needed. The associations have several hundred expectant drivers on their waiting lists.

Despite their survival over the years, the taxis have never been able to discard the stigma of involvement in criminal or paramilitary activity. They serve areas of the city normally regarded as sectarian ghettoes, and are widely believed to be subject to the influence of paramilitary groups. Some drivers have been murdered, and cabs have sometimes been involved in terrorist incidents. There is a belief that revenues are siphoned off to paramilitaries, and rumours circulate about connections with racketeering, welfare benefit fraud, etc. Apart from this, shared taxi operation appears to have no legal basis in Northern Ireland. As in Great Britain before 1985, there was no law to permit it. Various public authorities, and the Citybus company which suffers from competition, are convinced the taxis are illegal, but no test case has been mounted. In the 1970s there were attempts by the police to remove them on the grounds that they were unsafe and uninsured. In 1978 the public inquiry report of the Belfast Urban Area Plan recommended that black taxis should be "eliminated" due to their competition with bus services. Citybus claims to lose £2 million in revenue to the taxis each year, and this is probably an underestimate.

Since about 1980 the Department of the Environment for Northern Ireland has been more willing to tolerate the shared taxis as some degree of regulation has been imposed. Under the Road Traffic (Northern Ireland) Order 1981, owner-drivers are required to obtain PSV and hackney carriage licences; roadside spot-checks are frequent, and insurance is rigidly enforced. In 1986 the DoE(NI) revised taxi licensing by issuing green discs for 'public' taxis and yellow discs for private hire taxis (the term 'hire-car' is not used in Northern Ireland). This allows drivers with unsatisfactory records to be denied licences. Contrary to popular opinion, this did not fully legalize the Belfast operations as there is still no clarification on the issue of passenger sharing at separate fares. Northern Ireland was excluded from the 1985 Transport Act which authorized such services on the British mainland, and hence the largest and most long-established system in the UK is still under a legal cloud - an irony which is well appreciated.

Despite remaining doubts in some quarters, the shared taxis are clearly a permanent feature of public transport in Belfast and must be taken seriously. It is reasonable to believe that most of the operation is legitimate; it now carries up to 15% of all public transport trips in the city, and has greatly enhanced the mobility levels of disadvantaged population groups. While they have had a negative impact on buses in most parts of the city, it is interesting to note that on the Shore Road route to Rathcoole and the University campus (figure 12.2) the Ulsterbus company responded to competition by improving services. This seems to accord with the spirit of deregulation exhorted on the other side of the Irish Sea. That ambivalent attitudes still exist however, is attested by the complete omission of shared taxis from the latest review of the Belfast Urban Area Plan (Lavery 1988).

It is less widely known that shared taxis are not confined to Belfast but are widespread throughout Northern Ireland. 'Black' taxis on the Belfast model also operate in Londonderry, being set up in 1981 as a job creation venture, but run by a subsidiary of the West Belfast Taxi Association known as 'Derry Taxis'. With 37 vehicles, there are three basic routes serving catholic/nationalist housing estates, running into a city centre terminus. These are in direct competition with Ulsterbus services, which have not started from the disadvantages experienced in Belfast of being forced off the streets by rioting, and are therefore able to take effective counter-measures. Bus services have been improved in response, so that frequencies and fares are comparable to the taxis and comfort is superior; the taxis are quicker and more easily serve residential streets. Londonderry taxis are also tainted with paramilitary connections. Lavery (1986c) reports the system to be inefficient with poor customer service, especially in comparison with Belfast, and forecasts its demise.

Elsewhere in Northern Ireland, taxis are saloon cars rather than London-type cabs, and most will allow informal passenger sharing and divide up the fare. Students have been found to be a lucrative market: 2000 one-way trips per day are made by taxi between the University of Ulster's Jordanstown campus and central Belfast. Similarly, local car operators ferry students from Portstewart to the University's Coleraine campus in competition with buses. An undesirable element in this case is the flagrant poaching of passengers from bus stops a few minutes ahead of the bus' arrival. If shared taxis are to become respectable, such practices should be stopped.

(b) Patronage and Economics

An estimated 130,000 people inhabit the wards served by the Belfast shared taxis. In a sample of wards surveyed by Lavery (1986a,b), car ownership rates varied from 50% down to a mere 15% of households (city average 42%). Male unemployment rates range from 18% to 56%. Hence the dependence upon a cheap and plentiful service can be envisaged. Total passenger numbers are difficult to estimate, but might be 14 million per year in all. This compares with almost 30 million on Citybus services, and thus the taxis constitute as much as 32% of *road* public transport journeys. An estimate from a different source is 15% of *all* public transport trips (Lavery 1986a,b, 1988). This astounding level is matched only in Third World cities. On specific taxi routes, the market share is much higher: on the Falls Road (West Belfast) perhaps about 70% of trips are by taxi and only 30% by bus, and on the Shankill Road (North Belfast) over 50% are by taxi.

Only limited user surveys have been done. The new Ardoyne service in 1985 presented an opportunity to gauge the increase in mobility: from an admittedly small sample, 41% said the taxis had increased their mobility standards, with the average increase being 400%. A survey on the Shankill Road found that 23% of passengers would have been unable to make the trip without the taxis, and a further 17% would make fewer trips. The social value of the operation cannot be in doubt.

Drivers are responsible for meeting their own costs, although vehicle purchase, insurance and maintenance may be facilitated by their associations. A second-hand FX4 taxi may be obtained for £1550 (in 1985), whereas new or rebuilt models would cost £10,000-£14,000. Typical yearly running costs for such a vehicle have been worked out at £6220, of which two-thirds is due to fuel. Insurance can be £600-£1000 p.a. Drivers are allowed to keep all their own earnings and may work as many hours as they wish. Fares are standardised to match those on the buses. A six-day week at 10-12 hours per day might produce a weekly income of £300. A crude calculation based on this produces a total income for the shared taxi business in Belfast of about £4 million in 1985 (Lavery 1986a,b).

12.3 SHARED TAXIS AND HIRE-CARS AFTER 1985

Under the 1985 Transport Act, the taxi and hire-car industries underwent a deregulation of their operating conditions at least as fundamental and far-reaching as on the buses. Taxi licensing was made mandatory for all district councils in England and Wales, although it is still optional in Scotland. Prior to this, taxi licensing was carried out by 85% of district councils in England and Wales, and hire-car licensing by 70%. These licensing authorities were no longer able to impose idiosyncratic conditions, and could not limit numbers unless they had evidence that there was no unmet demand (i.e. the abolition of quantity licensing as on the buses). Most significantly, taxis and hire-cars were no longer limited to the 'exclusive hire' tradition but could now permit passengers making separate journeys to share the vehicle, paying separate fares. Hence this extended to the whole country (but not Northern Ireland) what had previously been possible only by extremely cumbersome procedures (see section 12.1). Such a policy is clearly compatible with the ideological basis of deregulation in general, but was also founded upon the conclusions of recent research by the TRRL (Coe 1986) that the taxi market was expanding and no longer dependent upon the custom of the wealthy alone. The government saw taxis as an under-developed business, needing only the relaxation of rules to release "new opportunities" (Department of Transport 1984). Little notice was taken of the possibility that, as a mode with a higher quality of service, the lower fares permitted by sharing would present unfair competition for the buses.

Shared use of taxis and hire-cars under the new system is possible by three mechanisms.

(i) *As a general provision*, where all passengers have booked in advance, on an informal or occasional basis.
(ii) *Under a specific local authority scheme*, where taxis may be hired on a separate fares basis without advance booking from an 'authorised place', usually a designated taxi stand.
(iii) *With a 'special restricted' operator's licence*, whereby a taxi operator may run a 'local service' within the meaning of the Act, along the same lines as a stage bus.

The first two options could be referred to as 'shared ride' mode, and the last as 'taxibus' mode. Following the Stourport scheme and the Belfast system, the 'taxibus' model of a frequent service along a fixed route would seem to have the greatest chance of success. The Belfast experience has, perversely, had little or no impact on the British mainland, largely due to inaccurate perceptions of Northern Ireland. The other modes resemble the occasional demand or area-wide services that proved relatively unsuccessful in rural areas. This has been borne out in practice.

A related consideration was the potential of a newly liberated taxi industry to create employment. There were now few remaining barriers to new entrants to the field, and hence there was scope for existing firms to expand and for new taxi businesses to set up. At a time of high unemployment this seemed an ideal opportunity, for example, for a redundant manual worker with his own car to go into business on his own. Government publicity was aimed at encouraging this, and taxi licences were reported to be changing hands at exorbitant prices.

By May 1987 under a quarter of district councils in England and Wales had decided to limit taxi numbers (previously 88%), and 17% were licensing taxis for the first time. Hence a sharp increase in licensed vehicles was to be expected. Based on a sample of authorities, Finch (1988) estimated an increase in numbers of 10% between 1985 and 1986; prior to this the 'natural' rate of increase was about 2% per year. A TRRL survey of district councils revealed that in 40% of cases most licence applications were made from the taxi trade, in 36% most were from the private hire sector and in 4% most were from newcomers. Out of the taxi trade applications, 59% were from drivers wishing to start their own businesses and the rest from firms seeking expansion.

Arrangements for taxi sharing, however, were less popular than the simple expansion of firms and licensed vehicles. Initial reluctance to set up 'shared ride' schemes (option (ii) above) stimulated the DTp to attempt its own demonstration projects. A scheme was authorized in June 1987 at London Heathrow Airport based on a special rank for shared taxis, but quickly foundered on the antipathy of cab drivers and the indifference of the public. Similar schemes at Waterloo and Paddington railway stations met a similar fate. Despite this setback, the DTp (as licensing authority in London) then authorized shared rides from all taxi ranks in the capital.

Upon the implementation of deregulation in autumn 1986, the pattern of new local service registrations by taxi operators was as shown in figure 12.3. These were extracted from the TRRL's Bus Registration Index by Finch et al (1986). This list is not definitive and may be an underestimate due to some taxi operators being classed as PSVs and some being 'non-commercial'. Almost 60 services were registered at this date, in the name of 37 operators, two of which were associations of several independent firms. The map pattern clearly differs from that of 1985, pre-deregulation, shared taxi services (figure 12.1). There is a greater degree of clustering, with some firms offering various routes in the same locality, as at Stockton/Darlington, Northallerton, South Lancashire/Manchester, Crawley and Southend. They are very largely town-based, with 70% of mileage being urban. However, the uncertainty endemic among the bus sector after deregulation also pervaded the taxi industry, and a number of these registered services did not commence operation or were withdrawn. By May 1987 the TRRL database revealed 99 local services registered by 76 operators (England and Wales only). Remarkably, it was found on verification that only 44 of these were actually running! Reasons given for this varied from the inability to get licences to the over-estimation of demand, but it would seem that many new ventures were unprofessionally planned and marketed (Finch 1988).

The term 'taxibus' is now coming into common usage, regardless of whether a minibus or car is used, referring to the relatively fixed routes and regular departures. 77% of registered services were to be run by 4/5 seat saloon cars or taxis, and the remainder by small minibuses with 8 seats (sometimes called 'microbuses'). The preference for traditional bus-type services is shown by the

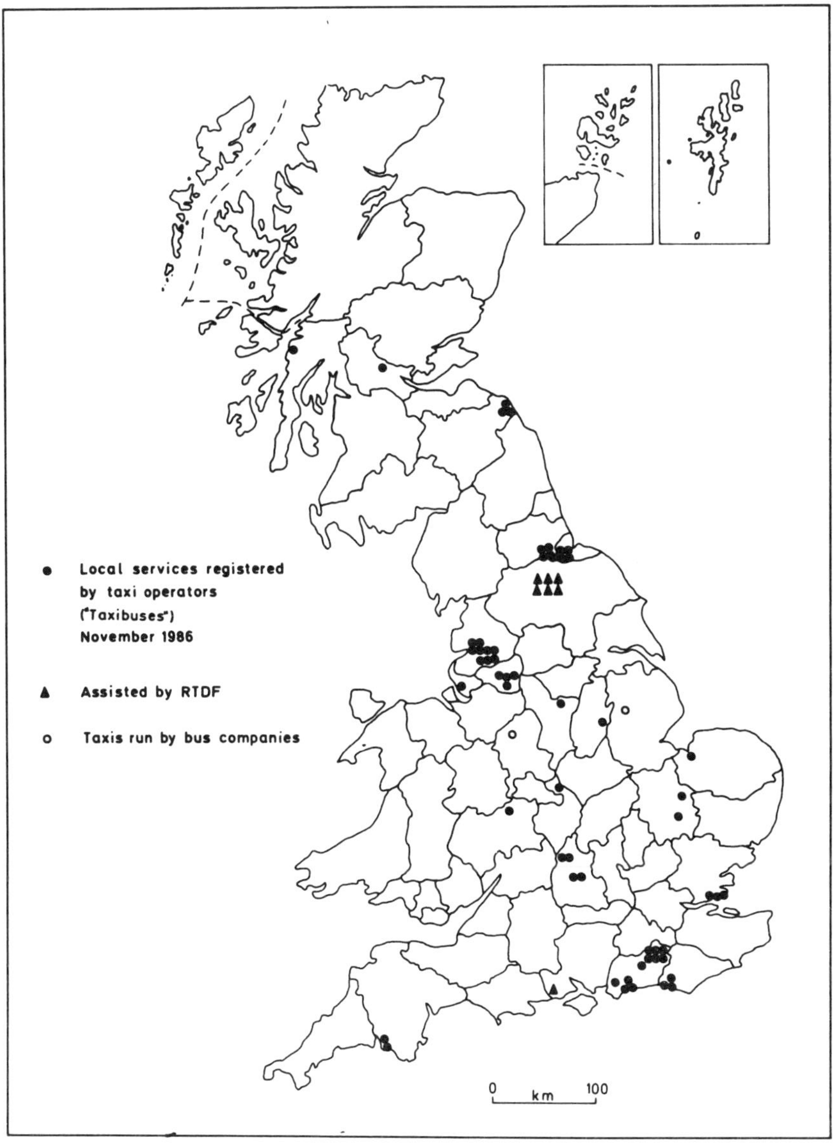

FIGURE 12.3 Post-deregulation 'taxibus' services 1986.

finding that 84% were planning to operate a fixed route only, and 12% with diversions from a fixed route. Taxis seem willing to sacrifice their normal advantage of D/R operation for more predictable, regular custom, but they can avoid the obligations of the traditional bus service in two ways. The first is the use of hail-and-ride (53%), and the other is to run for only part of the day (25%), or for specific journeys only such as the peak hours (38%), or a night service only (8%).

Rural shared taxi schemes may be making a comeback with the assistance of the Rural Transport Development Fund. Figure 12.3 shows only the first ventures at Northallerton (six services) and New Milton, Hampshire. One year on, by November 1987, there were 11 such grant-aided projects, whether licensed as taxis or as PSVs (see figure 14.2 below). These are spread throughout the country, with a minor cluster of three schemes in Cornwall. They tend to use 6-8 seat vehicles or larger minibuses, and are characterized by more diverse service patterns than in towns, being less dependent on regular routes but filling a multi-service role with D/R diversions or dial-a-ride. It remains to be seen whether these will outlast the RTDF.

In addition to the 99 commercially registered services in May 1987 there were also at least 14 known to be subsidized (Finch 1988). Most local authorities are willing to consider tenders from taxi operators, but find a lack of interest among the latter. Other councils believe taxibuses have insufficient capacity to be included in the tendering process.

The 'shared ride' mode, under which passengers may share a taxi for any journey if hired from an 'authorized place', has been unpopular and made no progress. By mid-1987 no local authority schemes of this sort were operational, as district councils were deferring consideration of the issue.

Finally, there is the complementary phenomenon of taxi vehicles being operated by bus companies - giving a new meaning to the term 'taxibus'. London-type FX4 taxis were acquired by the Potteries Motor Traction company in 1982, converted to carry seven passengers, and used as PSVs in the Stoke-on-Trent area under the name 'Flexibus' (section 5.1). The next best known scheme was effected by the Lincoln City bus company, which acquired 20 FX4s for use as small buses. From October 1986, the 'Lincoln Limos' have operated routes from the city centre bus station to various suburban termini. In mid-1987 they were reported to be carrying 5500 passengers per week. Such a trend may proceed to the extent of bus companies buying up taxi firms and integrating them into their own operations. The ex-NBC company Devon General, for example, has done this. Experiments may see the use of shared taxis or taxibuses perhaps in combination with minibuses (chapter 5) as a flexible response to different operating environments or passenger demands, or as feeders to corridor bus routes.

Chapter 13

THE RAILBUS

13.1 RAILBUS DEVELOPMENT

The railbus constitutes the only possible claim to a UCM that could be made by the rail sector. As the fixed nature of the rail network precludes any attempt at more flexible operation, and as 'community' involvement is limited to decorating local stations, then the 'unconventional' possibilities are confined to the technical field.

Railbus development can be seen in the context of two rather different but overlapping contexts. The first is the 'rural rail problem', which as a sub-set of the 'rural transport problem' contains the usual elements - small and dispersed demand, unprofitability - conducive to UCM solutions. The second is British Rail's major policy of 'DMU replacement', which aims to phase out the large fleet of diesel multiple units, now elderly, unreliable and uneconomic, in favour of more modern and efficient trains. DMUs, dating back in concept to the Modernization Plan of 1955, have become the standard stock for the least favoured parts of the network; BR had 3000 of them. They are not, however, confined to rural lines, but operate throughout BR's Provincial Sector - i.e. outside the Inter-City and South East commuter networks - and consequently run in a range of environments including urban. In the rural context, the problem may be seen in environmental and social terms (from the consumer's viewpoint), but the solutions offered are technical in nature.

Although a definition of a rural railway cannot be entirely unambiguous, it was estimated that in 1980 about 60 BR passenger services could be described as mainly rural in character (Hodge 1981). These accounted for about 1100 million passenger-kms (roughly 3.5% of the network total), and in 1980 attracted revenue of £36 million at an operating cost of £67 million (an operating ratio of 54%). Rural lines are not only catering for a minority market, but one that is particularly difficult to serve by a fixed inflexible axis like a railway line. Those presently existing are of course the relics of a much larger and denser network which has been gradually closed down over historical time, the main spate of closures occuring in the aftermath of the Beeching Report of 1963. Contraction followed by bus substitution has been the traditional response to low usage and high costs, but this is no longer acceptable as an automatic reaction to problem cases for various reasons, notably a greater awareness of the social value of rural lines (Hillman & Whalley 1980), the inadequacy of bus alternatives, the contribution of 'branches' to the 'trunk' network, and increasing popular protest. From 1970 very few stretches of line have actually been closed, and from 1974 successive transport ministers have reaffirmed their opposition to substantial cuts in the network. Nevertheless, this has not inspired confidence, and the political climate is such that rural lines appear to be constantly under threat. Closure scares flare up frequently (Whitelegg 1987).

In these circumstances the main question surrounding rural lines is how they can be kept going at minimal cost. Mere maintenance of the existing service would not be sufficient, as this sector was characterised by obsolete and deteriorating equipment (primarily the DMUs) and infrastructure, urgently in need of renewal. The policy had always been that new investment in such lines could not be justified economically (contrary to the view of rail supporters among the public that investment should be made as a deliberate policy of attracting new markets for rural lines, e.g. tourism). Infrastructure maintained on a grudging "make do and mend" basis would inevitably deteriorate to the point of restricting train operations and compromising safety standards (seen by some as part of a cynical ploy to force line closures by neglect). In 1980 it was estimated that up to 5000 out of 35,000 kms of track would be unusable by the end of the decade. Investment for infrastructure renewal was unavoidable, the issue being to identify the minimum standard required for safe and efficient operation at appropriate levels of service, and hence achieve a 'low cost rural railway' with simpler and cheaper technology (Hodge 1981).

Essential elements which could be replaced in a simpler form are (a) track - lighter rails and/or single line; (b) signalling - simplified systems possibly based on radio control of trains; (c) level crossings - automatic open crossings replacing barriers; (d) stations - basic and unstaffed (this has already been done); (e) rolling stock - modern, lightweight and cheaper to operate, replacing the DMUs. The concern of this chapter is with the trains themselves, but it should be noted these are part of an overall low cost system. There are interactions, for example: lightweight trains should prolong track life and eventually permit relaying with lighter rails; radio train control could be integrated with automatic level crossings; unstaffed stations require on-train fare collection, but one-man-operation could be sustained with some kind of automatic ticketing system.

To put this in perspective, improved rolling stock has played only a small part in the debate on the future of rural railways. More attention has been devoted to establishing appropriate financial arrangements, management structures, cooperation among organizations, improved marketing, and negative measures such as bus alternatives (e.g. Policy Studies Institute 1981). In the latter reference - conference proceedings - the term 'railbus' was applied to replacement or connecting bus services! A few years later, after railbuses had been introduced, a joint BR-local authority working party produced a report on the same subject in which only one recommendation out of 25 supported the use of modern lightweight rolling stock (Association of County Councils 1984).

Over the network as a whole, the DMU replacement policy had a false start. It was decided to refurbish the DMU fleet so that units from the late 1950s could continue into the 1980s and put off the evil day when a replacement would be needed. This breathing space was largely wasted with delays and technical problems, but eventually produced, as a first attempt, the Class 210. This was a diesel-electric multiple unit, designed to a high performance specification for commuter routes, but was too expensive and inappropriate for most Provincial Sector lines. It was the reaction to this which began the search for a lightweight design suitable for most routes including the rural, and resurrected the 'railbus' concept (Ford 1981, 1984).

The basic idea of producing a light rail car by mounting a bus body on a rail underframe is by no means new. There have been various designs of 'railbus' or 'railcar', sometimes exotic but usually unsuccessful, in service from time to time on the British rail system, the last of which had been withdrawn in 1968. Some of these are illustrated in Ford (1981). Numerous other examples can be found from railway history in Ireland and continental Europe. Better prospects for renewing the idea in the 1980s were offered by recent technical advances and the availability of mass-produced modular bus bodies.

There followed a spate of experimental designs for new trains, most in the early years adopting the railbus concept. The apparent confusion of types was partly a developmental progression, testing various facets of design and technology, but partly also due to some indecision over roles. Previous DMU types had been extremely versatile, working in many different environments, and the dilemma now was whether to produce a prototype capable of meeting most of these diverse demands, or several different models, such as 'heavy' and 'light' versions, each suited to its own niche in the market. The following types emerged (described in more detail in Hodge 1981, Ford 1981, 1984).

(a) LEV1 (Leyland Experimental Vehicle)
This was a railbus in its most literal form, with a single Leyland National bus body mounted on a simple rail underframe and a cab at each end (figure 13.1). It was produced in 1978 jointly by BR Research Division and Leyland Vehicles.

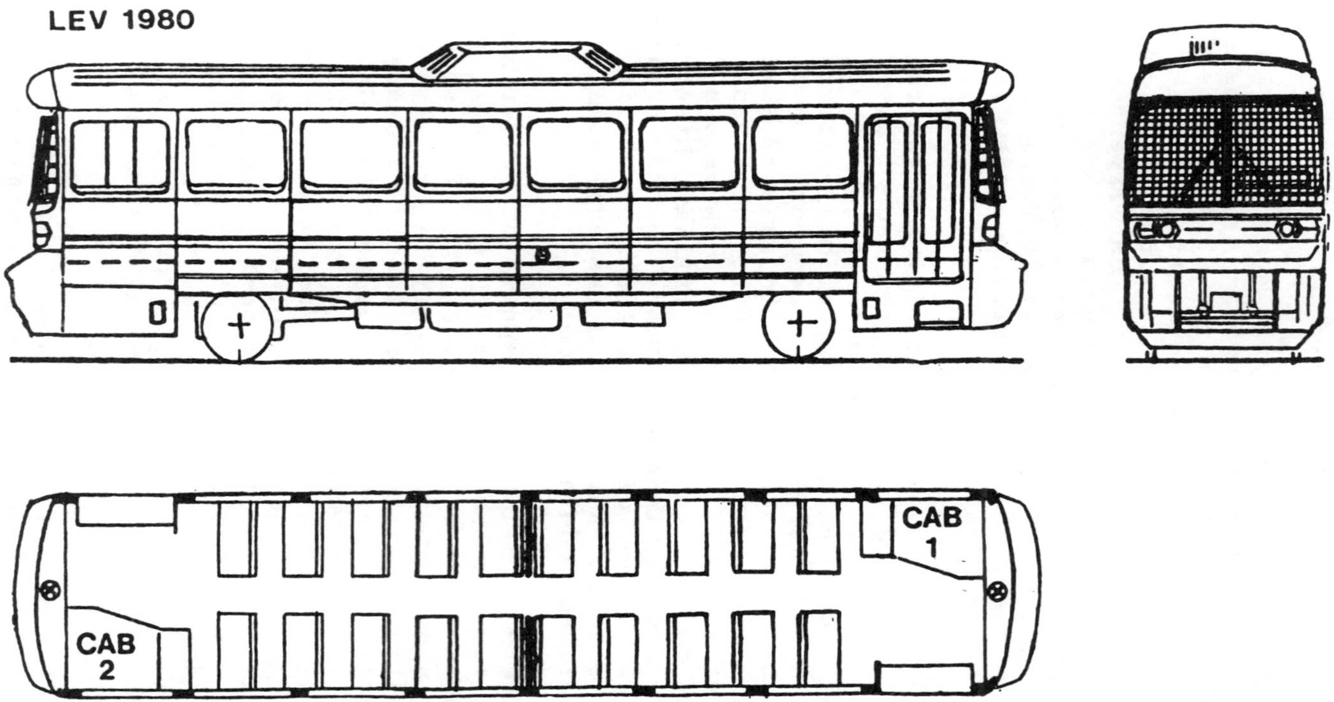

FIGURE 13.1 The Leyland Experimental Vehicle (railbus) (Hodge 1981, (c) British Rail).

Although research trials were undertaken, notably on the East Suffolk line, it was officially stated that its main purpose was to evaluate bus industry construction techniques, and that it was unlikely to be the eventual DMU replacement. LEV1 could hardly be taken seriously in this role, for although it was certainly cheap and light (16 tonnes against a two-car DMU's 58 tonnes), it was small (42 seats) and could not possibly cope with peak traffic even on rural lines. Passenger reactions during trials were favourable in respect of internal appearance, ride quality and noise, but probably this could only be sustained for journeys of up to two hours. Results were sufficiently positive to justify proceeding with a development programme for lightweight vehicles.

(b) **BRE-Leyland Railbus**
The next two models to appear were developed simultaneously but resulted in quite different products. This reflected some degree of rivalry between BR's research and engineering divisions, the former preferring the lighter railbus theme and the latter more traditional 'railway' design. Following LEV1, BR commissioned an extended unit based on the LEV concept. British Rail Engineering and Leyland Vehicles produced what was thought to be the definitive railbus, also known as 'R3', in 1981. This was a 56-seat vehicle, only 20 tonnes, with the required low fuel consumption and maintenance needs. Also regarded as merely a step in the development process, R3 did undergo passenger trials and market research showed that it scored more highly than a classic DMU, LEV and the Class 140, with respect to noise and ride quality. R3 was eventually sold to Northern Ireland Railways for service on two branch lines (see below). There was much interest in the export potential of such units, and a very similar model, sometimes known as LEV2, was sold to the United States for evaluation.

(c) **Class 140**
Intended to be a prototype lightweight train designed to BR business specifications for passenger service, and not as a mere experiment, the Class 140 turned out to be rather different from the railbus series. Although it used the same

modular body construction as LEV and R3, it was fitted with conventional BR front ends and driving cabs and had less resemblance to a bus. The body was strengthened, and other fittings were of 'rail' rather than 'bus' standard. The two-car unit carried 102 passengers at higher speeds, but it was more expensive and at 47 tonnes much heavier than the other prototypes. Passenger trials were conducted on a greater scale, on a variety of lines, but consumer reactions were much less favourable than for the lighter alternatives. In its defence, BR maintained that the 140 was not just for rural branch lines but more suited to the heavier duties formerly done by DMUs such as conurbation services.

The amount of testing of the three main products that had emerged by 1981 was variable, and not done in a manner allowing direct comparisons under equivalent conditions. A comprehensive demonstration project was planned for the East Suffolk line, which would have tested all the elements of the 'low cost rural railway' - infrastructure as well as train types (Hodge 1981). Although some improvements were made here, the integrated programme did not happen. Also, there might have been an opportunity to try out the new train technologies on re-opened lines, under a 1981 Amendment to the 1962 Transport Act which waived the normal closure procedure for experimental services. Other problems, however, prevented any use of this.

(d) Class 141

Criticism of the Class 140 led to a reversion to the 'railbus' concept, with the re-insertion of bus-type fittings. Heavy re-engineering of the National bus body was found to be no longer necessary. The resulting Class 141 is a two-car unit with a seating capacity of 94 (figure 13.2). Unlike its predecessors, this was not intended to be a prototype, as it was felt that sufficient experience had been gained with the R3 and 140 to design a new unit for production. Expenditure for the construction of 20 Class 141 train sets was authorized by the Department of Transport and these were delivered in 1983. As the first production model, this has come to be regarded as

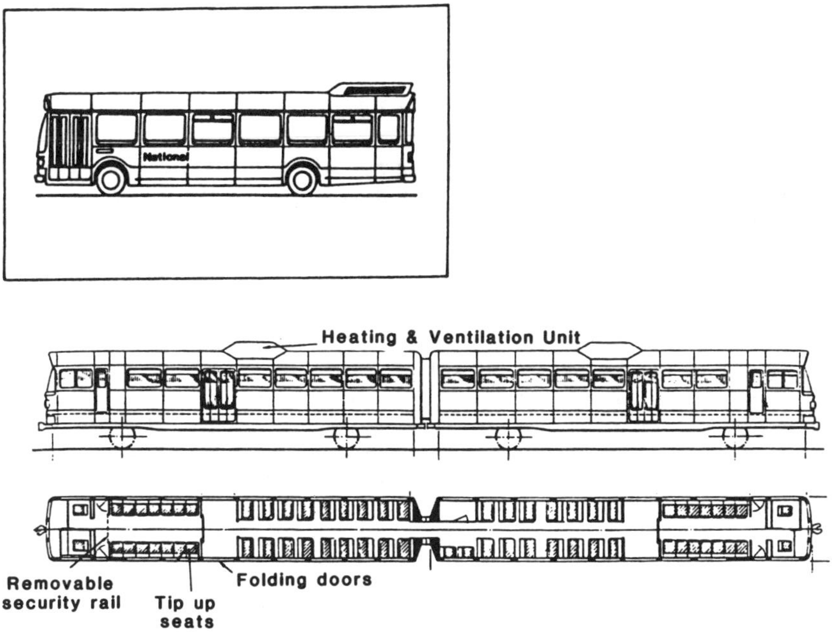

FIGURE 13.2 The Class 141 railbus.
(Inset) Leyland National (not to scale).
(Modern Railways, Jan 1983, (C) Ian Allan Ltd.).

the 'standard' railbus, although by no means the end of the line. The 141 acts as its own prototype, and reservations were noted with respect to weight and internal layout for consideration in later versions.

13.2 IMPLEMENTATION AND AFTER

The most likely customers for DMU replacements were the Passenger Transport Executives, who manage rail services in the conurbations and were the greatest users of the life-expired stock. West Yorkshire PTE had taken the most interest in the railbus development programme and had agreed to take some of the new units when they went into production (Abbott 1983). Hence, the first 20 Class 141s went to West Yorkshire under a leasing agreement for a two-year period (1984-6). They were in service in the Leeds/Huddersfield/ Wakefield area, and crossing the PTE boundary to Doncaster and York. It might seem anomalous that lightweight, railbus technology had been developed in the context of the problems of rural lines, yet the first production series went into service only in urbanized West Yorkshire. The need for stock renewal was perhaps more urgent in the PTEs, while the superior acceleration and braking of the 141s was more suited to shorter distance stopping services and could also accommodate the re-opening of intermediate stations. On the Leeds-Harrogate line reduced overall journey times permitted a more frequent service with the same number of trains. No opportunity arose for urban/rural comparisons with the railbus.

But this is also symptomatic of some continuing confusion over the appropriate roles for 'light' and 'heavy' train versions. For West Yorkshire PTE, the 141 was still too expensive and 'over-engineered' even for urban routes, there were reservations about several design points, but most important was the lack of capacity to handle urban peak demand. As a bus derivative, the train cars had a 2+2 seating arrangement, compared with 2+3 in the old DMUs. These points reflect the PTE's lack of influence, as customer, in the design process. However, it was generally agreed that the 141s performed well in service and did achieve the vital objective of cutting fuel and maintenance bills.

It is arguable how far one should take the railbus story, as it was but a component of the search for an effective DMU replacement. The typical "bus on rails" was the LEV, while the classic production railbus was the Class 141. Later models have retreated from the railbus concept, although for good railway operating reasons. First, the capacity restriction on the 141s had to be dealt with. BR had invited tenders for a further 75 two-car improved railbuses, and the contractors were able to produce a 'wide-bodied' version, accommodating 2+3 seating, without abandoning the Leyland National body pressings which were essential to the low cost. This was designated Class 142. Another bus builder, Alexanders, got part of the contract and in a similar way produced Class 143. These were successful models, and further substantial orders were placed. Although with the same pedigree, it is debateable whether these latter types retain enough 'bus' features to keep the 'railbus' label. In any event, the marketing men had disowned 'railbus' and renamed the whole series - Class 141/142/143 and the slightly different 144 - as 'Pacers'.

Meanwhile, another school of thought was convinced that the question of DMU replacement was best answered by rebuilding the classic DMU in a modernized form (Ford 1984, 1986). This was expected to be more suitable than the railbus for high density urban work in the PTEs, and able to cover a higher proportion of the range of service types formerly undertaken by the DMUs. With three cars per unit to provide the necessary capacity, and avoiding the mistakes of the Class 210 and 140, these new generation DMUs were built very quickly in 1983/4. Class 150/151 were not at all founded on the 'railbus' ethos, and thus no further details need be given here save to note that they were dubbed 'Sprinters', went into production and entered service in 1986 on numerous lines around the country, and were very successful and popular.

West Yorkshire PTE, the authority which accepted the original tranche of Class 141 railbuses, now operates a wide range of Pacers and Sprinters of different types (known collectively as 'Metrotrains'), and has succeeded in 'lightening' its formerly heavyweight local rail system. The original 20 Class 141s were being refurbished in 1988 and adapted so that they can couple with the newer Pacer units.

The only other true railbus to go into service was the single R3 unit which was acquired by Northern Ireland Railways. This has operated on the Belfast-Larne and Coleraine-Portrush branch lines, but has not been a success and has run only intermittently. Apart from mechanical problems, it has attracted complaints about overcrowding due to NIR's inability to predict occasional heavy loads, even in the off-peak. Latterly, it was used for ferrying maintenance gangs up and down the track.

It is also interesting to note that some of the problem rural lines which brought the problem to a head, especially the longer ones such as the Cambrian lines in Wales, have been provided with Sprinters. Although this has had some external funding from development agencies, the improvement in service quality is far beyond what might have been envisaged in the DMU days. On rural lines in general, there is still a need for a comprehensive application of all elements of the 'low cost rural railway'.

If the railbus is identified primarily with the Class 141, then the railbus era has been and gone, and it can be seen in retrospect as but a stage in a continuing development progression from the old DMU to a modern light rail system. It could be maintained that it is the "bus on rails" concept that is really 'unconventional', and that the newer types are dependent upon technical refinements of traditional train design. As a technology which aims at simplification rather than enhanced performance, and one which recalls earlier ideas, the railbus has parallels with 'appropriate technology' of the Third World (Nutley 1988c). The analogy does have some relevance for the UCM sector.

Chapter 14

EVALUATION AND POLICY ROLE

14.1 COMPARATIVE EVALUATION - ECONOMICS AND ACCESSIBILITY

The awesome task of evaluating all UCM types in a comparable manner has to be faced. Apart from the obvious problem of sheer complexity, the vital pre-requisite is to obtain a common basis for comparison. It is quite impossible to answer the inevitable and apparently simple question as to which is 'best'. This of course depends entirely on the role which UCMs are being asked to play, on the specific problem they are being asked to solve, on the environment in which they are expected to operate. A straightforward comparison of up to 20 modal types would be not only obscured by complexity but also undermined by futility. Only with prior definition of the problem context can the range of feasible UCM types be narrowed down, and these compared with respect to their ability to solve the problem.

It is not surprising that no author has attempted to undertake a full economic comparison of all UCM types. Occasionally, and much more realistically, there have been attempts to compare two or three modes which seem applicable to a single environment or problem context. On this basis, the total set of UCMs is divisible into more coherent sub-groups.

(i) High intensity conventional bus
PSV minibus
Environment - metropolitan/urban, high density corridor routes.
Comparative evaluation - Glaister (1985), Bly & Oldfield (1986).

(ii) Medium intensity conventional bus
PSV dial-a-ride
PSV minibus
Urban/suburban/new town, medium density, feeder/ distributor-town centre.

Martin (1978b), Tunbridge & Hale (1978), Tunbridge (1980).

(iii) Low intensity fixed-route conventional bus
D/R diversions with conventional bus
Community bus
Mixed/rural, low density, inter-town or country-town.
Balcombe (1980a).

(iv) Medium intensity conventional bus
Shared taxi or taxibus
Urban/suburban, feeder or medium density corridors.
Greening & Jackson (1983).

(v) Organized car sharing
Car pooling
Urban commuting, cf. 'ridesharing' in USA.
Green (1978), Vincent & Wood (1979).

(vi) Shared hire-car or taxi
Social car scheme
Rural/remote rural; trip purpose or destination may be restricted.
Watts (1980).

(vii) Postbus
School bus carrying public
Passenger/freight service
Mixed/rural, country-town; dual-purpose modes aimed at rationalization of total transport resources, institutional constraints but resource cost savings.
Lugton (1980).

(viii) Semi-formal social car scheme
Informal lift-giving
Hospital car and SSD car services
Rural/remote rural, country-town; all CT modes restricted to 'essential' trips.
Balcombe & Dredge (1980).

(ix) 'Welfare' transport, health and social services
CT minibus pooling and group hire schemes
Volunteer-run car schemes
Dial-a-ride for elderly and handicapped

All environments but mainly urban; comparison of CT and statutory services for persons with special needs.
Bailey & Layzell (1983), Sutton (1985, 1987).

(x) PSV dial-a-ride, 1970s
CT dial-a-ride, 1980s
Metropolitan/urban, but dispersed demand; comparison of D/R modes with different philosophies and different client groups.
Bailey (1984).

(xi) Dial-a-ride for elderly and handicapped
Subsidized taxi scheme
Metropolitan/urban, dispersed demand; D/R modes catering for general mobility needs of persons with handicaps.
Heraty (1984), Potter & Heiser (1987).

Relevant economic data can be found in chapters 3-12, and none will be repeated here. The only attempt to compare more than a few modes is Martin (1978a), but this is purely economic and not related to environmental or problem contexts. Much of the valid comparative work, other than the purely theoretical, stems from the RUTEX programme, as this has been the only significant opportunity to obtain consistent data for a large number of modes and services, in similar environments (Transport & Road Research Laboratory 1980). A major objective was matching UCM type with specific local problem characteristics. In reality, pragmatic factors play a large part in deciding which modes are actually implemented. Nevertheless, a model was attempted by Balcombe (1980a) in which demand density (passengers per sq km) was used as a crude surrogate for 'environment' and related to operating characteristics of D/R modes. Theoretical work, however, plays virtually no role in UCM planning.

After RUTEX, the TRRL attempted to assess the suitability of the newly introduced modes by hypothesizing the transfer of each one to the other experimental sites, and estimating how it would fit the pattern of demand there. Several alternatives were compared hypothetically with the mode actually chosen for each area. Results are tabulated by Balcombe (1980b), and the general conclusion for these relatively remote rural areas was that most RUTEX services could have been substituted by a social car scheme, which would have been cheapest, and many also by a shared hire-car. Many assumptions are made in this kind of exercise, such as transferability of results from one area to another, and the practicality of implementation.

Very rarely is there an opportunity to compare the economic performance of several modes operating in the same environment, where demand patterns may be thought broadly similar, and also where there is a reasonable sample of each mode so that individual peculiarities may be averaged out. Some data which comes close to these requirements are presented in table 14.1. The two adjacent counties of Dyfed and Powys had between them in 1981/2, nine postbus services, fifteen area-wide social car schemes and a wide range of private sector stage buses, in a low density region of widely dispersed settlement. The table confirms distinct economic profiles for the three modes.

Private stage buses have by far the greatest average cost, but much higher revenues reflecting considerably higher passenger loadings, but nevertheless these still leave much greater deficits than the two UCMs. This stage bus sample includes a couple of short town services and inter-town links as well as the more common village-town shopping journey, but these are all representative of the range of service types in the area. The comparatively more specialized nature of the two UCMs is apparent. It is interesting that the average postbus mileage is twice that of stage buses, reflecting the former's 5-6 day a week frequency, and hence cost/mile and deficit/mile are low. Postbus costs are lower than this frequency would suggest due to the subtraction of unavoidable mailvan costs. Costs per mile for the three modes are typical of rural conditions in the early 1980s, as are the operating ratios. Omitting frequency and mileage, cost/trip on postbuses is much higher than on stage buses, and hence passenger loadings are low relative to distance covered.

TABLE 14.1 **Comparative performance indices: Dyfed and Powys 1981/2 (source: raw data from Dyfed CC and Powys CC).**

	Postbuses	Social Cars	Private Stage Buses
Sample size (no. of services)	9	15	21
Av. cost £	1710	774	4566
Av. revenue £	1191	163	2049
Av. deficit £	519	611	2517
Av. mileage	21977	4772	10342
Av. no. trips/year	1657	208	6656
Av. catchment pop.	1300	7620	1540
Operating ratio (rev./cost)	0.69	0.21	0.44
Cost/mile (pence)	7.78	16.21	44.15
Deficit/mile (pence)	2.36	12.80	24.33
Cost/trip (pence)	103.19	372.11	68.59
Deficit/trip (pence)	31.32	293.75	37.81
Trips/person/year	1.274	0.027	4.322

Social car schemes are quite distinctive: the large catchment area population reflects the mode's orientation towards diffuse trip demands over a wide area. Low costs, trip numbers, revenues and deficits are indicative of the low service levels offered, or restricted, by social cars. Endorsing conclusions made elsewhere about this mode, the cost/trip and corresponding deficits are very high indeed, yet the low level of service means that overall costs and deficits are acceptable, and make this option appear cheap. The contrast between cost/trip and cost/mile means passenger numbers are very low relative to distance covered, reflecting the taxi-like quality. Overall, differences in trip rates indicate the general purpose nature of stage buses and the exclusivity and 'essential' nature of social cars, with postbuses in between.

The lesson of this is that although the environmental and general problem contexts appear similar, each mode has its own very distinct characteristics. They are not competing with each other, so the question of which is 'best' is immaterial, but are orientated towards specific patterns of demand in different localities. They complement each other; each has a useful place. The question becomes how well they are satisfying the particular needs of specific places, and whether this could be done better in any other way. Simplistic indices such as 'cost per trip' are inadequate. A more localized scale of analysis is required, and much more attention should be paid to articulating the precise nature of problems or 'needs' and to defining the benefits attributable to any mode or policy.

Value for money is the inevitable and imperative aim of any service provider, whether it is the private sector seeking the greatest income from its investment or the public sector seeking to fulfil its obligations to obtain maximum benefits for the taxpayer. In public transport the subsidy of the one by the other confirms this obligation, but also exposes the contradiction between providing services of a type and scale shaped by commercial criteria and those of a type and scale regarded by public authorities as necessary for the population to whom it is responsible.

In one sense the consumer population may constitute a 'market' for the service, levels of which may be determined by the expressed 'demand' for it, but this sort of purely economic/ commercial approach has long been shown, in many environments around the world, not to produce a satisfactory overall level or standard of that service. This is implicitly to appraise the service from the consumer's viewpoint, and acceptance that consumer 'need' for a certain level of service may have a stronger claim to fulfilment than that which could be sustained by the market is not without some ideological basis. As the entire field of study under review in this book is founded on such a viewpoint and is accepted, in varying degrees, in most parts of the world, it is not felt necessary to justify this approach. Suffice it to say that transport is a derived demand, a means to an end, and not a product from which the consumer gains any direct personal benefit, and therefore there is no question of the consumer being expected to pay the real cost as a result of his exercising free choice. The need to obtain a reasonable range of goods and services which may only be available at locations beyond walking distance of home means that, for the non-car user, some consumption of public transport is essential for the pursuit of a normal way of life.

The role of transport then becomes the provision of *access* to distant facilities, raising the utility of places, but of course adding to the effective cost of goods and services eventually consumed. It is possible to argue from this that public transport is almost a public service in the welfare sense, and therefore should be provided by the state on a similar basis to health, social services and education. Levels of service provided on the basis of need, and adequate subsidy to sustain this, with a commitment to stability and continuity, would be the desirable consequence. Some European countries (of which the UK is not one) do operate public transport on the basis of a similar commitment.

It is not intended to pursue a polemic on the role of public services, but to indicate the consequences for the *evaluation* of transport in the context under discussion. While a comparison of 'costs' and 'benefits' is ultimately inescapable, the conventional balance sheet of expenditure and income or even the broader form of cost-benefit analysis has its ideological bias. 'Value for money' does not preclude 'value' being assessed in more relevant, subjective or intangible ways. The 'benefits' of a transport service must be interpreted in terms of the job which society is asking transport to do, that is, to provide access to distant locations. Accessibility can be measured (although there are many different methods), and the additional accessibility provided by any service constitutes the crucial 'benefits' of that service. Especially in the contexts of rural areas and of minority groups such as the handicapped, the 'transport problem' which provoked the growth of UCMs and other policy measures can nearly always be interpreted as a *lack of accessibility*.

This has already been referred to at various points above; there is a mountain of empirical evidence from rural case studies on problems of accessibility, mobility and economics (Wilmers & Moseley 1975, Hillman et al 1976, White 1978, Oxley & Seaton 1979, Moseley 1979, Halsall & Turton 1979, Nutley 1983, Cloke 1985). There may be an analogy here with the position of the infirm or handicapped in an urban area where conventional public transport, although apparently abundant, is effectively useless due to unsympathetic design or operational factors; realistically, accessibility to facilities might be as poor as in the countryside. It was mentioned above in connection with RUTEX (section 1.4) that traditional measures of benefit such as value of travel time are irrelevant in these circumstances, and that RUTEX could be criticised for not taking a problem-solving approach and therefore not appreciating the importance of accessibility.

It is important to understand that this concept is a more abstract one that goes further than obvious visible benefits such as number of passengers carried. Accessibility differs from mobility or trip rates, and also from expressed demand or actual service patronage. The point can be clarified by focusing on the misleading use of the term 'demand'. Transport demand or patronage of services is low in rural areas not because this is an accurate reflection of what the public wants or needs, but simply because service provision (frequency of buses) is low for economic reasons. It is the low levels of service which severely constrain the trip opportunities available to residents, and therefore bus patronage (apparent 'demand') is low. This of course has feedback effects for transport operators. However from the consumer's viewpoint, there is a real latent demand for a wider range of trip opportunities (access to facilities in distant locations) than can be provided by the existing transport system. It is this gap between the levels of accessibility *needed* and that which can actually be achieved which comprises the scale of the 'problem'. To merely equate low bus patronage with low demand is to argue the problem out of existence.

Accessibility is a measure of *opportunity*. It measures the degree to which it is *possible* for a group of people in village A to travel, under prevailing space, time and economic constraints, to town B in order to avail themselves of goods and services located there. It is not related to travel behaviour in that it does *not* predict how many people *will* travel from A to B.

The methodological problem then becomes how to establish a yardstick against which to measure the adequacy of existing services. How much transport *should* be available in these circumstances? For comparative evaluation, a normative standard is required, measured in accessibility terms. This would allow measurement of the standards achievable under the prevailing transport system relative to the optimum 'needed' by the population - the shortfall will be immediately apparent. Further, any policy input, such as the introduction of new services (or withdrawals), can be assessed in terms of its effect on access levels, and these 'benefits' can be offset against costs.

As far as UCMs are concerned, this approach can be used to evaluate the cost-effectiveness of all the modal types, and the potential of each to improve access levels in local areas. A decision on which mode is 'best' is unlikely to be simple! Similarly, one can compare the cost-effectiveness of UCMs with that of conventional transport, and also with 'non-transport solutions' which might be implemented in the longer term (see below).

14.2 COMPARATIVE EVALUATION - CASE STUDIES

(a) Method

Although there are a large number of relatively simple, aggregative measures of accessibility, that required here needs to be articulated at the local scale in order to evaluate individual services and impacts on local communities, and at a personal level in order to differentiate the needs and lifestyles of various social groups. The method owes much to that pioneered by Moseley et al (1977) in East Anglia.

The optimum standard (100% accessibility) is achieved where the whole population has an adequate level of access to every one of a set of destination facilities ('functions') defined to be socially necessary. These could include: workplaces (towns as employment centres), shopping (the village grocery store, as well as nearby towns), health (doctor, dentist, chemist, hospital for treatment and for visiting), administration or personal business (post office, bank, local authority and government welfare offices), and leisure (from the village pub or community centre through libraries and cinemas to the facilities of large towns). Work, shopping and leisure might be disaggregated according to the local urban hierarchy on the basis that larger towns have a greater range of opportunities. Setting standards like this constitutes a normative view of 'need' (Koutsopoulos 1980).

The demand for these functions is deemed to originate in a set of villages within a defined study area. Village populations are broken down by social group, e.g. working people, housewives, elderly, children - but further breakdown might produce insignificant subsamples. Each group is necessarily divided into car owners and non-car owners. The possibility of a linkage between village populations and distant functions (most commonly in a small town) being acceptable requires pre-defined standards of frequency (days per week) that the function is 'needed' and minimum stay time at the destination.

Every potential connection between social groups in villages and relevant functions is subjected to a test procedure to determine whether the locations of supply and demand, transport links, and the defined standards can be coordinated in time and space. For non-car owners, one first asks whether the function is located in the home village and whether its opening hours are compatible with each social group's free time. In most cases a journey by public transport will need to be sought in order to connect with the nearest suitable outlet of the function. Questions then incorporate the possibilities of making the whole return journey within the group's free time, after allowing for the necessary stay time at the destination, and of making that journey at the necessary frequency. A major impediment to access being secured is the case of certain functions being barred to working people and schoolchildren through being unavailable outside work or school hours. In fact, this 'time clash problem' is the only significant barrier to the accessibility of car owners, who have a simpler test procedure to pass, and who predictably emerge with much higher access levels. More details of this 'accessibility test' can be found in Nutley (1981, 1983).

As a result of these procedures, accessibility is measured by the number of social group-function contacts which are proved accessible, or weighted by the population of each group, in each case as a percentage of the total desirable contacts. Results can be expressed by village or by area, by social group, or by type of function, and can be represented by a variety of graphic and cartographic means. Policy appraisal can be achieved by analysis before and after implementation, and it would normally be possible to distinguish the contributions of individual services. Similarly, the evaluation of alternative options or modes can be achieved by comparison of the accessibility produced under each option with that existing at present.

(b) Evaluation of RUTEX

This method has been applied to the evaluation of the Welsh RUTEX schemes which took place in Dinefwr District of the County of Dyfed in 1977-9 (see section 1.4). The survey population was confined to nine villages surrounding the small market town of Llandovery. The method is difficult to apply to dispersed populations but this can be eased by including quite small nucleations with populations as low as 30. It is a great advantage of the method that questionnaire surveys are *not* required, although of course they could be useful for supplementary information. Fieldwork is necessary to establish the availability of local services and details such as doctors' surgery hours, etc.

In this case the technique requires a parallel 'before' and 'after' study, so that the accessibility effects of RUTEX can be deduced. The particular UCMs introduced here were:

(i) One postbus, based in Llandovery, operating two routes in rural areas to the north and south of the town; the timetable was complex with each route having a service on four weekdays and alternate Saturdays; on two weekdays there were morning and afternoon services which allowed rural dwellers to make a return trip to town.
(ii) School contract buses run by local independent companies under Section 30 permits; four routes connecting the villages with Llandovery schools and operated twice every weekday in school term.
(iii) A social car scheme throughout the RUTEX area, administered by the WRVS in local sub-areas; the usual restrictions of 'urgent trips' apply.

See Dinefwr Rutex (1979) for details of these. In addition there were complementary adjustments to timetables by the main local bus company and British Rail.

Accessibility for car users averaged 73%, and it was assumed that this group derived no benefit from public transport innovations. Prior to RUTEX, the existing public transport provision bestowed an average access level of 46% for non-car users. The car/non-car gulf is clear, but it is interesting that subsequent case studies elsewhere in rural Wales showed much *lower* non-car access rates. Three of the nine villages are sited on a main valley route conveying a reasonable bus service and, crucially, a railway line; these tend to be the larger settlements and retain more local functions. The average results are biased by this, so that while these places have 50-60% non-car accessibility, the outlying villages with only market day buses and very few functions score 0-16%. It is the latter which RUTEX was primarily aimed to assist.

Table 14.2 gives the basic results on a social group basis. The most disadvantaged group is not the elderly as is commonly supposed, but non-car owning working people. They are prevented from using many facilities due to their being unavailable outside work hours; much conventional public transport and nearly all of RUTEX was confined to these times. The same principle applies to a lesser degree in the case of schoolchildren. It is the 'housewives' and 'elderly' groups who have higher non-car accessibility (through fewer time constraints) and who gain most from RUTEX (as these are nearly all weekday off-peak services). The social car scheme poses some difficulties as the method really demands a timetable, but the 'urgent trips' requirement is effectively limiting its use to the same two groups and for medical journeys. Making this assumption, the RUTEX gains can be re-calculated.

Benefits are therefore selective. While RUTEX has raised the overall non-car accessibility rate by an apparently modest 5.5%, this gain accrues almost exclusively to the housewives

TABLE 14.2 Dinefwr RUTEX: accessibility by social group ((c) Nutley 1983).

% accessibility

	Car	Non-car owners		RUTEX	with social car	
	Owners	pre-RUTEX	post-RUTEX	gain	post-RUTEX	gain
Econ. Active Males	63	33	33	<0.5		
Econ. Active Females	63	33	33	<0.5		
Housewives	99	61.5	70	8.5	75	13.5
Elderly	89	51	52.5	1.5	61	10
Schoolchildren	-	41	41	0		
Total	73	46	48	2.5	51	5.5

and elderly with increases of 13.5% and 10% respectively. Also, these gains *have* accrued to the smaller, remoter villages, with four achieving access gains between 2.5% and 12.5% (the others, zero). In terms of destination functions, RUTEX has secured access gains of up to 20% for a dozen facilities located in nearby towns, mainly for the 'housewives' group.

Cost-benefit evaluation can be conducted as shown in table 14.3. Although there might be difficulties, it is possible to distinguish the contributions of each mode. The benefit/cost ratio is simple and makes no attempt to convert 'benefits' to monetary values. In absolute terms, 'units of accessibility' measure the numbers of people achieving access to additional functions. 'Costs' are operating deficits only, and exclude 'setting-up costs' and the capital costs of new vehicles in this first year of operation. Hence costs of the postbus and school bus seem absurdly low, even at 1978 prices; the infinity ratio for the latter becomes hypothetical. Greater benefits attributable to the school buses reflect its regular daily timetable (the method discounts its absence in school holidays), whereas the postbus timetable proved too complex. The contrast with the social car is fascinating. Even with restrictions on eligible population and destinations, its D/R operation means that it is *potentially* available for *all* eligible trips at any reasonable time. Therefore the accessibility it provides (regardless of the actual number of passengers) can be greater than that of other modes. Despite this, the 'personal service' of the social car incurs a high cost per trip and thus the total cost is very much higher (NB. no fares were charged at this date). It might be reiterated that actual passengers carried play no part in this equation, although of course they influence the size of deficits (usually negatively, but social car costs increase with the numbers of trips). Accessibility can be seen as a measure of *opportunity*, and the ratio reflects the value of keeping these opportunities open.

More detailed results of this analysis can be found in Nutley (1981, 1983).

TABLE 14.3 Dinefwr RUTEX: accessibility benefits and costs ((c) Nutley 1983).

Initial units of accessibility = 6852

	Benefits (additional units of accessibility)	Costs (1978 prices) £	Benefit/Cost Ratio (units of accessibility per £)
Postbus	115	78	1.47
School bus	224	0	∞
Re-scheduling by private operator	20	0	∞
Social car	477	1744	0.27
Total	836	1822	0.46

Total desirable units of accessibility = 14964

*One unit of accessibility = one person accessible to one function.

(c) Other Studies

Other attempts to measure accessibility by this kind of methodology have been rare, and its application to the assessment of UCMs even rarer. Two studies are worth mentioning.

Kilvington & McKenzie (1985) used a very similar technique to predict local accessibility problems in Breckland District, Norfolk. Having done this, it would be possible to assess the contribution made by the twelve community car schemes recently inaugurated in the district. Only two were in fact studied in detail, but even these reflected wide differences in pre-existing facilities and access levels, and the range of trips permissable by the car schemes. Accessibility analysis played no part in planning the new services. In one parish where existing accessibility was fairly high, the number of 'observed problems' relatively low, and eligible community car trips very restrictive, the car scheme 'solved' 7% of previous access problems. In the other parish where existing transport was much poorer, 'observed problems' higher, and the range of eligible car trips more generous, 20% of previous access problems were made good. Activities facilitated were overwhelmingly medical in each case.

Sutton (1985) conducted a similar study of the Soar Valley community bus in Nottinghamshire. After its introduction in 1979, the community bus had encouraged a 48% increase in bus patronage by local residents, and a questionnaire survey indicated that this satisfied 33% of latent demand. An accessibility analysis on similar lines to those above showed that before introduction 21% of the adult population experienced 'access problems' with respect to at least two facilities. The Soar Valley CB neatly complemented existing buses in terms of facilities served, such that access problems then remained for only 9%. Four out of five affected settlements showed a significant improvement. Of 25 defined destination functions, seven were made much more accessible by the CB, and only five still caused significant access problems after its appearance.

(d) All Modes Compared

Studies undertaken *post facto* like those cited above are able to assess only those UCMs actually in operation in the area concerned. It would hardly ever be possible therefore to evaluate more than three different modes in the same area (as in the RUTEX study). The possible performance of other modes would have to be hypothesized. Any public authority contemplating the solution of accessibility problems by the introduction of UCMs would benefit from an evaluation exercise which covered all the modal types likely to be of use in the area. A more valid comparison would be offered if all types were tested in the same environment and addressed to the same set of problems. The vital questions are: which of the wide range of UCMs makes the most cost-effective contribution to prevailing problems, and also can UCMs perform better than conventional buses or other conceivable approaches?

Such an exercise would have to be hypothetical, and involve planning the operation of successive UCMs in a specific area with quantified access deficiencies. Only one study of this sort has been done (Nutley 1985). An area was chosen in the southern part of Radnor District in the Welsh county of Powys. Demand for access to services was deemed to originate from fourteen villages in the area; these are generally small, most with populations under 100, and with very few facilities of their own. Four small towns surrounding the area are the main service centres. At the time of the survey in 1981 these were connected by a basic level of bus services which ran along the main roads but did not penetrate the study area. Four of the villages were close enough to the arterial routes, but the remainder had 'market day only' services or no public transport at all. The crucial problem was in gaining initial access to the 'outside world' network. Using the same methodology as described above, the overall accessibility level in 1981 was 50.9%, this being an average of 73.6% for the car users and only 32.6% for non-car users. The latter is itself an average, for a few villages close to the main roads are larger with more facilities, achieving 40-54%, while others are muc weaker with 0-10% non-car accessibility. Differences among social groups are similar to those in the RUTEX study.

The basic approach taken was to hypothesize the introduction of additional transport services in this area, designing networks and schedules in such a way as to make good existing access deficiencies, to measure the additional accessibility procured and then to relate this to costs. Forty of these alternative scenarios were worked out, falling into three categories: (i) providing more conventional buses; (ii) implementing UCMs; (iii) planning 'non-transport solutions' (see below).

(i) Bus solutions

In the situation of considering improvements to local accessibility it is probably best to start with *conventional buses*. The main point of interest here is the trade-off between access provision and increasing cost. How much accessibility can be provided before costs become unacceptable? What would it cost (and is it feasible) to secure 100% accessibility? Is there any particular level of service which is more cost-effective than others?

Initially, the approach taken was to design bus networks at successively higher levels of service, corresponding to the frequencies at which certain destination functions were defined to be needed, under the methodology. In planning route locations, certain guidelines were adhered to: they took the form of village-town, or town-town passing through villages en route; they were kept as short as possible consistent with linking a reasonable number of villages; all fourteen survey villages were provided, as far as possible, with the same level of service under any one plan; three or four separate service routes were necessary to cover the whole area; and links were made with all of the surrounding towns under most plans. It is very important that hypothetical networks are accompanied by worked out timetables, to check the feasibility of the plan and to allow access to functions to be computed accurately. Therefore, as far as possible, route systems were designed so that start and return times were compatible with social groups' free time, and arrival and stay times coincided with facilities' opening hours. Also, connections were organised, wherever possible, with existing public transport along the peripheral major roads, in order to acquire access to more distant places (the farthest places needed for this case study were Hereford, where a major hospital is sited, and Cardiff, the regional centre).

Costing of these hypothetical bus routes was done by applying the average deficit incurred by a sample of routes run by independent bus companies in adjacent areas. In all cases the existing local transport services were assumed to be removed, although this led to a saving of only about £20 per week.

Thirteen 'bus' plans were actually tested, but for present purposes it is sufficient to give the results of a few of these (table 14.4) to express the variations obtainable (full results can be found in Nutley 1985). Plans K, N and Q represent progressive increases in frequencies, from weekly to daily, and succeed in raising overall access levels by 1.5-9% and non-car levels by 3-16%. Even a daily service is well short of 100% accessibility, and also there is a significant jump in costs (and fall in benefit/ cost ratios) at higher than once per week frequencies. This is because each route, schedule and service level is aimed at a particular sub-set of defined functions. The next stage therefore was to cumulate these plans successively, in order to build up accessibility to as high a level as possible. Saturation coverage of buses is comprised in Plan V, cumulating all routes with service levels between one and seven days per week. Although inevitably high cost, this achieves the highest access levels with non-car users almost reaching the level of car users (71-73%). This equality indicates that the limit has been reached, for the remaining shortfall of 27% is due to the 'time clash problem', whereby certain functions are denied to some groups through being available only in work/school hours. This impasse can only be breached by the extension

TABLE 14.4 South Radnor case study: transport options (source: Nutley 1985).

Plan		Access level (%) Overall	Non-car	Cost (1982/3 prices) £/week	Benefit/Cost Ratio
A	Initial situation (1981)	50.9	32.6		
	Conventional buses replacing existing local transport				
K	frequency 1/wk., off-peak	52.5	35.4	10	46.0
N	frequency 5/wk., Mon-Fri for work journeys	55.5	40.8	235	5.8
Q	frequency 6-7/wk., daily trips to hospital and Mon-Sat evenings	59.8	48.6	300	8.8
V	cumulated services with frequencies 1-7/wk.	72.4	71.5	850	7.6
W	combined services with frequencies 2 x 1/wk (off-peak + Sats) & 6/wk (off-peak)	61.0	50.8	230	13.1
	Unconventional modes replacing existing local transport				
X1	social car for medical trips	57.9	45.1	10	207.1
X2	social car for medical trips + conventional bus 2 x 1/wk.	65.6	59.2	60	73.2
U1	school buses carrying adult passengers	50.5	31.9	-10	n/a
U2	community bus 'minimum' service	55.0	40.0	10	122.4
U3	community bus 'maximum' service	58.9	47.1	60	39.9
U4	'courier' service, multi-route	52.4	35.2	155	2.8
U5	'courier' service, single route	53.6	37.5	185	4.4
P1	postbus, two routes	51.9	34.3	20	14.3
P2	postbus, two routes + conventional bus 2 x 1/wk	58.6	46.4	75	30.4
P3	postbus, two routes + social car for medical trips	58.6	46.4	25	91.2
U6	community bus 'minimum' service + social car for medical trips	62.6	52.9	20	167.9

of facility opening hours, which will be considered below. Plan W is an attempted compromise, achieving significant access gains with intermediate service levels and costs, and a rather better benefit/cost ratio (table 14.4).

(ii) UCM solutions

The cost-effectiveness of various UCMs can now be compared with this 'conventional' solution, and it is possible that the idiosyncracies of UCMs could be tailored to plug some of the difficult access gaps exposed by the analysis. For example, there is the problem of providing suitable transport to medical facilities, on the basis that access could theoretically be required at any time, while in practice trip *rates* for health purposes are very low. Some sort of D/R mode seems appropriate, and accordingly a *social car scheme* was proposed. For the South Radnor study area, Plan X1 suggests a car service able to cater for all medical trips (but not for other purposes) required by non-car owners, and available at all reasonable times. Its great flexibility raises the non-car access rate by 12.5%, but cannot avoid the time clash problem. It is very cheap to operate and generates a very impressive benefit/cost ratio (table 14.4). This is because, for this mode, costs depend on the number of car trips. A sample of similar schemes in adjacent areas produced an average rate of 0.0006 trips/person/week which, applied to the study area, gives an estimate of only 1.5 trips/week; the average cost per trip on equivalent schemes was £4-5.

As access rates are constrained by the exclusion of non-medical trips, such a plan could be usefully complemented by a basic level of conventional buses to broaden the range of functions accessible. Plan X2 added to the social car two bus routes of weekly frequency, which raised access levels to 65% overall (59% non-car) with a very good benefit/cost ratio.

In South Radnor, as in many rural areas, the map of *school bus* routes is much more extensive than the map of stage carriage services. This under-exploited resource might be adapted to carry adult passengers. Five of the existing school contract routes were thought to be potentially useful for this purpose, although in some cases a larger vehicle would need to be deployed. From the access point of view, problems stem from their absence in school holidays and weekends, from the fact that return trips are too long for the elderly (the methodology imposed a journey duration limit on this group), and that the return trip is too early for working people. If existing local transport is removed before implementing this policy (Plan U1), then accessibility actually *falls*. Its only redeeming feature is cheapness, as the valid costs are those in addition to transporting schoolchildren, meaning in practice the extra running costs of larger vehicles. Adult trip rates and revenues were based on those for school bus services included in the Welsh RUTEX (Dinefwr Rutex 1979).

A *community bus* for the study area was then hypothesized. The most intriguing aspect was the means of serving the whole area with just a single vehicle. It was assumed to be based in one of the central villages. Two strategies were devised: Plan U2, a 'minimum' service running on four days per week and with 15% mileage available for the more lucrative hires and excursions; and Plan U3, a 'maximum' service running throughout the week with only 2.5% mileage available for private hires, etc. The first of these was designed to ensure that every village had a mid-day service into town on one weekday, with a Saturday working planned to make connections with existing transport to more distant centres. Plan U3 adds to this a variety of trips aimed at specific urban facilities. As would be expected, the second plan produces higher access (by 7% for non-car owners) but at higher cost, although benefit/cost ratios are reasonable. It is most unlikely that, with only one vehicle, the non-car access rate could be raised above the 47% attained here. Costings were based upon those for similar schemes in Devon and Clwyd, using trip rates experienced by these, and higher revenues from private hires (Devon Rutex 1978, Clwyd 1978).

Plans U4 and U5 suggest a *'Courier'* service for the study area, based upon the well-known

Border Courier system in southern Scotland (Borders 1981). Again, only a single multi-purpose vehicle would be realistic. As in the Borders, its main task was assumed to be the conveyance of medical supplies between doctors' surgeries (located in the small towns) and the major regional hospital (in this case, Hereford), although it might take different routes for passengers' benefit. The main problems for present purposes stem from this emphasis on goods carriage, as this cannot be credited in the evaluation of accessibility, and route patterns severely constrained by the location of surgeries and hospitals are of little use in effecting personal access to everyday services.

The scheduling problem is tackled in Plan U4 by a multi-route system, so that every village has a service on at least two days per week, and in (U5) by working a long and circuitous single route, to link as many villages as possible, at a daily frequency (part of this might be D/R). This latter is slightly more successful, but neither plan can really be said to produce any worthwhile improvements in accessibility, while the costs, based upon those of the Borders scheme, appear fairly high.

Possibilities of implementing *postbuses* also face various constraints. The main difficulty in planning is that postbus services must not deviate from the routes and times already used by the Post Office for dealing with the mail. Eight mailvan routes cover the South Radnor study area, and none of these serves more than three of the survey villages (see figure 7.5). Detailed examination concluded that only two routes, tapping six villages, might hold possibilities for useful conversion to postbuses. Lack of correspondence between morning delivery and afternoon collection runs would reduce the useful accessibility that the postbus could provide for some villages; restricted Saturday mail operations have a similar effect. For these and other reasons (e.g. mail schedules causing the arrival in town after the surgery has closed), access gains attributable to Plan P1 are very small.

The postbuses obviously need to be supplemented. Plan P2 adds a basic level of conventional buses (two different weekly services). Plan P3, noting the complementary nature of the two modes, combines postbuses with the social car scheme. Coincidentally, access levels achieved are identical (46.4% non-car, and 58.6% overall), although the distribution of benefits is different, and the latter plan is cheaper. Costing of postbuses was done on the basis of averages from all such services in adjacent areas, which gave an operating deficit (over and above the cost of running mailvans) of 1.5 pence per kilometre (limited data are published in Powys 1981).

A great many combinations of modes and service types could be conceived, although experience of the evaluation procedure should suggest the most promising. The final 'unconventional' plan, (U6), pairs up two UCMs which complement each other in terms of access possibilities and which also have encouraging benefit/cost ratios. First, the community bus in (U2) shows reasonable access gain for little net cost, while operating selective stage-type services; second, the social car from (X1) is the best way (through being demand-responsive) of catering for those trip types which may be urgently needed at any time, but which are made only rarely. Results are most favourable, with very reasonable access scores (52.9% non-car, and 62.6% overall) at low cost, and with an impressively high benefit/cost ratio.

(iii) Non-transport solutions

Alternative scenarios considered so far have concentrated on 'transport solutions' to what are commonly perceived to be 'transport problems'. An 'accessibility' interpretation however brings in the land use element, in that the 'problem' becomes that of the spatial separation of residences (demand) and location of facilities (supply). In this broader sense, there are other means of bringing people closer to necessary functions or activities. According to Moseley (1979) the transport option is only the most obvious or immediate. Alternatives would be: to bring services to the people in remote areas by the 'mobile service option'; to rationalise the rural settlement pattern by concentrating population in larger 'key villages' with all basic facilities; or to encourage a dispersed pattern of small service units in all villages. These could be called *'non-transport solutions'*. Another would be the extension of facility opening hours.

Although these may seem unrealistic and only attainable in the longer term, if at all, they should be included in the analysis for the sake of hypothetical comparison. Selective results are given in table 14.5, but further details can be found in Nutley (1985).

First, one might postulate *additional village facilities*. Plan E assumes that all survey villages have the basic complement of post office, small shop, public house and village hall/community centre (NB. primary schools are omitted as it is assumed statutory school transport is adequate). This raises non-car accessibility by only 4.5%. If, in addition, all villages were provided with a bank, doctor's surgery and chemist (Plan G), it is raised by only 12.5%. Car owners also benefit slightly from this type of policy. Primarily, these figures can be no higher because the great majority of the necessary functions are of the 'urban' type; but also, time-clash problems remain, even for facilities located on one's doorstep. It is unfortunate that no reliable cost data could be found, and costings are therefore crude estimates, but clearly this is not a cost-effective option (table 14.5).

The problem of time conflicts, which has been found such an inhibiting factor, could theoretically be solved by the *extension of facility opening hours*. This would mean public service outlets open on weekdays extending into the evenings or Saturdays. For Plan X3, a range of shopping, health and administrative functions (in the towns) was assumed to have suitably extended opening times. But it was immediately apparent that such provisions are of little use to non-car owners without suitable public transport

TABLE 14.5 South Radnor case study: non-transport options (source: Nutley 1985).

Plan		Access level (%) Overall	Non-car	Cost (1982/3 prices) £/week	Benefit/Cost Ratio
A	initial situation (1981)	50.9	32.6		
	Non-transport solutions: local service outlets				
E	provide all villages with post office, shop, pub, community centre	54.8	37.2	2020	0.5
G	provide all villages with post office, shop, pub, community centre, bank, chemist, doctor	61.8	43.0	12380	0.2
	Non-transport solutions: extend opening hours of town facilities				
X3	extend opening hours of shops, health, admin. functions	63.9	35.4	2450	1.6
X4	extend opening hours of shops, health, admin. functions + conventional bus 2 x 1/wk.	71.9	49.8	2500	2.5
X6	extend opening hours of shops, health, admin. functions + conventional bus 2 x 1/wk. + social car for medical trips	78.8	62.4	2510	3.3
	Non-transport solutions: mobile services				
M1	mobile post office, shop to all villages	51.7	32.2	195	1.1
M3	mobile post office, shop, bank, peripatetic doctor/ chemist to all villages	53.4	35.4	825	0.9
M4	mobile shop, peripatetic doctor/chemist to all villages + conventional bus 2 x 1/wk.	61.6	50.2	775	4.1

to connect with the new business hours. Plan X3 benefits mainly the car owners, whose accessibility rises to 99%. To remedy this, Plan X4 adds a basic level of conventional buses, and this does succeed in raising non-car access to 50%. Adding a complementary social car service (Plan X6) produces the highest overall access rate at 78.8% but, due to the bias towards car owners, the non-car rate is less impressive at 62.4%. Costing is again crude due to lack of data, but costs must be high, outweighing the value of benefits.

Mobile or peripatetic services, whereby movement is made by the supplier and not the consumer, have a long history in the countryside (and to some extent in town) and are today more widespread than normally given credit for (Moseley & Packman 1983). These might conceivably be regarded as 'unconventional modes', although not within the current definition.

Mobile services may be postulated in the study area, assumed to be based in the surrounding towns and operating circuits of convenient groups of villages, and scheduled as far as possible in accordance with conditions defined by the method. Plan M1 provides a mobile shop and post office to every village currently lacking, while (M3) provides in addition a mobile bank and peripatetic doctor/chemist. As with additional 'fixed' village facilities, the main problem is that relatively few functions can realistically supplied by this means, and therefore the access gains are disappointing. (M4) adds the basic level of conventional buses, and this succeeds in raising overall accessibility above 60% (non-car 50%), with a slightly better benefit/cost ratio. Costings were again based on very sparse information.

While mobiles are certainly cheaper than establishing more 'fixed' village services, it would appear that these are little more cost-effective than any other type of 'non-transport solution'. This makes sense intuitively, as a service provided to a village from a van constitutes one function, while a bus into town makes available the whole range of functions located in the town.

(iv) Conclusions
Having tested forty planning options, a decision on the 'best' for the study area is by no means simple. For example, if the objective were simply to provide maximum accessibility, one would find that in most cases this can only be done at high cost, with a fairly low benefit/cost ratio. A 'minimum cost' objective would simply produce very little extra accessibility. Cost-effectiveness, as measured by the benefit/cost ratio, tends to be high in cases where the costs are relatively low, and therefore access gained tends to be modest. There are further problems concerning equity in the distribution of benefits (Nutley 1985).

These contradictions are evident if one lists plans in rank order of access gain, in which case most high scorers would be ruled out by cost, or in rank order of benefit/cost ratio, in which case most high ratios are associated with mediocre access levels (tables 14.4 & 14.5). Looking for a compromise, only two score well on both criteria - (X2), the combination of social car scheme plus a basic level of conventional buses, and (U6), the community bus 'minimum' service backed up by a social car scheme. (P3), the postbuses plus social car, might have merit as a third candidate. Access gains by these are significant, but still nowhere near 100%, for it would seem that this is the best that can be achieved at reasonable cost (say, £100 per week as a realistic limit for public support).

Detailed conclusions must of course be dependent upon the assumptions of the evaluation and planning methods, the specification of the plans tested, and the characteristics of the study area. Despite this, it is possible to draw general conclusions of wider applicability. It is highly probable, for example, that:

$1. more village functions or mobile services cannot be substitutable for public transport;
$2. accessibility is severely inhibited by time conflicts regardless of transport policy, while

the extension of opening hours is unlikely to be the best solution;
$3. the access provided by conventional buses shows diminishing returns to increased service levels, with the most cost-effective level appearing to be a combination of one off-peak and one Saturday service per week;
$4. UCMs by themselves are unlikely to lead to very high access levels, but if fairly modest gains are acceptable then they might be cost-effective.

On UCMs, it is clear that the social car is most successful when assessed in this way. It differs from all other modes in being demand-responsive and not being obliged to keep to a timetable. It is ideal for rare but important trips such as for medical purposes - the above methodology required *daily* access to doctors and hospitals on the grounds that illness might occur at any time, but in practice health trip *rates* are very low. Such situations require a mode of communication which is constantly available, but which keeps costs down by operating only when the potential need is actually felt by consumers. Hence, accessibility (a measure of potential) is relatively high and overall costs low, despite high cost per trip. The problem is precisely this restriction to 'urgent' trips, usually interpreted in a medical sense. Social cars available for any trip purpose are unknown, and it is extremely doubtful if they could handle a much higher level of demand, even if they are very localized. A social car scheme is therefore best used to complement some other transport, either conventional buses or another UCM, and it was such combinations that proved most successful in the above exercise. It is possible that this might also apply to other D/R CT modes of the 'welfare' type, but there would be many cases where the complementary mode could not be used by handicapped persons. Use of CT modes for general mobility, especially in urban areas where most operate, would have many implications for accessibility and cost - the above methodology has not been applied in such a context.

Other UCMs suffer from idiosyncratic constraints and therefore, by themselves, make only modest contributions to non-car accessibility. Rather than attempting to discriminate among them, it is more appropriate to ask each community to decide which level of accessibility it finds acceptable (and the standards set by the methodology may, in retrospect, have been too high), and then to develop a suitable mode on pragmatic grounds.

As a postscript, it might be noted that two years after the South Radnor case study a new postbus service was introduced in the area between Builth Wells and Painscastle (see figure 7.5).

14.3 THE STATUS OF UCMS IN 1985

The foregoing detailed accounts of individual modes have included estimates of the numbers of services and their geographical distribution in 1985. This is obviously a convenient date as the eve of deregulation. The implications of the new policy have been briefly considered at various points above, but this book must conclude with an assessment of the current status of UCMs, their contribution to the overall public transport sector, and their future role.

(a) Geographical Distribution

Inevitably, specific modes are usually considered in isolation from others, and inter-modal comparison is difficult. Apart from the economic and accessibility comparisons attempted above, it might be useful to look at the geographical distribution of all types to obtain an overview of national coverage. Maps have been presented to show the locations of the most common modal types, and hence a 'national picture' could be derived from a combination of figure 7.3 (postbuses), figure 10.3 (community buses), figure 11.1 (social cars) and figure 12.1 (shared taxis), all referring to 1985. A composite map of all modes might seem to be necessary, but is difficult to construct and even more difficult to interpret for the familiar reasons: services by different modes

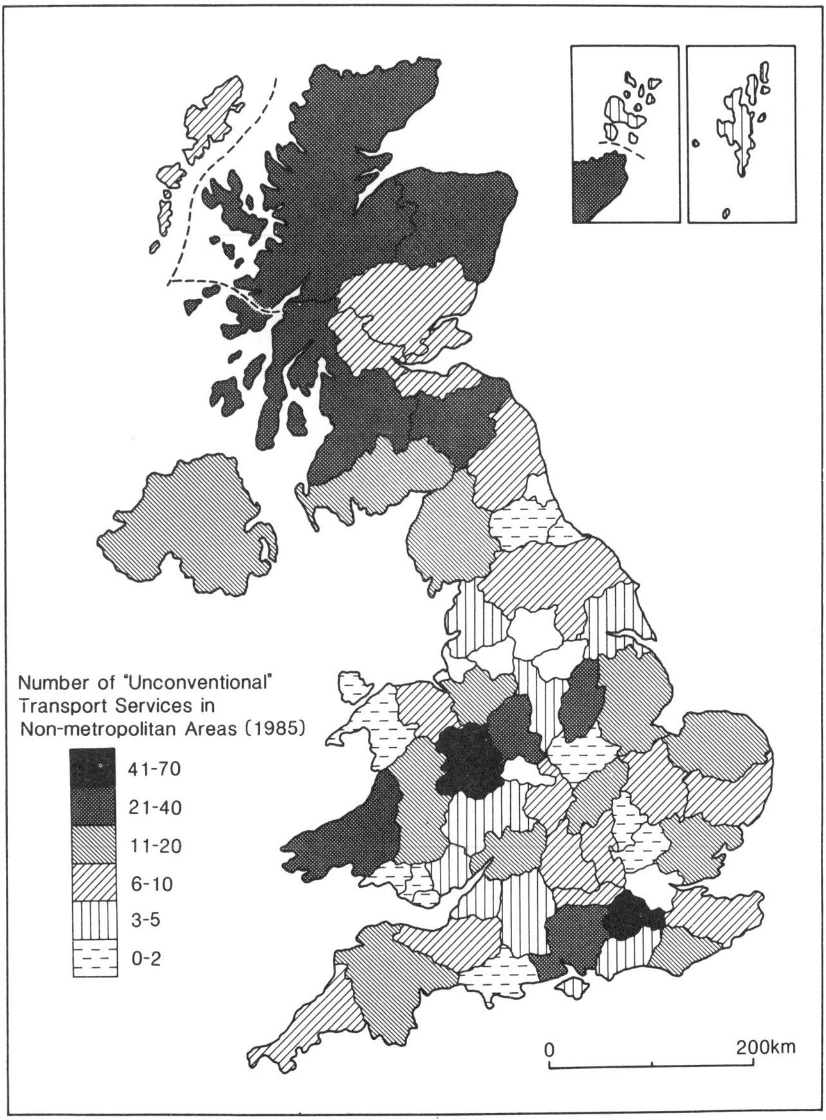

FIGURE 14.1 **Aggregate distribution of modes of all types 1985.**

cannot be easily compared; problems of definition abound; services are organised at different scales in different areas. Figure 14.1 should therefore be treated with great caution and regarded only as a guideline to the variation that exists. UCM totals are obtained by combining numbers of services by the above four modes, plus dial-a-rides, and some easily identifiable 'others'. The main omission is school buses. Many county totals are dominated by social cars, and this mode presents the most problems of definition and enumeration. Highest scores in Shropshire and Surrey are due to large numbers of car schemes, but these reflect the parish level of organisation and the availability of complete information. In other areas we are informed that car schemes cover the whole county, but without accurate numbers an arbitrary value has been used; a similar expedient has been applied to the unknown number of illicit shared taxis in Northern Ireland. As mentioned above, similar problems bedevilled the construction of figure 9.1 on 'welfare' CT schemes. It should be remembered that there is some overlap between the latter map and figure 14.1, as 'community' modes are common to both.

The aggregate distribution does give a fair impression of the extent of UCM development by 1985, and confirms the significant expansion that has taken place, when compared with equivalent data for 1979. There is at first sight no simple pattern to the composite map, and the maps of individual modes (figures 7.3, 10.3, 11.1, 12.1) are a little more revealing; the latter have been discussed, in a rural context, in Nutley (1988a). First, there seems to be an extraordinary complementarity between the dominance of postbuses in Scotland and the dominance of social cars in (mainly southern) England and Wales; both are reflected in the composite map. Otherwise, certain counties are characterised by clusters of services of the same mode: clearly it is local authority policy and not geography which explains this.

Some local authorities appear to be more active than others in promoting UCMs. Before county policies are examined, further qualifications must be made. If some counties or regions appear to be inactive, then it may be because they have few 'problems' with respect to mobility and accessibility, or conventional public transport is believed to be adequate. Conversely, counties/regions with many UCMs may be lauded for their initiative, but this may only reflect a situation where existing transport was so poor that alternatives were urgently needed. This endorses the importance of evaluating UCMs relative to the travel needs of the population and the standard of all other local service facilities, as done in the above accessibility exercises.

Low scoring areas on figure 14.1 include largely urbanized counties (Durham, Lancashire, Cleveland, Avon, the Glamorgans) where one might not expect to find UCMs (as opposed to CT). Rural, low scoring counties are rather enigmatic (Leicestershire, Dorset). It is conceivable that some local authority areas appear weak on the maps because they have developed transport strategies based on modes not included here. Some areas might get by with an informal network of lift-giving arrangements, or rely on social services transport. Others have based policies around the rationalisation of conventional bus networks (e.g. Hertfordshire). Most important in this respect, however, is the integration of stage carriage with school services, and some authorities have centred their public transport policies almost entirely on this (e.g. Gwynedd).

Otherwise, local preferences for particular modes do seem to be indicated by the maps referred to: Powys, Northumberland and the whole of rural Scotland for postbuses; Dyfed, Shropshire, Lincolnshire and Cumbria for social cars; Norfolk, Northamptonshire and Devon for community buses; North Yorkshire for shared taxis, etc. Are there any reasons for this?

Simple correlations of modal types with environmental or social conditions are probably futile. Examination of PTPs and TPPs reveals little sign of local authorities having conducted any systematic evaluation of alternative modes, either of the 'social needs' type or traditional cost-benefit appraisal. Community needs are certainly taken into account, but in a subjective and *ad hoc* way, such as reacting to local representations, 'trying out' new services as and when the budget allows, or encouraging the community itself to adopt 'self-help' measures. It would appear that modes which emerge on the ground have arisen through pragmatic adjustment of the possibilities under financial and institutional constraints - the 'art of the possible'. Having found a mode which appears to succeed under local conditions, the tendency is to continue a successful formula by replicating the original in other areas. A neighbourhood effect may then appear, whether the initiative comes from the authority or the community.

A wide gulf therefore exists between the means of implementation (or non-implementation) in practice and the evaluation and planning techniques that might have been used. Operational pragmatism seems to be the main factor in deciding what happens on the ground. One should not therefore assume that map patterns are to any great extent an accurate reflection of need. The degree to which, at a national scale, the UCM/CT sector satisfies travel needs is unknown, unless the appropriate analyses are done in all areas. For example, many PTPs report an absence of interest among local communities for setting up voluntary transport schemes: the point has already been made that self-help transport may reflect the degree of local expertise as much as need, and it is possible that those communities where needs are greatest lack the necessary skills and experience to fend for themselves (Banister & Norton 1988).

The other major blank on the map is Northern Ireland, where the reason for the lack of 'unconventional' activity is the absence of any suitable legislation to encourage it. With one small exception, none of the stream of permissive legislation passed in Britain since 1977 was extended to Northern Ireland. Explanations must be conjectural: perhaps the fact that the quasi-monopolistic bus company in the province -

Ulsterbus - has consistently covered costs and not made subsidy demands like the 'territorial' bus companies in Britain, has led to some complacency. Ulsterbus has traditionally been opposed to 'experiments'. The voluntary sector has been concerned for the handicapped but not for self-help 'public' transport. The exception is a little-known Order in Council from 1981 which allows payment for private lift-giving. Immediate needs seem to be for a system of minibus permits, community bus permits, seedcorn funding like the Rural Transport Innovation Grant, and a legitimation of the controversial shared taxis. Draft legislation was expected in 1989.

(b) **Market Share**

As with any other transport mode or sector, its role and status at any time could be judged by the market share it holds. This hardly seems appropriate to UCMs, as they are by nature not commercial or competitive enterprises, but quite the opposite, i.e. very largely aimed at mopping up pockets of demand which have been neglected by the market. It is axiomatic, therefore, that the proportion of public transport trips accounted for by UCMs is inevitably small, and probably very small. This could be extremely difficult to work out, for the usual reasons. One estimate was made by Cumbria County Council, who found that patronage on postbuses, social cars, a community bus and a shared taxi in the county together comprised about 1% of all public transport trips (Cumbria CC, personal communication). This low figure is not surprising, and may be typical of other counties, and begs the question of whether the importance of UCMs has been exaggerated.

There are several answers to this. First, the fact that this is a highly disparate sector with (arguably) up to 20 different modes means that considerable time and effort has to be spent in describing and analysing each one, and this may give the impression that the sector is more important than it really is. However, this misses the point. The great majority of UCMs were intended as local solutions to local problems, and they must be locally appraised. Thus, 1% of the county aggregate might constitute 100% of public transport trips in places which previously had no services at all.

Thirdly, the UCM/CT sector might constitute much more than 1%, depending on what is included. Familiar and traditionally understood UCMs such as postbuses and community buses are additional transport resources deliberately designed as a 'minimum' expedient at local scales. Other UCMs are adaptations of conventional transport, not so locally confined, and if more widely encouraged could be deployed over whole networks. If the new generation of PSV minibuses is regarded as 'unconventional', then this is thought to comprise 10-15% of the national bus fleet. It is conceivable that many existing buses could be adapted to a Flexibus (partly D/R) mode, to combine passengers with freight, or to operate as a 'multiple service bus'. Such adaptations might be easier to implement and benefit more people than attempting to set up completely new infrastructure for a local community bus. Taking advantage of recent legislation, a greater proportion of the taxi market will be on a 'shared' or 'taxibus' basis. If the long-overlooked 'welfare' CT sector is included then the 'market' increases enormously: persons needing special transport arrangements to cope with physical mobility problems might comprise 10-25% of the population; the average county might have 100-150 voluntary transport organizations. It is possible therefore that the 1% of the market occupied by specific visible UCMs might expand up to, perhaps, 50% if *all* likely modes were included and all possibilities for adaptation were taken up. However, this is not to say that such a scenario would necessarily be the best means of meeting the transport needs of any area.

(c) Local Authority Attitudes

Through the medium of PTPs and TPPs, which are official documents, one can get a valid impression of attitudes towards UCMs and their role in overall public transport provision. Local authorities are virtually unanimous on the limited role of such expedients. The following quotations are typical.

> "A network of supplementary transport to fill the gaps in the regular system is...being considered for introduction in cases where social needs cannot be fulfilled using existing public transport resources." (Dyfed PTP 1979/80).

Others are even more cautious:

> "Where appropriate, and particularly in outlying rural areas, to give consideration to the provision and use of alternative means of achieving a reasonable level of service in the most effective and economic manner, including investigation of, and where appropriate, assistance to less conventional forms of public passenger transport." (Hampshire PTP 1985/6).

The supplementary nature of UCMs is emphasised by forbidding competition with other transport:

> "Unconventional services should not be in conflict with conventional services nor abstract traffic from them; they are only likely to suit a limited number of small areas of the County." (West Sussex PTP 1980/81).

Many authorities discourage the idea of UCMs replacing conventional buses:

> "Unconventional public transport services, e.g. minibuses, community buses, rural car services and postbuses would appear to have an increasing role to play as car ownership increases and conventional services become more uneconomic. Each of these, however, has its disadvantages and few of them can completely replace the conventional bus service." (Gwynedd PTP 1979/80).

Positive attitudes are nearly always tempered with caution:

> "The Regional Council has always encouraged and promoted unconventional solutions to the Region's transport problems where these appear cost-effective..... to a limited degree they also now constitute a small part of the Strategic Network. By their nature however these solutions tend to offer only a

> restricted level of service.... These services are therefore no substitute for a conventional regular service..." (Borders TPP 1986/7).

While some are fairly negative:

> "It is the view of the County Council that schemes of this nature can make only a very small contribution to the public transport network in the County." (Avon PTP 1979/80).

On the potential contribution of the voluntary sector, some authorities have taken the lead in promoting action, but others insist on the main initiative coming from the community itself. Contrast the following:

> "The Council will give further encouragement to self-help schemes such as the Langwathby Community Minibus in suitable areas, giving financial support where funds are available."
> "It is the County Council's policy to establish the Voluntary Social Car Service throughout the rural areas of the County." (Cumbria PTP 1985/6).

> "The County Council is always willing to consider financial support and advice for new schemes and experiments, but initiative and organisation for schemes must also come from the local communities." (Cheshire PTP 1985/6).

> "The County Council remains keen to promote community transport but recognizes that such solutions cannot be imposed from without; they must grow out of local commitment, and few communities are prepared to make that commitment while they have any chance of retaining a professionally run bus service, however infrequent." (Devon TPP 1985/6).

In a depressingly high number of cases, local authorities report a lack of initiative or will among community organisations:

> "This type of scheme [a voluntary car] depends very much on the goodwill of local organisers and voluntary drivers. Previous attempts to introduce this type of scheme met little local enthusiasm...." (Gwynedd PTP 1979/80).

> "Unfortunately recent experience indicates a general lack of will, even from rural communities, to become involved in self-help schemes." (Kent PTP 1985/6).

An interesting point is the degree to which councils display changing attitudes, or the perception of changing requirements, between the late 1970s and mid-1980s. Only a few show clear signs of accepting change.

> "...Nevertheless, government proposals contained in the Transport Bill (1985) may mean that unconventional solutions will have a greater part to play. Further initiatives are therefore being taken to ensure that such solutions are not overlooked." (Cheshire PTP 1985/6).

Compare the following quotes from the same county:

> "...as yet there has not appeared to be significant evidence of local demand or justification for the promotion of community bus and car sharing schemes." (Northumberland PTP 1979/80).

> "...there are nevertheless other forms of public transport, such as the Post Bus or Community Bus, that may be attractive in certain circumstances. It is therefore County Council policy to ...encourage the development of other modes of transport in rural areas where such schemes can be shown to be more cost-effective than conventional services." (Northumberland PTP 1984/5).

Again, changing attitudes from one county over six years:

> "...no constructive efforts have so far been made by any rural community in Essex towards

> initiating a broad-based type of self-help scheme..."
> [the Council] "...is no longer disposed...to underwrite costly experiments such as dial-a-bus and park-and-ride which in its experience have failed to achieve any significant change of mode or generate a substantial increase in ridership." (Essex PTP 1979/80).
>
> "The County Council welcomes the powers conferred upon it under [the 1981 Act] and will be willing to exercise them in respect of unconventional schemes, such as scheduled shared taxi services, should it consider them appropriate to meet the needs of any isolated rural community. It will similarly be prepared to consider making funds available to permit the introduction of community bus services under the provisions of..." [the 1981 Act]. (Essex PTP 1985/6).

(d) **Present and Future Role**

In general, there is no doubt that by the landmark year of 1985 UCMs were universally perceived to be confined to a local, supplementary role. Economic and accessibility analysis has endorsed their suitability for this. As throughout 1980s Britain, consensus on public services provision is no longer tenable, and all sectors have been expected to reappraise their resources, operations and role. In the light of upheavals in all sectors of transport, there can be no confidence that UCMs can retain their hard-won niche and resist pressures for change. All prevailing trends in the later 1980s suggest that a different role might be thrust upon them.

(i) Continuing innovation and adoption of new modes.
(ii) Rise of the 'self-help' movement.
(iii) Sequence of permissive legislation, up to and including the 1985 Act, aimed at facilitating new types of service.
(iv) Continuing decline of conventional public transport, believed by many to be perpetuated by deregulation, and compounded in rural areas by the steady loss of other local facilities.

(v) Financial pressures on local authorities to enforce 'least cost solutions'.

Taken together, the unavoidable inference to be drawn from these trends is that UCM/CT growth might be expected to *replace* traditional public transport. This must be the main issue at present. Is the 'unconventional' sector now expected to adopt a wider role and begin to replace conventional public transport, as state support is withdrawn? If so, can they succeed in doing this? Such issues have not yet been officially acknowledged. There has certainly been a trend for local authorities, having had transport service cuts forced upon them, to feel obliged to consider UCMs as low cost substitutes - a role for which they were not designed.

It has never been seriously suggested, even hypothetically, that a complete network of UCMs could adequately take over from conventional buses the job of providing for the everyday travel needs of the entire non-car population. Could this happen? PSV minibuses would come nearest, if they are regarded as 'unconventional'; it was suggested above that 'multiple service' and 'flexible' buses and passenger/goods integration (public/schools integration already having gone as far as possible) could play a much greater part. Most other modes would remain localised, although additional 'schemes' could always be encouraged. But, what *is* the policy? After deregulation, there is no policy but laissez-faire. As expressed in the White Paper (Department of Transport 1984) the government view seems to be, assuming the success of the overall deregulation strategy, that UCMs will expand but remain supplementary. However, the encouragement given to UCMs in the White Paper and in the subsequent Act implies that deregulation will not succeed everywhere. The proposal for a Rural Transport Innovation Grant stemmed from the admission that "A free market will help but cannot on its own solve..." [the rural problem].

Similar ambivalence surrounds official attitudes to community transport. Self-help is encouraged, as an easy option for the state, in that potentially expensive local problems are taken off its hands (although there may still be significant public support costs and administrative aid). Again, it implies deregulation and the free market might fail in some circumstances. It was pointed out earlier that this could be seen as a cynical ploy by government to withdraw from its traditional responsibilities while at the same time applauding local initiative. Considerable responsibility attaches to the local volunteer organizers, particularly with respect to the crucial test of maintaining community support and finance in the medium and longer term. If a community bus fails, then the conventional bus is unlikely to return. Banister (1982) indicates that self-help could be a transitional stage before the complete demise of public services, and in this case, failure could be blamed on the community itself with officialdom seeing this as evidence that there is "no demand".

14.4 NEW MODES SINCE 1985

(a) Deregulation

The period from 1986 to the time of writing in 1988 is of course dominated by the early stages of the newly deregulated bus system (not applicable in London or Northern Ireland). A basic outline of the policy principles and likely issues was given in chapter 1 and, as the place of UCMs within the overall strategy is still very minor, it is not necessary to enter into a detailed discussion of the changes brought about under the new regime. It is intended merely to search for evidence on the role of UCMs. The change has certainly brought the expected uncertainty and unpredictability into the bus sector, and although this may lessen with time, the network under the new system will probably remain less stable than under the old. It is anticipated that the situation will not settle down until at least 1990, and it may not be until then that firm trends become clear.

It is expected that the number and diversity of UCMs will continue to increase, but the relationship with the new policy could be either positive or negative. Positive encouragements to innovation are:

(i) Legal changes, permitting certain types of operation not previously possible, primarily shared taxis and 'taxibus' services.
(ii) The purely commercial stimulus of reduced bureaucracy, more competition and (after privatization of the NBC and municipal bus companies) new private sector entrants, producing *inter alia* 'unconventional' minibus operations.
(iii) The Rural Transport Innovation Grant, providing central government funds for new rural services of an 'innovatory' nature.

Negative factors in the appearance of UCMs refer to the problem exposed in the previous section, namely that deregulation will cause gaps to appear in unprofitable areas which could only be filled by some kind of (possibly inadequate) UCM. Where routes are 'non-commercial' and the local authority does not see the need or does not have the resources to support a tendered service, or where the subsidy requirement is too high, or where part of a subsidised route loses traffic to the predatory actions of a more powerful competitor, then local communities may be deprived of public transport. UCMs initiated under (i), (ii), (iii) above may have found new markets, but they may also be acting as replacements under such a scenario, and therefore not constitute any real addition to overall transport resources. Only detailed local case studies could establish this.

Also, there was concern that existing UCMs, possibly developed and supported by local authorities and communities over several years, might be undermined by commercial services abstracting traffic and will themselves be subject to competitive tender. In response to lobbying on this, postbuses and community buses were granted five years (1986-91) exemption from tendering (under SI 1985/1921). Other voluntary modes would not normally be subject to tender, and those UCMs which are adaptations of conventional buses would have to register or tender in the usual way. The only UCMs definitely known to have been killed off by deregulation are the Bassetlaw Community Bus and the Horncastle Bus Club.

Deregulation is being extensively monitored by TRRL and results of the initial stages have been reported in a generalised form. Summaries of new service registrations up to October 1986 have been provided by Finch et al (1986) and Balcombe et al (1987), while a review of progress under the new system up to November 1987 has been made by Balcombe et al (1988). An intermediate report (Rickard et al 1988) is a distillation of the interim results of half a dozen on-going research projects commissioned by the TRRL and the Scottish Development Department, which analyse the effects of deregulation in a range of non-metropolitan case study areas. The general conclusions are that the early stages of deregulation have been generally successful, with little dramatic change to networks, little disruption and considerable cost savings to operators and local authorities. Despite this fairly bland conclusion, the reports do include many examples of problems involving registration and tendering strategies, service changes and withdrawals, competition, fare inconsistencies, information and reliability, but which are presented merely as temporary local difficulties. Time will tell, but some observers have attributed this uncritical stance to the TRRL's lack of independence. Other reviews (e.g. Knowles et al 1988) concentrate on change and dislocation, local variations and uncertainties, and comparison with the original ideals behind the policy.

Interest in the potential for further UCM development demands an examination of the extent of 'conventional' service withdrawals under the new conditions. Amid the controversy surrounding the new policy was the prediction that commercial services would compete in areas with highest traffic potential, and abandon routes which were unremunerative, as cross-subsidy is no longer possible. Survival of 'non-commercial' routes then depends entirely on whether the local authority's budget is large enough to cover all the necessary tenders. Where previously subs-idized services register as commercial and reduce the subsidy bill, then it might be possible to support all those remaining. It should be remembered, however, that under the 1983 Transport Act, the minister may limit any expenditure on public transport and thus subsidized services will remain at the mercy of government caprice.

Alarmist fears of the complete collapse of rural services have certainly not been justified. Reports of service reductions in the countryside are fairly numerous but idiosyncratic, and cannot be generalised or quantified at an aggregate level. These cases are submerged in national average figures, which indicate that commercially registered services constituted 70-80% of mileage previously operated (Balcombe et al 1987). As would be expected, among the TRRL/SDD case study areas there was much greater variation: the lowest was 20% in rural north Powys; a sharp urban/rural split was recorded in Clwyd (average 52%) and Lancashire; rural Scotland, against the expected trend, shows higher rates, between 50% in Dumfries & Galloway and 70% in Highland Region; apparently rural west Wiltshire returned 79% commercial registrations (Rickard et al 1988). Given these higher than expected rates, local authorities were generally left with sufficient resources to support the non-commercial services they felt necessary. In Mid-Wales there was virtually no change to the network, as the councils had over previous years built up the most efficient set of subsidized routes (Knowles et al 1988). There is very little information on services abandoned, either dropped by commercial operators or not offered for tender, save an admission that frequency reductions occurred in early mornings, evenings and Sundays and there was "substantially reduced coverage of rural and suburban areas" (Balcombe et al 1987). Of the study areas, some losses were reported in north Powys. But the overall conclusion is that the new system "has resulted in the maintenance of the level of the pre-deregulated network..." (Rickard et al 1988).

At first sight, there is little scope here for the generation of new UCMs to fill gaps in the deregulated network. Many new services are run by minibuses instead of traditional vehicles, but many other modes such as taxis and community transport have a different licensing procedure or do not have to register, and are therefore "outside the system" as far as many of these studies are concerned. It is not yet possible to test the "UCM replacement" thesis advanced earlier.

Continuing uncertainty suggests that service losses and low cost or CT replacements might appear at any time in the future. For example, the surprisingly high rate of commercial registrations conceals some dubious situations indicative of a general fragility in the industry. There is evidence that uneconomic routes were registered as commercial as operators strove to protect their territory and avoid competitive tendering; future de-registration is possible. Cost-cutting by operators has been achieved mainly by staff redundancies and unethical employment practices causing the reduction of wage levels; these are 'once only' savings. Many tenders have been deliberately under-priced to gain contracts. Such trends could lead to increases in future subsidy demands, while the much-heralded savings in county transport budgets may simply invite those budgets to be cut by government. While there are short term grounds for claiming success, deregulation has done nothing for stability and confidence, and it is in this respect that UCMs will retain a potential replacement role.

(b) Innovation

Unfortunately, PTPs were discontinued by the 1985 Transport Act, but in the new circumstances there are two sources of data on UCMs. The TRRL's Bus Registration Index is a complete database of post-deregulation services, from which apparently 'unconventional' ones might be extracted. Many UCMs are excluded from registration, and only services run by taxi operators and commercial minibuses could be identified by Finch et al (1986). Just after implementation of the policy in October 1986, almost 60 services were registered by taxi operators (mapped in figure 12.3), and these have since expanded further (Finch 1988). This is clearly a major gain when compared with the previous difficulties of licensing a shared taxi service. All are of the 'taxibus' type rather than the shared use of conventional taxis. There have also been subsidised taxibus services contracted by local authorities (more details have been given in chapter 12). Although it is difficult to be precise about numbers of vehicles or routes, there were almost 200 locations known at this time to have mini- or midibuses. At the end of 1986 there were about 3000 PSV minibuses on stage services, a figure which has since doubled to reach an estimated 10-15% of the market (see chapter 5).

In the TRRL/SDD case study areas, most references to 'innovation' also concern minibuses. They are present, or even common, in many towns in all but one of the English and Welsh study areas, but not yet in the Scottish. Some bus companies report plans to convert significant proportions of their fleet to small vehicles. There was only one report amongst the study areas of a new taxibus service (in Wiltshire) (Rickard et al 1988).

The other source of data on new UCMs is the list of projects assisted under the Rural Transport Innovation Grant, which was set up under the 1985 Act to fund the development of new local services in rural areas. In England, this is known as the Rural Transport Development Fund and administered by the Development Commission, to the value of £1 million. In Scotland and Wales the RTIG is administered by the Scottish and Welsh Offices with budgets of £170,000 and £100,000 respectively. It is mainly intended as a capital grant, especially for new vehicles, but is also available for other 'setting up' costs such as driver training and publicity, the appointment of local transport advisers or brokers, and even contributions to initial running costs. The eligibility conditions are that it should facilitate a *new* service, not competing with other transport, which has local support and will benefit rural communities. Strangely, the specification does not define "innovation" and it is unknown whether any application has been turned down for being too 'conventional'!

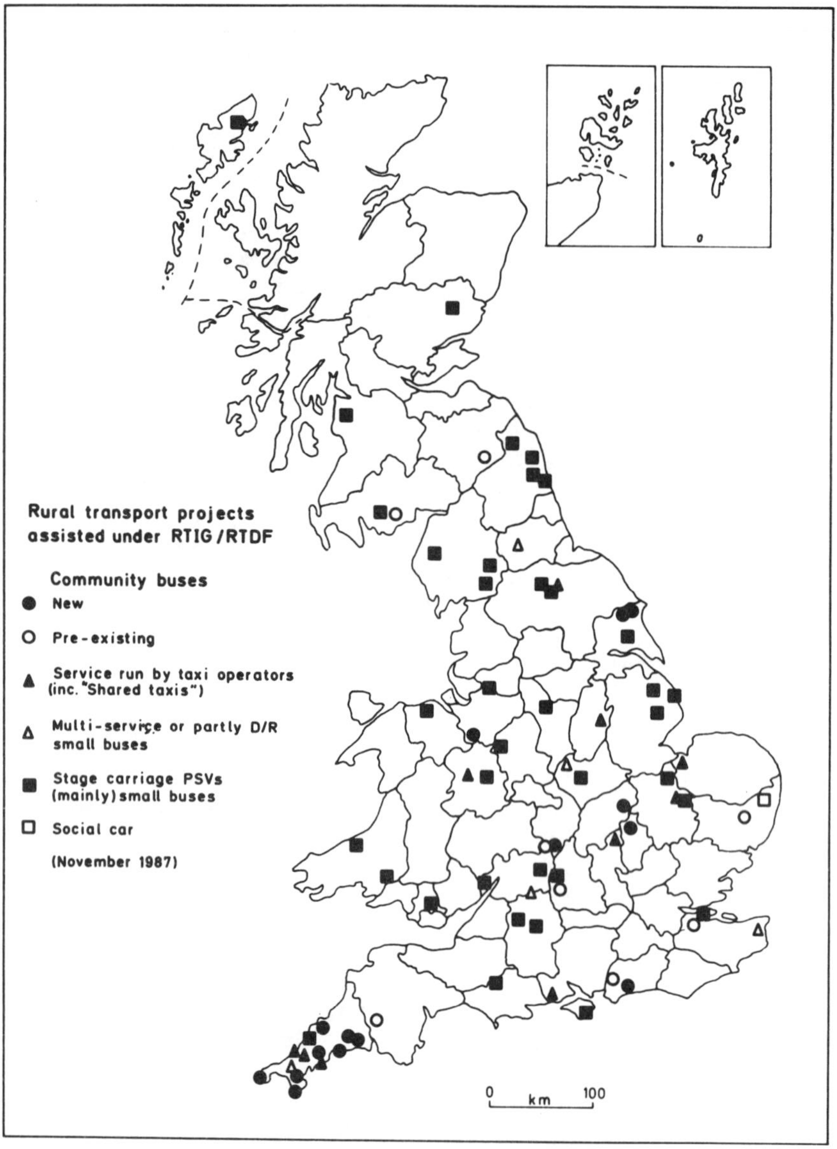

FIGURE 14.2 Rural Transport Innovation Grant: projects assisted 1987.

Assisted projects up to the end of 1987 are shown on figure 14.2. It is immediately apparent that not all of the new services count as UCMs under the definition employed here. Out of the 75 schemes shown on the map 35 appear to be ordinary PSVs, although most are using small vehicles. There are 15 newly established community buses, of which eight were set up in Cornwall as a result of a joint initiative by the County Council and local voluntary workers. In addition, another eight community bus associations have been allowed to purchase a second vehicle. The eleven services offered by taxi firms are all of the semi-stage 'taxibus' type. Another five PSV buses or minibuses demonstrate some degree of D/R and/or multi-purpose operation, such as attendance to health and education needs, and with wheelchair facilities. Grants for 'infrastructure' were given to 18 counties for the appointment of transport advisers or brokers; a few smaller grants were given for local surveys and publicity. Exceptionally, grants have also been made for the re-opening of railway stations on the Leeds-Settle-Carlisle line.

In the geographical sense, the major anomaly is the lack of interest in Wales and Scotland. Overall, the rate of progress by the end of 1987 is on the face of it quite impressive. Strictly, however, if one were to count only the completely new services operated by modes within the UCM definition, then only 32 out of the 75 on the map would qualify.

Appraisal of these developments must refer back to the question posed earlier on the role of the UCM sector. Do these new services constitute a real increase in overall transport resources? Are they catering for long term outstanding needs? Or, are they just filling in gaps exposed by deregulation? Answers to these could only be obtained by a series of detailed case studies on the local effects of the new policy.

(c) Prospect

Current trends indicate that the expectations prior to the 1985 Act that UCMs will increase are being verified. They are still a small but probably a growing proportion of total public passenger transport by road. On their role, however, it would seem that certain types at least are unlikely to remain supplementary. Attempts to predict where the post-deregulatory confusion might be leading would not be helped by over-generalizing the disparate array of modes described above as a single sector with a uniform role. There are three main subdivisions.

In a competitive climate composed of independent and newly privatized bus companies, those UCMs which are adaptations of conventional buses - multi-purpose or multiple service buses, D/R diversions through to fully fledged dial-a-rides, free shoppers' buses and passenger/freight combinations - would seem to have no impediment to further expansion. There are no legal obstacles and no special licences are required. In environments of the type which spawned most UCMs - low, dispersed or specialised demand - there might be less or no commercial stimulus but such adaptations may be the only way of capturing a viable traffic load. Whether or not the spate of new urban minibuses is regarded as being within the UCM definition, its occurrence certainly testifies to some spirit of innovation. Many of these flexible forms have appeared with the assistance of the RTIG and, although it is uncertain how many examples subsist without this grant aid, it is evidence of a willingness to innovate which should spread more widely. The spread of 'taxibus' services supports the trend, despite the need for special licences. At least in urban environments, certain UCM types could succeed, become established and even 'conventional' in time. A danger is that such experimentation will be complacently ascribed to the benefits of competition, with no concern shown to the instability produced by deregulation, and no apparent need for any coherent central policy based on the transport needs of consumers.

An intermediate group is those modes which are inhibited from taking advantage of "new commercial opportunities" (if they exist in the areas concerned) through their dependence on institutional cooperation. Postbus development seems to have nowhere further to go, due to the constraints of linking two different service types. Combining passenger traffic with public sector institutional demands by means of school buses or more ambitiously through an ESCORT- or Courier-type system could, and perhaps should, become normal in 'non-commercial' areas, although any commitment to satisfying travel needs might well be outweighed by the demand for resource savings through rationalization.

It is in the voluntary (CT) sector where UCMs are under the greatest threat of having an unsuitable role thrust upon them. In the absence of any policy which guarantees a minimum level of service for meeting needs in all areas, the local community itself is the last resort. All 'non-commercial' areas are vulnerable to being left exposed by a deregulated free market system or by local authority budgets inadequate to subsidize services. The abandonment of political consensus on state support for public services (not just transport) has been made on the assumption that private or voluntary initiative exists to fill any gaps, an assumption that is rarely sustained at local level. Hence, community bus and social car schemes reflect localities with the requisite organizational skills rather than needs, and highlight a social/spatial polarization between those who *need* help and those who *can* help. The traditional CT sector (providing mobility for the elderly and handicapped) is squeezed between increasing demands, contracting state support for public health and welfare facilities, unreliable sources of funding, and ambivalence over the role of the volunteer. In many ways, the UCM/CT role within transport is a microcosm of 1980s Britain.

REFERENCES

ABBOTT J. (1983) The Class 141 railbus: the reaction of the Passenger Transport Executives. *Modern Railways*, 40, 69-71.

ACHESON N. (1988) *The use and availability of minibuses in the Northern Ireland voluntary sector*. Belfast: Northern Ireland Council for Voluntary Action.

ANON (1986) The minibus story. *Bus Fayre*, 9, 68-94.

ASSOCIATION OF COUNTY COUNCILS (1984) *Review of rural railways*. London: ACC.

ATHERTON G. & TAYLOR J. (1988) *Getting to school by bus. Safety, supervision and standards of service*. Glasgow: Scottish Consumer Council.

AWDAS D. (1978) Rural transport coordination. In CRESSWELL, 1978, op cit, 119-138.

BAILEY J.M. (1979) Social transport based on 'need'. *Surveyor*, 154 (Sep. 20), 14-16.

BAILEY J.M. (1984) Is there a role for dial-a-ride in Britain? *Proceedings of Seminar J, PTRC Conference*, 153-168.

BAILEY J.M. & APPLEBY L. (1986) A comparative study of special transport for disabled people in Britain, *Contractor Report* 13. Crowthorne: Transport & Road Research Laboratory.

BAILEY J.M. & LAYZELL A.D. (1983) *Special transport services for elderly and disabled people*. Aldershot: Gower.

BALCOMBE R.J. (1979) The rural transport experiments: a mid-term review, *Supplementary Report* 492. Crowthorne: Transport & Road Research Laboratory.

BALCOMBE R.J. (1980a) Demand-responsive bus services in rural areas. In TRANSPORT & ROAD RESEARCH LABORATORY, 1980, op cit, 40-60.

BALCOMBE R.J. (1980b) Summary and conclusions. In TRANSPORT & ROAD RESEARCH LABORATORY, 1980, op cit, 94-103.

BALCOMBE R.J. & DREDGE A.S. (1980) Community transport. In TRANSPORT & ROAD RESEARCH LABORATORY, 1980, op cit, 61-78.

BALCOMBE R.J. et al (1987) Bus deregulation in Great Britain: a review of the opening stages, *Research Report* 107. Crowthorne: Transport & Road Research Laboratory.

BALCOMBE R.J. et al (1988) Bus deregulation in Great Britain: a review of the first year, *Research Report* 161. Crowthorne: Transport & Road Research Laboratory.

BANISTER D. (1981) Self-help supersedes public transport. *Geographical Magazine*, 53, 725-728.

BANISTER D. (1982) Community transport for rural areas: panacea or palliative? *Built Environment*, 8, 184-189.

BANISTER D. et al (1984) Towards needs-based transport planning. *Traffic Engineering & Control*, 25, 372-375.

BANISTER D. (ed) (1985) *Rural transport and planning: bibliography with abstracts*. London: Mansell.

BANISTER D. & NORTON F. (1988) The role of the voluntary sector in the provision of rural services - the case of transport. *Journal of Rural Studies*, 4, 57-71.

BEESLEY M.E. & GLAISTER S. (1985) Deregulating the bus industry in Britain - a response. *Transport Reviews*, 5, 133-142.

BENDIXSON T. (1977) *Instead of cars*. Harmondsworth: Penguin.

BLY P.H. & OLDFIELD R.H. (1986) Competition between minibuses and regular bus services. *Journal of Transport Economics & Policy*, 20, 47-68.

BONSALL P.W. (1980a) Car pooling in the USA: a British perspective, *Supplementary Report* 516. Crowthorne: Transport & Road Research Laboratory.

BONSALL P.W. (1980b) A survey of attitudes to car sharing: a database for microsimulation, *Supplementary Report* 563. Crowthorne: Transport & Road Research Laboratory.

BONSALL P.W. et al (1983) Ridesharing in Great Britain: performance and impact of the Yorkshare schemes. *Transportation Research*, 17A, 169-181.

BORDERS REGIONAL COUNCIL (1981) *Border Courier service: final monitoring report*. Newtown St Boswells: BRC.

BOVY P.H. & KRAYENBUHL V. (1979) Introductory report. In ECONOMIC RESEARCH CENTRE, 1979, op cit, 5-50.

BROWN C. & BARNES A. (1985) Experience with ESCORT. In *Twelfth Seminar on Rural Public Transport*, edited by P.R. WHITE, pp 16-25, Polytechnic of Central London.

CAMBRIDGESHIRE COUNTY COUNCIL (1977) *Rural midibus project phase 1*. Cambridge: Cambs. C.C.

CARPENTER T.C. (1977) Postbuses in Scotland. In TRANSPORT & ROAD RESEARCH LABORATORY, 1977a, op cit, 157-168.

CHARNOCK D.B. et al (1983) Public transport coordination in shire counties - the East Sussex approach. *Traffic Engineering & Control*, 24, 471-476.

CHILDS R.J. (1978) Nottinghamshire's rural transport experiments: the Bassetlaw Community Bus. In *Seventh Seminar on Rural Public Transport*, edited by P.R. WHITE, pp 32-58, Polytechnic of Central London.

CLOKE P. (ed) (1985) *Rural accessibility and mobility*. St. David's Univ. Coll., Lampeter: IBG Rural Geography Study Group.

CLWYD COUNTY COUNCIL (1978) *New ways forward: alternatives in public transport in Clwyd*. Mold: CCC.

COE G.A. & FAIRHEAD R.D. (1980) Rural travel and the market for public transport in Rutex areas. In TRANSPORT & ROAD RESEARCH LABORATORY, 1980, op cit, 3-18.

COE G.A. & JACKSON R.L. (1981) The taxi and private hire car industries in England and Wales, *Laboratory Report* 1011. Crowthorne: Transport & Road Research Laboratory.

COE G.A. (1985) Taxibus: a shared taxi service in Stourport, *Supplementary Report* 841. Crowthorne: Transport & Road Research Laboratory.

COE G.A. (1986) The taxi and hire car industry in Great Britain in 1985, *Research Report* 68. Crowthorne: Transport & Road Research Laboratory.

COUSINS S. (1979) *Green cars- a brief guide to shared car ownership*. Milton Keynes: Alternative Technology Group, Open University.

CRESSWELL R. (ed) (1978) *Rural transport and country planning*. Glasgow: Leonard Hill.

CROCKFORD E.D. (1974) The Harrogate "Chauffeur Coach" service, *Traffic Engineering & Control*, 15, 764-769.

DAVIES H.E.H. (1987) The East Sussex transport broker: a preliminary report, *Research Report* 88. Crowthorne: Transport & Road Research Laboratory.

DEPARTMENT OF TRANSPORT (1977) *Transport policy*, Cmnd 6836. London: HMSO.

DEPARTMENT OF TRANSPORT (1978) *A guide to community transport*. London: HMSO.

DEPARTMENT OF TRANSPORT (1979) *Innovations in rural bus services. The Government's response to the Eighth Report from the Select Committee on Nationalised Industries*, Cmnd 7743. London: HMSO.

DEPARTMENT OF TRANSPORT (1984) *Buses*, Cmnd 9300. London: HMSO.

DEVON RUTEX WORKING GROUP (1978) Rural transport experiments: the Exe Valley Market Bus, *Supplementary Report* 427. Crowthorne: Transport & Road Research Laboratory.

DEVON RUTEX WORKING GROUP (1979) Rural transport experiments: the mid-Devon lift-giving scheme, *Supplementary Report* 525. Crowthorne: Transport & Road Research Laboratory.

DEVON RUTEX WORKING GROUP (1980a) Rural transport experiments: the Taw Valley car service, *Supplementary Report* 570. Crowthorne: Transport & Road Research Laboratory.

DEVON RUTEX WORKING GROUP (1980b) Rural transport experiments: South Exmoor car service, *Supplementary Report* 579. Crowthorne: Transport & Road Research Laboratory.

DICKINSON D.W. (1977) Stage bus diversion schemes. In TRANSPORT & ROAD RESEARCH LABORATORY, 1977a, op cit, 201-219.

DINEFWR RUTEX WORKING GROUP (1979) Rural transport experiments: the Welsh schemes, *Supplementary Report* 507. Crowthorne: Transport & Road Research Laboratory.

EASTERN COUNTIES OMNIBUS COMPANY (1975) *Rural midibus proposals*. Norwich: ECOC.

ECONOMIC RESEARCH CENTRE (1979) *ECMT Round Table 40: Paratransit*. Paris: European Conference of Ministers of Transport.

ENNOR P.D. (1977) Public transport in rural areas: a review. In TRANSPORT & ROAD RESEARCH LABORATORY, 1977a, op cit, 129-155.

FARRINGTON J.H. (1985) Transport geography and policy: deregulation and privatization. *Transactions, Institute of British Geographers*, NS 10, 109-119.

FEARNSIDE K. (1980) Experience of lift-giving schemes in Shropshire. In *Ninth Seminar on Rural Public Transport*, edited by P.R. WHITE, pp 7-22, Polytechnic of Central London.

FINCH D.J. et al (1986) Innovative public transport services, *Working Paper* WP(TP) 21. Crowthorne: Transport & Road Research Laboratory.

FINCH D.J. (1988) The impact of the 1985 Transport Act on the taxi industry, *Research Report* 148. Crowthorne: Transport & Road Research Laboratory.

FORD R. (1981) The DMU dilemma. *Modern Railways*, 38, 345-349.

FORD R. (1984) DMU replacement- can BR ever get it right? *Modern Railways*, 41, 121-125.

FORD R. (1986) The DMU replacement story. *Modern Railways*, 43, 315-320.

FULFORD C. (ed) (1987) *Hackney Community Transport annual report 1986*. London: Hackney CT.

GALLOP K.R. (1975) Demand responsive diversions on a rural bus service. In *Fourth Seminar on Rural Public Transport*, edited by P.R. WHITE, Polytechnic of Central London.
(Updated version in: WHITE, 1978, op cit, 73-94).

GLAISTER S. (1985) Competition on an urban bus route. *Journal of Transport Economics & Policy*, 19, 65-81.

GLASSBOROW D.W. (1978) Rural bus services -an assessment of current experiments. In CRESSWELL, 1978, op cit, 161-168.

GRAMPIAN REGIONAL COUNCIL (1985) *Community transport in Grampian*. Aberdeen: GRC.

GREEN G.R. (1978) Car-sharing and car-pooling -a review, *Supplementary Report* 358. Crowthorne: Transport & Road Research Laboratory.

GREENING P.A.K. & JACKSON R.L. (1982) Minibus pooling in Great Britain: legal and economic aspects, *Supplementary Report* 726. Crowthorne: Transport & Road Research Laboratory.

GREENING P.A.K. & JACKSON R.L. (1983) Shared taxi operation: cost considerations, *Supplementary Report* 793. Crowthorne: Transport & Road Research Laboratory.

GRIGG A.O. (1987) An assessment of the East Sussex coordinated rural transport project (Escort), *Research Report* 112. Crowthorne: Transport & Road Research Laboratory.

GRIMMER M. (1979) Dial-a-ride systems in Great Britain. In ECONOMIC RESEARCH CENTRE, 1979, op cit, 65-86.

GWILLIAM K.M. et al (1985) Deregulating the bus industry in Britain - the case against. *Transport Reviews*, 5, 105-132.

HALL D.R. (1988) The changing nature of minibus technology and deployment in UK public transport strategies, with particular reference to the Tyne & Wear area. In *Transport technology and spatial change*, edited by R.S. TOLLEY, pp 31-78. Stoke-on-Trent: IBG Transport Geography Study Group.

HALSALL D.A. (ed) (1982) *Transport for recreation*. Edge Hill College, Ormskirk: IBG Transport Geography Study Group.

HALSALL D.A. & TURTON B.J. (eds) (1979) *Rural transport problems in Britain: papers and discussion*. Univ. Keele: IBG Transport Geography Study Group.

HARKELL G. (1988) *London dial-a-ride review '88*. London: London Dial-a-Ride Users Association.

HARRIS M. (1978) A bus for the community. *Modern Transport*, 2, 74-79.

HERATY M.J. (1984) Monitoring demand-responsive transport for disabled people: the Readibus example, *Supplementary Report* 828. Crowthorne: Transport & Road Research Laboratory.

HILLMAN M. et al (1976) *Transport realities and planning policy*, PEP No. 567. London: Political & Economic Planning.

HILLMAN M. & WHALLEY A. (1980) *The social consequences of rail closures*, PSI No. 587. London: Policy Studies Institute.

HODGE P. (1981) Low-cost rural railways and the BR-Leyland Railbus. In *Tenth Seminar on Rural Public Transport*, edited by P.R. WHITE, pp 7-20, Polytechnic of Central London.

HOLDING D.M. (1983) The Swiss postal bus system- a comparison with British rural bus operation. In *Eleventh Seminar on Rural Public Transport*, edited by P.R. WHITE, pp 44-74, Polytechnic of Central London.

HUNTER D. & SMYTHE S. (1988) Efficiency debate. *Community Transport Magazine*, 7(1), 30-31.

JACKSON R.L. (1982) Minibus pooling: the Ilford experiment, *Supplementary Report* 739. Crowthorne: Transport & Road Research Laboratory.

JACKSON R.L. (1984) The role of taxis and private hire cars in Great Britain, *Laboratory Report* 1135. Crowthorne: Transport & Road Research Laboratory.

KAYE D. (1986) A community bus 10 years on. *Buses*, 38, 74.

KAYE D. & KITE R. (1983) *Community buses*, Unpublished MS.

KILVINGTON R.P. & McKENZIE R.P. (1985) A technique for assessing accessibility problems in rural areas, *Contractor Report* 11. Crowthorne: Transport & Road Reasearch Laboratory.

KIRBY R.F. et al (1975) *Paratransit: neglected options for urban mobility*. Washington DC: The Urban Institute.

KNOWLES R.D. (ed) (1985) *Implications of the 1985 Transport Bill*. Univ. Keele: IBG Transport Geography Study Group.

KNOWLES R.D. et al (1988) The age of the bus - or is it? *Geographical Magazine* 60(6), 36-41.

KOUTSOPOULOS K.C. (1980) Determining transportation needs. *Traffic Quarterly* 34, 397-412.

LAVERY H.I. (1986a) Public transport by shared taxi- the Belfast experience. *Conference paper*, Transport Operations Research Group, Univ. Newcastle.

LAVERY H.I. (1986b) Black Taxis - the shared taxi service in Belfast. *Omnibus Magazine*, 357, 39-43.

LAVERY H.I. (1986c) *Derry Taxis- the way ahead for the next five years*. Unpublished MS, Dept. Transport Studies, Univ. Ulster, Jordanstown.

LAVERY H.I. (1988) *Shared taxis and the Belfast Urban Area Plan 2001- evidence to the Public Inquiry*. Dept. Transport Studies, Univ. Ulster, Jordanstown.

LLOYD F.J. (1977) Dial-a-ride in the UK: a case study. In TRANSPORT & ROAD RESEARCH LABORATORY, 1977a, op cit, 99-113.

LUGTON J. (1980) Dual purpose transport. In TRANSPORT & ROAD RESEARCH LABORATORY, 1980, op cit, 79-93.

LUGTON J. & MACNIVEN D. (1979) The Scottish rural transport experiments. In PLANNING EXCHANGE, 1979, op cit, 3-22.

McCRACKEN D.M. & MILLAR G.I. (1974) A positive approach to school traffic. In *Third Seminar on Rural Public Transport*, edited by P.R. WHITE, Polytechnic of Central London.
(Also in WHITE, 1978, op cit, 58-72).

McLAREN A.D. (1979) Running a community bus service. In PLANNING EXCHANGE, 1979, op cit, 37-38.

McLAUGHLIN B. (1987) Rural policy into the 1990s - self help or self deception. *Journal of Rural Studies*, 3, 361-364.

MARTIN P.H. (1977) The Harlow dial-a-bus experiment: comparison of predicted and observed patronage, *Supplementary Report* 256. Crowthorne: Transport & Road Research Laboratory.

MARTIN P.H. (1978a) Comparative assessment of unconventional bus systems, *Supplementary Report* 387. Crowthorne: Transport & Road Research Laboratory.
(reprinted in TRANSPORT & ROAD RESEARCH LABORATORY, 1977a, op cit, 115-128).

MARTIN P.H. (1978b) The costs of operating dial-a-bus, minibus and conventional bus services, *Supplementary Report* 409. Crowthorne: Transport & Road Research Laboratory.

MASEY H. & HOPKIN J.M. (1986) Transport for children, elderly and disabled people. First impressions on the effects of the 1985 Transport Act, *Working Paper* WP(TP)25. Crowthorne: Transport & Road Research Laboratory.

MAWFORD R. (1976) Alternative forms of rural bus operation. In *Fifth Seminar on Rural Public Transport*, edited by P.R. WHITE, Polytechnic of Central London.
(Also in WHITE, 1978, op cit, 185-207).

MEADOWS T. & OXLEY P. (1988) Joining up. *Community Transport Magazine*, 7(5), 37-40.

MILEFANTI D.C. (1978) Government Rural Transport Experiments 1977. In CRESSWELL, 1978, op cit, 169-178.

MINISTRY OF TRANSPORT (1961) *Rural bus services: Report of the Committee*. London: HMSO.

MINISTRY OF TRANSPORT (1963) *Rural transport surveys: Report of preliminary results*. London: HMSO.

MINISTRY OF TRANSPORT (1965) *Rural bus services: Report of local enquiries*. London: HMSO.

MITCHELL C.G.B. (1974) Dial-a-ride in north America: report of a study tour in December 1973, *Supplementary Report* 95. Crowthorne: Transport & Road Research Laboratory.

MOSELEY M.J. (1979) *Accessibility: the rural challenge*. London: Methuen.

MOSELEY M.J. et al (1977) *Rural transport and accessibility* (2 vols). Norwich: Univ East Anglia.

MOSELEY M.J. & PACKMAN J. (1983) *Mobile services in rural areas*. Norwich: Univ. East Anglia.

MOYES A. (1977) Recreational transport in England and Wales: some recent developments. *Geography* 62, 209-212.

MOYES A. (1987) *An analysis of statistics from the Dyfed CC/WRVS Country Cars Service, Aberaeron scheme, 1984/5*. Unpublished MS, Dept. of Geography, Univ. Coll. Wales, Aberystwyth.

MURRELL R. & NEWPORT R. (1985) Minibuses in urban areas: straw in the wind or flash in the pan? *Conference paper*, Transport Operations Research Group, Univ. Newcastle.

NATIONAL CONSUMER COUNCIL (1978) *Rural rides. Experiments in rural public transport- a consumer view*. London: NCC.

NATIONAL COUNCIL FOR SOCIAL SERVICE (1977) *Rural transport: new developments in rural transport provision*. London: NCSS.

NATIONAL NORFOLK (1977) *Village bus service- the first year report*. Norwich: National Bus Company.

NICKSON J. (1981) The post bus faces a difficult future. *North 7 - The Highlands & Islands Magazine*, 45, 15-17.

NORFOLK COUNTY COUNCIL (1976) *Community bus experiment*. Norwich: NCC.

NORTH YORKSHIRE RUTEX WORKING GROUP (1979) Rural transport experiments: the Ripon Flexibus, *Supplementary Report* 491. Crowthorne: Transport & Road Research Laboratory.

NORTH YORKSHIRE RUTEX WORKING GROUP (1980a) Rural transport experiments: the Northallerton Hospital Transport Service, *Supplementary Report* 552. Crowthorne: Transport & Road Research Laboratory.

NORTH YORKSHIRE RUTEX WORKING GROUP (1980b) Rural transport experiments: Colsterdale car service, *Supplementary Report* 589. Crowthorne: Transport & Road Research Laboratory.

NORTH YORKSHIRE RUTEX WORKING GROUP (1980c) Rural transport experiments: Hackforth and district car sharing service, *Supplementary Report* 594. Crowthorne: Transport & Road Research Laboratory.

NUTLEY S.D. (1981) *The evaluation of accessibility levels in rural areas*. Cardiff: Welsh Office.

NUTLEY S.D. (1983) *Transport policy appraisal and personal accessibility in rural Wales*. Norwich: Geo Books.

NUTLEY S.D. (1985) Planning options for the improvement of rural accessibility: use of the time-space approach. *Regional Studies*, 19, 37-50.

NUTLEY S.D. (1988a) "Unconventional modes" of transport in rural Britain: progress to 1985. *Journal of Rural Studies*, 4, 73-86.

NUTLEY S.D. (1988b) "Unconventional modes" of transport in the United Kingdom - a review of types and the policy context. *Transportation Research A*, 22A, 329-344.

NUTLEY S.D. (1988c) "Unconventional modes" - appropriate technology? In *Transport technology and spatial change*, edited by R.S. TOLLEY, pp 1-30. Stoke-on-Trent: IBG Transport Geography Study Group.

OLDFIELD R.H. & BLY P.H. (1985) Using smaller buses in Central London, *Research Report* 31. Crowthorne: Transport & Road Research Laboratory.

OPEN TECH TRANSPORT PROJECT (1986) *Your guide to the 1985 Transport Act*. Cardiff: Transport Publishing Projects.

ORGANISATION FOR ECONOMIC COOPERATION AND DEVELOPMENT (1979) *Transport services in low density areas*. Paris: OECD.

ORRISS H.D. (1976) Community transport. *County Councils Gazette*, 69, 84-85.

ORSKI C.K. (ed) (1975) Special issue on Paratransit. *Transportation*, 4 (4).

OXLEY P.R. (1977) Dial-a-ride in the U.K. -a general study. *CTS Report* 12, Centre for Transport Studies, Cranfield Inst. of Tech. (reprinted in TRANSPORT & ROAD RESEARCH LABORATORY, 1977a, op cit, 43-77).

OXLEY P.R. (1980) Dial-a-ride: a review. *Transportation Planning & Technology*, 6, 141-148.

OXLEY P.R. & SEATON R. (eds) (1979) *Mobility in rural areas. Conference proceedings*. Centre for Transport Studies, Cranfield Inst. of Tech.

PETTIT T. (1987) Dyfed country cars. *Community Transport Magazine*, 6(5), 19.

PHAROAH T. (1987) Shared cars: key to reducing traffic. *Town & Country Planning*, 55, 240-242.

PLANNING EXCHANGE (1979) Rural Transport Experiments, *Forum Report* 17. Glasgow: Planning Exchange.

POLICY STUDIES INSTITUTE (1981) *The future of rural railways*. London: PSI.

POTTER S. (1987) *Who uses dial-a-ride?* London: London Dial-a-Ride Users Association.

POTTER S. & DAVIS D. (1986) *On bus survey of dial-a-ride users*. London: Federation of London Dial-a-Rides.

POTTER S. & HEISER B. (1987) *The development of dial-a-ride and other forms of transport for people with disabilities in London*. London: London Dial-a-Ride Users Association.

POWYS COUNTY COUNCIL (1981) *Powys Post Buses*. Llandrindod Wells: PCC.

RHYS D.G. & BUXTON M.J. (1974) The rural transport problem: a possible solution. *Town & Country Planning*, 42, 555-558.

RICKARD J.M. et al (1988) Deregulation of local bus services: interim results from non-metropolitan study areas, *Research Report* 139. Crowthorne: Transport & Road Research Laboratory.

RIGBY J.P. (1979) A review of research on school travel patterns and problems, *Supplementary Report* 460. Crowthorne: Transport & Road Research Laboratory.

RIMMER P.J. (1980) Paratransit: a commentary. *Environment & Planning A*, 12, 937-944.

ROWLANDS P. (1984) Swap-body buses - a rural answer? *Transport Engineer* (October), 67.

SCOTTISH RUTEX WORKING GROUP (1979) Rural transport experiments: Blackmount services, *Supplementary Report* 446. Crowthorne: Transport & Road Research Laboratory.

SCOTTISH RUTEX WORKING GROUP (1980) Rural transport experiments: Stair service, *Supplementary Report* 618. Crowthorne: Transport & Road Research Laboratory.

SCOTTISH RUTEX WORKING GROUP (1981a) Rural transport experiments: Ruralink service, *Supplementary Report* 665. Crowthorne: Transport & Road Research Laboratory.

SCOTTISH RUTEX WORKING GROUP (1981b) Rural transport experiments: South Ayrshire Hospital Transport Service, *Supplementary Report* 666. Crowthorne: Transport & Road Research Laboratory.

SELECT COMMITTEE ON NATIONALIZED INDUSTRIES (1977-78) *Eighth Report. Innovations in rural bus services*, HC 635. London: HMSO.

SKEWIS W.I. (1963) Here's a model system for rural areas. *Bus & Coach* (January), 32-33.

SLEVIN R. & COOPER A.E. (1973) Minibus and dial-a-ride: initial experiences with the Abingdon experiment. *Traffic Engineering & Control*, 14, 586-589.

SOUTHDOWN MOTOR SERVICES (1977) *Cuckmere Community Bus report*. Brighton: SMS.

SPELLER B.E. & MITCHELL C.G.B. (1975) The Harlow dial-a-bus experiment, *Supplementary Report* 127. Crowthorne: Transport & Road Research Laboratory.

STARK J. (1979) Development of postbus services within Wales and the Marches. In WELSH OFFICE, 1979, op cit, 16-19.

STONHAM P. (1985) Minis could raise maxi problems for local buses. *Surveyor* (15 August), 10-13.

SUTTON J.C. (1983) A minibus for the community. *Community Transport Quarterly*, 2(1), 20,43.

SUTTON J.C. (1985) *Social policies and transport services: an investigation into secondary and tertiary modes of passenger transport in the UK*. PhD thesis, Trent Poly., Nottingham.

SUTTON J.C. (1987) Community transport services in the United Kingdom: history, organization and recent developments. *Specialized Transportation Planning & Practice*, 2, 325-350.

TAYLOR J. & THOMPSON C. (1986) *Starting up. Running a community minibus. A basic guide.* Manchester: National Advisory Unit for Community Transport.

TRANSNET (1987) *Vehicle pooling schemes. Seminar papers from a one-day training session organised by Transnet and the LCTA.* London: Transnet.

TRANSPORT & ROAD RESEARCH LABORATORY (1977a) Symposium on unconventional bus services: papers, *Working Paper* WP(PT) 23. Crowthorne: TRRL.

TRANSPORT & ROAD RESEARCH LABORATORY (1977b) Symposium on unconventional bus services: summaries of papers and discussions, *Supplementary Report* 336. Crowthorne: TRRL.

TRANSPORT & ROAD RESEARCH LABORATORY (1980) The rural transport experiments: proceedings of a symposium, *Supplementary Report* 584. Crowthorne: TRRL.

TRANSPORTATION RESEARCH BOARD (1980) Paratransit 1980. *Transportation Research Record* 778.

TRANSPORTATION RESEARCH BOARD (1985) Transportation for elderly and handicapped persons, paratransit, and ridesharing. *Transportation Research Record* 1018.

TRANSPORTATION RESEARCH BOARD (1986) Innovations in ridesharing. *Transportation Research Record* 1082.

TUNBRIDGE R.J. (1980) A comparison of optimal minibus, dial-a-bus and conventional bus services, *Laboratory Report* 928. Crowthorne: Transport & Road Research Laboratory.

TUNBRIDGE R.J. & HALE D.M.J. (1978) Comparative assessment of demand responsive and fixed route bus services, *Laboratory Report* 847. Crowthorne: Transport & Road Research Laboratory.

TUNBRIDGE R.J. & MITCHELL C.G.B. (1977) The preliminary design of many-to-few dial-a-bus services, *Laboratory Report* 789. Crowthorne: Transport & Road Research Laboratory.

TURNER R.P. & WHITE P.R. (1987) NBC's urban minibuses: a review and financial appraisal, *Contractor Report* 42. Crowthorne: Transport & Road Research Laboratory.

TURNOCK D. (1977) The postbus: a new element in Britain's rural transport pattern. *Geography*, 62, 112-118.

VINCENT R.A. & WOOD K. (1979) Car-sharing and car-pooling in Great Britain: the recent situation and potential, *Laboratory Report* 893. Crowthorne: Transport & Road Research Laboratory.

WALDEN-JONES J. (1979) The use and value of voluntary car services in Wales. In WELSH OFFICE, 1979, op cit, 20-23.

WALMSLEY D.A. (1985) The use of minibuses in Great Britain, *Research Report* 6. Crowthorne: Transport & Road Research Laboratory.

WARRINGTON A. (1987) Taking the plunge. *Community Transport Magazine*, 6(2), 23-29.

WATTS P.F. (1976a) The operational characteristics of the Carterton dial-a-bus service, *Laboratory Report* 694. Crowthorne: Transport & Road Research Laboratory.

WATTS P.F. (1976b) The Harlow dial-a-bus service: some operational statistics, *Supplementary Report* 225. Crowthorne: Transport & Road Research Laboratory.

WATTS P.F. et al (1978a) British postbuses - a review, *Laboratory Report* 840. Crowthorne: Transport & Road Research Laboratory.

WATTS P.F. et al (1978b) British postbuses - service details, *Supplementary Report* 366. Crowthorne: Transport & Road Research Laboratory.

WATTS P.F. (1980) Rural shared hire-cars: a comparative assessment of their potential. In TRANSPORT & ROAD RESEARCH LABORATORY, 1980, op cit, 19-39.

WATTS P.F. (1984) The potential for postbus development, *Laboratory Report* 1091. Crowthorne: Transport & Road Research Laboratory.

WATTS P.F. (1985) Coordinated public transport in East Sussex: County Rider services 823 and 825, *Research Report* 7. Crowthorne: Transport & Road Research Laboratory.

WELSH OFFICE (1978) *Rural transport: a symposium 1977*. Cardiff: Welsh Office.

WELSH OFFICE (1979) *Supplementary transport: a symposium 1978*. Cardiff: Welsh Office.

WHITE P.R. (ed) (1978) *Rural public transport. A selection of papers from Rural Public Transport Seminars held at the Polytechnic of Central London 1972-1976 with bibliography*. Transport Studies Group, Polytechnic of Central London.

WHITELEGG J. (1984) Transport policy, fiscal discrimination and the role of the state. *Political Geography Quarterly*, 3, 313-329.

WHITELEGG J. (1987) Rural railways and disinvestment in rural areas. *Regional Studies*, 21, 55-63.

WILKES P.F. & MADGETT J. (1977) The Norfolk community bus I & II. In TRANSPORT & ROAD RESEARCH LABORATORY, 1977a, op cit, 169-189.

WILMERS P. & MOSELEY M.J. (eds) (1975) Rural public transport and planning, *Discussion Paper* 4. London: Regional Studies Assoc.

WILMERS P. et al (1981) Self-help. *The Planner*, 67, 59-74.

WINFIELD R.C. (1978) The use of school bus services by the public in Dyfed. In WELSH OFFICE, 1978, op cit, 131-132.

Appendix 1

LEGISLATION RELEVANT TO UNCONVENTIONAL MODES AND COMMUNITY TRANSPORT 1968 TO 1985

TRANSPORT ACT 1968

[Main purposes: bus nationalization, sets up NBC and SBG;
bus subsidies, fuel tax rebate, new bus grant authorized;
urban: PTEs set up in conurbations;
National Freight Corporation set up;
railways: subsidies for specific lines]
**UCMs
'Section 30 permits' allows LEA contracted school buses to carry fare-paying adult passengers on spare seats, and not need PSV standards;
also available for small passenger vehicles up to 12 seats, not needing PSV.**

POST OFFICE ACT 1969

[Main purpose: to change the status of the Post Office from a government department to a public corporation]
**UCMs
PO vehicles allowed to carry passengers;
postbuses eligible for local authority subsidy, fuel tax rebate and new bus grant.**

LOCAL GOVERNMENT ACT 1972
LOCAL GOVERNMENT (SCOTLAND) ACT 1973

[Main purpose: to reorganize local government in GB]
**Transport
Non-metropolitan counties made responsible for public transport coordination and subsidy;
TPPs to be prepared annually, including policies and financial programme, as a bid for Transport Supplementary Grant.**

RAILWAYS ACT 1974

[Main purpose: to re-formulate the basis for govt. funding of railways, esp. 'Public Service Obligation' grant to uneconomic sectors]
**UCMs
PSO gave some security to experimental schemes on marginal rural lines.**

LOCAL GOVERNMENT (MISCELLANEOUS PROVISIONS) ACT 1976

[Amongst various non-transport items...]

**UCMs

Extends the 1847 Town Police Clauses Act to allow District Councils in England and Wales to license taxis and hire-cars and to enact bye-laws for their control.**

1976 DoE Consultation Document "Transport Policy"

1977 DTp White Paper "Transport Policy"

Contains encouragement for UCMs and announces RUTEX programme.

MINIBUS ACT 1977

**UCMs

'Minibus permits' allow minibuses (up to 16 seats) used exclusively for educational, religious or social welfare purposes, for the benefit of the community and without a view to profit, to operate for hire or reward without PSV licensing. An enormous benefit for CT organizations.**

PASSENGER VEHICLES (EXPERIMENTAL AREAS) ACT 1977

**UCMs

Allows the Sec. of State for Transport to relax licensing regulations in certain designated areas, so that experimental schemes can be tested. This was used to set up the RUTEX programme, but remained in force afterwards, e.g. was later used to permit certain shared taxi/hire-car schemes.**

INNER AREAS ACT 1978

[Main purpose: to rejuvenate the inner city with development schemes and govt. funding]

**UCMs

Under the Urban Programme, provided funding and volunteer workers for numerous CT and dial-a-ride services.**

TRANSPORT ACT 1978

[Main purposes: to set up a semi-formal planning system for public transport (whereas the TPP system had over-emphasized roads), by requiring non-metropolitan counties in England and Wales to produce PTPs, defining the 'needs' of the area;

railways: various provisions on freight transport]

**UCMs
Community buses: minibuses (8-16 seats) may operate stage carriage services, for the benefit of the community and without a view to profit, without PSV standards; road service licences may be obtained for stage routes by a simplified procedure.
Social cars: schemes approved by local authority and traffic commissioners may charge fares to passengers on a non-profit basis without becoming a PSV; limited advertising allowed.
Lift-giving: casual passengers might offer a contribution to drivers' costs, but no advertising.**

TRANSPORT ACT 1980

[Main purposes: deregulation of express bus services; relaxation of stage bus licensing; trial areas for complete deregulation;
National Freight Corporation privatized]
**UCMs
Lift-giving and social car schemes: fares may be charged on non-profit basis if agreed before journey, advertising allowed.
School buses may carry adult fare-paying passengers without permits; LEA buses may do same, and may run normal stage services when not needed for schools work.
Minibus permits available for larger vehicles (17+ seats) on similar conditions.**

PUBLIC PASSENGER VEHICLES ACT 1981

[Commonly known as the 'PPV Act'.
A 'consolidating' Act, combining all recent legislation on road passenger transport]
**UCMs
Reaffirms recent changes on minibus permits, school buses, community buses, car sharing, etc.
'Restricted' operator's licence for certain PSVs - could be used by taxi firms to run minibuses for contract hire.**

TRANSPORT ACT 1962 (AMENDMENT) ACT 1981

[Known as the 'Speller Act'.
Waives formal railway line closure procedures so that experimental rail services can be tried on closed lines or freight lines]
**UCMs
Potentially useful for railbus experiments.**

CIVIC GOVERNMENT (SCOTLAND) ACT 1982

[Amongst various non-transport items...]

**UCMs

Allows District Councils in Scotland to license taxis and hire-cars and to enact bye-laws for their control. **

LONDON REGIONAL TRANSPORT ACT 1984

[London Transport removed from control of GLC, to become 'London Regional Transport' under direct control of DTp;
encourages some bus routes to be put out to tender]

**UCMs

Allowing some private sector competition stimulated experiments with PSV minibuses on urban routes.**

1984 DTp White Paper 'Buses'

Lays down policy for deregulation of bus industry, and general encouragement for UCMs.

LOCAL GOVERNMENT ACT 1985

[Main purpose: abolition of the GLC and six English Metropolitan County Councils]

**Transport

PTEs to remain, controlled by reps. from boroughs;
encourages urban bus routes to be put out to tender;
private competition stimulated use of PSV minibuses.
Loss of GLC support a serious threat to London CT and DaR services.**

TRANSPORT ACT 1985

[Main purposes: complete deregulation of bus industry in GB, non-commercial routes to be put out to tender;
NBC subsidiary companies to be privatized;
Local authority and PTE bus fleets to be privatized;
Local authorities have no role in public transport planning or coordination; PTPs discontinued]

**UCMs

Taxis and hire-cars: sharing by passengers at separate fares legalized; this possible either informally where all passengers booked in advance, or under a local authority scheme from an 'authorized place' without booking, or as a semi-stage service under a 'special restricted' operator's licence, not counting as a PSV.
'Community bus permits' allow CBs to operate public stage services or otherwise for hire and reward on non-profit basis.

Minibus permits: modified rules, now also available for recreational use.
Grants: DTp may pay transitional rural bus grant, and Rural Transport Innovation Grant for UCMs and other new services;
CT defined as 'public transport' and may be grant-aided as such by local authorities; LRT may fund CT and DaR services in London.**

Appendix 2

DEFINITIONS OF MODES

D/R diversion (Flexibus)	Stage bus diverting from fixed route/ schedule on demand-responsive (D/R) basis, or operating flexible routes on similar basis.
Multiple service bus	Stage bus operating different routes for different types of traffic on different times of day or days per week.
Contract bus	Stage bus 'bought in' for specific service by agency without normal transport responsibilities, e.g. district, town or parish council.
Subscription bus (cooperative bus)	Stage bus contracted by local association and financed by subscriptions; no fares charged to members.
Free shoppers' bus	Stage bus paid for by local traders/ retailers to bring customers to shopping centre; no fares.
School bus	Bus contracted or owned by local education authority for transport of school pupils, also carrying fare-paying adult passengers.
Postbus	Vehicle operated by Post Office for mail collection and delivery services, adapted to carry passengers.
Community bus (CT)	Small vehicle owned and managed by committee of local volunteers, operating a variety of stage, D/R services, excursions and hires, with volunteer drivers (there are many variations).
PSV dial-a-bus	Public Service Vehicle operated in D/R mode according to telephone requests, using 'hi-tech' control equipment.
'Welfare' dial-a-bus (CT)	Vehicle run by community group in D/R mode for people with special needs, such as the elderly and disabled; usually with adaptations such as wheelchair lifts.
Social car (CT)	Private car run by volunteer owner/ driver to carry persons on essential trips impossible to make by other means; pool of cars coordinated by local committee.

Hospital car (CT)	Private car run by volunteer owner/driver to carry outpatients to hospital; organized either by health authority to supplement ambulance service, or as social car exclusively for hospital trips.
Lift-giving scheme (CT)	Informal arrangement by private car owners to give lifts, perhaps on regular basis, to other persons; with local advertising but no central organization.
Car pooling (CT)	Arrangement among car owners to use a single vehicle for regular trips (mainly the journey to work) on an alternate basis.
Shared taxi/hire-car	Taxi or hire-car authorized to admit passengers sharing the vehicle, at separate fares.
Passenger/freight service (Courier)	Stage bus modified to carry some freight, parcels, sundries; routes may be determined by regular freight delivery runs.
Demountable vehicle	Vehicle with interchangeable bodies, detachable from single chassis/cab unit, e.g. bus body interchangeable with freight van.
Railbus	Light rail vehicle (diesel multiple unit) consisting of bus body mounted on rail underframe.

(CT) = 'community transport' mode

Appendix 3

GLOSSARY OF INITIALS AND ACRONYMS

BR	British Rail
BROH	bus revenue operation hours
CB	community bus
CC	County Council (England & Wales)
CT	community transport
DaR	dial-a-ride
DGH	District General Hospital
DHSS	Department of Health and Social Security (Great Britain)
DMU	diesel multiple unit
DoE	Department of the Environment (Great Britain)
DoE(NI)	Department of the Environment (Northern Ireland)
D/R	demand-responsive
DRT	demand-responsive transport (transit)
DTp	Department of Transport (Great Britain)
ECOC	Eastern Counties Omnibus Company
ESCORT	East Sussex Coordinated Rural Transport
GLC	Greater London Council
GP	General Practitioner
HCS	Hospital Car Service
IT	information technology
LEA	local education authority
LEV	Leyland Experimental Vehicle
LRCC	Lincolnshire Road Car Company
LRT	London Regional Transport
MIND	the National Association for Mental Health
MSC	Manpower Services Commission
NAU	National Advisory Unit (for Community Transport)
NBC	National Bus Company
PHAB	Physically Handicapped and Able-bodied
PSV	Public Service Vehicle
PTE	Passenger Transport Executive
PTP	Public Transport Plan (England & Wales)
PTT	Post, Telephone, Telegraph (continental Europe)
RC	Regional Council (Scotland)
RTDF	Rural Transport Development Fund (England)
RTIG	Rural Transport Innovation Grant (Scotland & Wales)
RUTEX	Rural Transport Experiments
SBG	Scottish Bus Group
SDD	Scottish Development Department
SSD	Social Services Department (of local authority)

TPP	Transport Policies and Programmes
TRRL	Transport & Road Research Laboratory
UCM	unconventional mode
VT	Voluntary Transport
WRVS	Women's Royal Voluntary Service

INDEX

CARDIFF
UWCC LIBRARY